BARRON'S

HOW TO PREPARE FOR THE
COLLEGE BOARD ACHIEVEMENT TEST

CBAT MATH

LEVEL II

THIRD EDITION

Howard P. Dodge
The Wheeler School
Providence, Rhode Island

BARRON'S
Barron's Educational Series, Inc.
New York • London • Toronto • Sydney

©Copyright 1987, 1984, 1979 by Barron's Educational
Series, Inc.

All rights reserved.
No part of this book may be reproduced
in any form, by photostat, microfilm, xerography,
or any other means, or incorporated into any
information retrieval system, electronic or
mechanical, without the written permission
of the copyright owner.

All inquiries should be addressed to:
Barron's Educational Series, Inc.
250 Wireless Boulevard
Hauppauge, New York 11788

Mathematics Achievement Test questions selected from
*The College Board Achievement Tests—14 Tests in 13
Subjects*, College Entrance Examination Board, 1986.
Reprinted by permission of Educational Testing Service,
the copyright owner of the test questions.

Permission to reprint the above material does not
constitute review or endorsement by Educational
Testing Service or the College Board of this
publication as a whole or of any other testing
information it may contain.

Mathematics Achievement Test Questions: pages viii–x,
#1–15

Library of Congress Catalog Card No. 87-1237
International Standard Book No. 0-8120-3881-9

Library of Congress Cataloging-in-Publication Data

Dodge, Howard P.
  Barron's how to prepare for the College Board
achievement test, mathematics, level 2.

  Includes index.
  1. Mathematics—Examinations, questions, etc.
2. Mathematics—1961–   . I. Title. II. Title: How
to prepare for the College Board achievement test,
mathematics, level 2. III. Title: College Board
achievement test, mathematics, level 2.
QA43.D57 1987   510'.76   87-1237
ISBN 0-8120-3881-9 (pbk.)

PRINTED IN THE UNITED STATES OF AMERICA

789 100 9 8 7 6 5 4 3 2 1

CONTENTS

Introduction
iv

How This Book Is Structured   iv
How the Level II Test Is Structured   iv
Suggestions for Scoring Higher on the Test   v
Questions Students Ask About This Achievement Test   vi
Sample Math Level II Questions   viii

SECTION 1
DIAGNOSTIC TEST

DIAGNOSTIC TEST
1

Answer Key
6

Answer Explanations
7

SECTION 2
REVIEW OF MAJOR TOPICS

CHAPTER 1. INTRODUCTION TO FUNCTIONS
11
[1.1] Definition / 11
[1.2] Function Notation / 12
[1.3] Inverse Functions / 12
[1.4] Odd and Even Functions / 14
[1.5] Multi-Variable Functions / 15

CHAPTER 2. POLYNOMIAL FUNCTIONS
16
[2.1] Definition / 16
[2.2] Linear Functions / 16
[2.3] Quadratic Functions / 18
[2.4] Higher Degree Polynomial Functions / 19
[2.5] Inequalities / 24

CHAPTER 3. TRIGONOMETRIC FUNCTIONS
25
[3.1] Definitions / 25
[3.2] Arcs and Angles / 27
[3.3] Special Angles / 28
[3.4] Graphs / 29
[3.5] Identities, Equations, and Inequalities / 32
[3.6] Inverse Functions / 36
[3.7] Triangles / 38

CHAPTER 4. MISCELLANEOUS RELATIONS AND FUNCTIONS
41
[4.1] Conic Sections / 41
[4.2] Exponential and Logarithmic Functions / 45
[4.3] Absolute Value / 47
[4.4] Greatest Integer Function / 49
[4.5] Rational Functions / 49
[4.6] Parametric Equations / 51
[4.7] Polar Coordinates / 52

CHAPTER 5. MISCELLANEOUS TOPICS
55
[5.1] Permutations and Combinations / 55
[5.2] Binomial Theorem / 57
[5.3] Probability / 59
[5.4] Sequences and Series / 61
[5.5] Geometry / 64
[5.6] Variation / 69
[5.7] Logic / 70
[5.8] Statistics / 72
[5.9] Odds and Ends / 73

CHAPTER 6. SUMMARY OF FORMULAS
76

CHAPTER 7. ANSWERS TO PRACTICE EXERCISES
80
Answer Key / 80
Answer Explanations / 82

SECTION 3
MODEL EXAMINATIONS

Model Test 1
94
Model Test 2
99
Model Test 3
105
Model Test 4
110
Model Test 5
116
Model Test 6
121
Model Test 7
126
Model Test 8
131
Answer Sheets / 137
Answer Key / 143
Answer Explanations / 144
Index / 165

INTRODUCTION

HOW THIS BOOK IS STRUCTURED

This introduction has been designed to help you approach the Mathematics Level II Achievement Test with confidence. It describes the makeup and content of the test, gives you suggestions that will help you score as high as possible, and answers the questions most asked by students who take the test. Once you have read through the introduction, you are ready for the body of the book which is divided into three sections.

Section One is a diagnostic test covering the topics on the Level II examination. Each question has a solution. At the end of each solution is a number in square brackets that refers you to a place in this book where that topic is reviewed. If you do not get the answer, you can look up the proper section in the book either by checking the contents page or by looking at the heads at the top of each page. For example, if you get question 26 wrong, you are referred to [4.4]. Looking on the contents page, you can see that 4.4, "Greatest Integer Function," begins on page 49. Otherwise you could flip through the pages until you come to Chapter four and then check the heads on right-hand pages until you reach topic 4.

Section Two is a review of the topics covered by the Level II test. You can either read through it in its entirety or use it as a reference section whenever you have difficulty with a problem. This section consists of a discussion of the methods and formulas necessary to solve the many types of problems you will encounter. A fuller explanation of some of the more elementary topics will be found in *Barron's How to Prepare for College Board Achievement Tests/Mathematics Level I*. Since this is a review, very few derivations of formulas will be found. If you are curious about derivations, you should refer to your high school texts.

Section Three contains eight sample Level II examinations, each consisting of 50 multiple-choice questions. A solution is given for each problem. Also provided is a cross reference in square brackets that refers you to the appropriate review topic.

HOW THE LEVEL II TEST IS STRUCTURED

The Mathematics Level II test contains 50 multiple-choice questions, each with five possible answers, only one of which is correct. You will have one hour to complete the test.

Although the test is aimed at students who have had 3½ or more years of high-school mathematics, it is often taken by those who have had 3 years of strong mathematics courses. Regardless of the amount of mathematics you have completed, remember that this examination is prepared for talented students who have studied trigonometry and elementary functions in detail. If you have such a background, you will be better prepared for this test than for the Mathematics Level I Test (which contains several questions on Euclidean plane geometry) since the material tested will have been more recently covered in your courses.

According to the most recent College Board information, the questions in the Mathematics Level II Achievement Test are divided into five categories. (The percentages indicate the approximate weighting of each topic in the examination.)

18% – 1. Algebra
20% – 2. Geometry (coordinate geometry in two or three dimensions, transformations, synthetic geometry in three dimensions, and vectors)
    8% – a) Solid Geometry
    12% – b) Coordinate Geometry
20% – 3. Trigonometry (properties and graphs of the trigonometric functions, inverse trigonometric functions, identities, equations, and inequalities)
24% – 4. Elementary functions
18% – 5. Miscellaneous topics (sequences, series and limits, logic and proof, probability and statistics, and number theory)

The topics in the model examinations in this book have been weighted in the same way in order to give you a true experience of the breadth you will encounter in the actual examination.

Although this is an advanced test, there are several relatively elementary questions among the 50 asked. For example,

1. An equation of line ℓ could be
   - [A] $x=5$
   - [B] $x+y=5$
   - [C] $y=5$
   - [D] $y=x+5$
   - [E] $y=0$

2. The inequality $3-x<2x$ is equivalent to which one of the following?
   - [A] $x>1$
   - [B] $x<1$
   - [C] $x<3$
   - [D] $x>3$
   - [E] $x>\frac{3}{2}$

The answer to [1] is [A] since x must remain 5 regardless of the value of y. Problem [2] can be transformed into answer [A] by adding x to both sides and dividing by 3.

SUGGESTIONS FOR SCORING HIGHER ON THE TEST

There is no quick, shortcut method of preparing for achievement tests. The review should be planned several weeks in advance of the examination, and a regular time should be set aside each day for review and practice. This preparation will lead to an increase in knowledge, a deeper understanding of the material, and greater confidence when taking the actual Mathematics Level II examination. With such a wide variety of topics covered, no student is expected to have studied every one. Do not be surprised if a few questions appear which seem to make no sense. On the November 1982 Level II examination, for example, students who earned a raw score between 42 and 50 (out of a possible 50 points) received an 800—the highest possible scaled score.

Scoring

Your raw score is determined by the number of correct answers *minus* one fourth of the number of incorrect answers. This system means that if you do not know how to do a particular problem, you should stop to think before you guess blindly. If you can eliminate only one choice, it may not be worth guessing, but if you can eliminate more than one choice and you are guessing at four or more questions, it may be worth the risk. Remember that a raw score of as little as 42 points could give you a perfect 800. Note that this varies from year to year, depending on how other students score. One year a raw score of 40 could give you a scaled score of 780, but another year that same 40 could translate to 800. On one recent test given by the College Board, a 20 gave 610, a 30 gave 700, and a 40 gave 780.

To help you raise that score, the following lists summarize suggestions for this examination in particular, mathematics examinations in general, and any type of timed examination.

Suggestions for the Level II test
- Go through the entire test doing only those problems that can be done easily.
- Go back through the test doing those problems that seem to be on familiar material but will take a little time to figure out.
- Finally, if there is time, use the five answer choices as a guide to try to figure out the correct answer to the remaining questions OR check the answers on the problems that have previously been completed.
- Be particularly careful to mark the correct space on the answer sheet.

Suggestions for mathematics examinations
- Round off and estimate wherever possible. Minimize calculation.
- Look for the shortcuts which are built into many problems.
- Work in units (hours, ounces, square miles) consistent with those in the answer choices.
- Break down complex word problems and make computations one step at a time.

Suggestions for any timed examination
- *Budget your time.* Work swiftly: don't linger over difficult questions.
- *Guess intelligently.* Random guessing will probably not raise your score.
- *Read each question carefully.* Answer the question asked, not the one you may have expected.
- *Save the hardest questions for last.*

- *Mark answers clearly, accurately.* Check the numbering of your answer sheet often. Erase cleanly; leave no stray marks. Mark only one answer. You will be marked wrong if two spaces are filled in.
- *Change answers if you have a reason for doing so.* However, it is usually best not to change based on hunch or whim.

QUESTIONS STUDENTS ASK ABOUT THIS ACHIEVEMENT TEST*

What is an achievement test?

An achievement test tests your knowledge of a subject and your ability to apply that knowledge. Unlike an aptitude test, an achievement test is curriculum-based and is intended to measure how much you have learned rather than how much learning ability you have.

How is it used?

Colleges use achievement tests to predict an admission candidate's future success on the level of work typical at that college. Applicants to colleges use the tests to demonstrate their acquired knowledge in subject areas important to their future goals.

Some colleges, especially the most selective schools, have found achievement test results to be better predictors of success in related college courses than SAT scores and, frequently, the high school record. Because grading systems, standards and course offerings differ among secondary schools, a student's achievement test score may provide the college with more and better information about what the student knows than his or her high school transcript does.

Colleges also use achievement test results to place students in appropriate courses after they have been admitted.

High schools use achievement tests to examine the success of the school's curriculum and to indicate strengths and weaknesses.

Should I take a math achievement test?

Depending on the secondary school you attend and the college to which you apply, you may not have much choice. Most private colleges and many state universities require applicants to take three achievement tests, though often not specifying which ones. According to a counselor at a selective independent secondary school, "Ninety-nine percent of the colleges our students apply to require three achievement tests, so we ask all of our students to take them regardless of where they apply."

Many secondary school counselors believe that an application is improved by achievement test scores even when a college does not require them, because the tests allow the candidate to demonstrate proficiency in the area of his or her greatest strength.

Which tests you take will depend on your most successful courses in secondary school and your intended major in college. You should not take an achievement test in a subject unless you have been successful in it or have made a special effort to prepare for the test (and it is best if you satisfy both conditions).

Examine the description of the contents of the Level II test to see whether you have studied the topics tested.

Many secondary school counselors believe that the achievement tests most helpful to your application are an English test, a math test and a third test in your best third subject. If you intend to major in a science or in any other math-related field (engineering, economics, etc.), you definitely should take a math achievement test.

How important are achievement tests in college admissions?

College admissions committees generally regard achievement test scores as "confirming" evidence to be compared with high school grades, the level of courses taken, and SAT scores, Grades (especially because of grade inflation) and courses offered vary greatly from high school to high school (and even from teacher to teacher within the same high school). A grade in "intermediate math," for example, at a selective independent school may respresent something entirely different from the same grade in a course of the same name at a small rural public high school.

No other factor on an application is so uniform an assessment of all applicants as the standardized achievement test.

The scores are used not only as a measure of knowledge but also as a means of interpreting the high school record. Low achievement test scores accompanied by high grades, for example, may suggest an inflated grading system. High scores with average grades may

---

*Adapted from *How to Prepare for the College Board Achievement Tests: Mathematics Level I*, 3rd Ed., by James J. Rizutto, Barron's Educational Series, Inc., Woodbury, New York, 1982. Reprinted with permission.

mean the applicant is attending a more demanding high school that offers a challenging program to competitive students chosen for special abilities.

In general, test scores are more important if they disagree with the high school record and less important if they agree. They are also more important if they lie outside the range of scores of students normally selected by the college and less important if they fall within the range. And, finally, they are more important if they conflict with the scores on the SAT than if they conform.

In other words, if they don't say the same thing as the rest of the record, it's the new message that is important. If the new message is positive, it will help. But in the majority of cases, the achievement test scores confirm the rest of the data on the application.

When and where are achievement tests given?

Achievement tests are usually given in the morning of the first Saturday in November, May, and June and the last Saturday in January. There are hundreds of test centers around the country. Dates vary from year to year; those for the near future are given below. For other information on dates and test sites, write to the College Board Admission Testing Program for its current schedule.

To whom do I write for information and registration?

Write to:

College Board ATP
Box 592
Princeton, New Jersey 08541

or call whichever of the following is closer to you:

Princeton, NJ   (609) 771-7600
Berkeley, CA   (415) 849-0950

The College Entrance Examination Board, an association of colleges and secondary schools, is headquartered at 888 Seventh Avenue, New York, NY. To arrange for testing, use *only* the Princeton address.

What should I bring with me to the test?

You should bring your ticket of admission, some positive form of identification, two or more No. 2 pencils (with erasers, or bring an eraser), a watch (though test centers *should* have clearly visible clocks) and a simple twist-type pencil sharpener (so that you need not waste time walking back and forth to a pencil sharpener if your point breaks).

No books, calculators, rulers, scratch paper, slide rules, compasses, protractors or any other devices are allowed in the examination room.

SAT and Achievement Test Dates

| TEST DATES || REGISTRATION DEADLINES ||
NATIONAL	NEW YORK STATE	REGULAR	LATE
†October 10, 1987		September 18, 1987	September 30, 1987
November 7, 1987	November 7, 1987	October 2, 1987	October 14, 1987
December 5, 1987	December 5, 1987	October 30, 1987	November 11, 1987
January 23, 1988	January 23, 1988	December 18, 1987	December 30, 1987
*March 19, 1988	*March 19, 1988	February 12, 1988	February 24, 1988
May 7, 1988	May 7, 1988	April 1, 1988	April 13, 1988
June 4, 1988	June 4, 1988	April 29, 1988	May 11, 1988

† Only in California, Florida, Georgia, Illinois, North Carolina, South Carolina, and Texas.
* SAT only.

SAMPLE MATH LEVEL II QUESTIONS

This section contains directions and several questions from an actual Math Level II test, reprinted with the permission of the Educational Testing Service. These particular questions were chosen to familiarize you with the various types that might appear. They were also chosen because their answers might be found by using unorthodox or "non-routine" methods or by intelligent guessing—methods which your math teacher might frown upon, but which sometimes work with multiple-choice type questions. Each question is followed by a comment describing such an approach.

Obviously, if you understand the problem and can solve it directly, do so. These methods are only suggested to show you that there is often another way to look for an answer. The directions, questions, and comments follow.

Directions: For each of the following problems, decide which is the best of the choices given. Then blacken the corresponding space on the answer sheet.

Notes: (1) Figures that accompany problems in this test are intended to provide information useful in solving the problems. They are drawn as accurately as possible EXCEPT when it is stated in a specific problem that its figure is not drawn to scale. All figures lie in a plane unless otherwise indicated.

(2) Unless otherwise specified, the domain of a function f is assumed to be the set of all real numbers x for which $f(x)$ is a real number.

1. The set of all ordered pairs (x,y) that satisfy the system $\begin{cases} y=x \\ xy=1 \end{cases}$ is

   (A) $\{(-1, -1)\}$ (B) $\{(-1, 1)\}$ (C) $\{(1, 1)\}$
   (D) $\{(-1, -1), (1, 1)\}$ (E) $\{(-1, 1), (1, -1)\}$

Comment: An easy, straightforward problem. Just substitute x for y in the second equation: $x^2=1$. $x=\pm 1$. Don't jump at (B) as the answer. Make sure you ANSWER THE QUESTION—ordered pairs for x and y. Correct answer (D).

2. If k is an integer less than zero, which of the following is less than zero?

   (A) $-k$ (B) $-(-k)$ (C) $(-k)^2$ (D) $(k)^2$
   (E) $-(k)^3$

Comment: Try a specific negative integer for k. For example, let $k = -1$, and check the given choices until you find one that answers the question: (A) $-(-1) > 0$, (B) $-(-(-1)) < 0$. You found it. Go on to the next question.

3. When a certain integer is divided by 5, the remainder is 3. What is the remainder when 4 times that integer is divided by 5?

   (A) 0   (B) 1   (C) 2   (D) 3   (E) 4

Comment: Again, try a specific number that satisfies the conditions of the problem. For example, when 8 is divided by 5, the remainder is 3. $4 \cdot 8 = 32$, which, when divided by 5, leaves a remainder of 2. Correct answer (C).

4. Which of the following defines a function that will associate a positive integer y with each positive integer x so that x and y have the same tens' digit?

   (A) $y=10x$ (B) $y=11x$ (C) $y=100x$
   (D) $y=101x$ (E) $y=111x$

Comment: This time try a specific two-digit number, for example, 21. Answers (A) and (C) can be eliminated since they shift all digits. Simple multiplications of 21 by 11, 101 and 111 lead to the correct answer (D).

5. The solution set of $\dfrac{(x+1)^2}{x} > 0$ is

   (A) the empty set   (B) $\{x|x>-1\}$ (C) $\{x|x>0\}$
   (D) $\{x|x>1\}$ (E) $\{x|x \text{ is any real number}\}$

Comment: Think about the problem before "diving in" to solve a quadratic inequality. Notice that the top is always positive or 0. Therefore, the denominator, x, must also be positive. We must also rule out $x=-1$ which makes the numerator 0. The solution occurs when $x>0$. Correct answer (C).

6. In the figure, the bases of the right prism are equilateral triangles, each with perimeter 30 centimeters. If the altitude of the prism is 10 centimeters, what is the total surface area of the solid in square centimeters?

   (A) 100   (B) $\dfrac{250}{\sqrt{3}}$ (C) $100\sqrt{3}$
   (D) 300   (E) $50\sqrt{3} + 300$

Comment: Since the lateral sides are squares, 10 centimeters on a side, their area is 300. Equilateral triangles have a $\sqrt{3}$ in their area, which must be added to the 300. (E) must be the correct answer.

7. If $f(x) = \dfrac{1}{x}$, which of the following could be the graph of $y = f(-x)$?

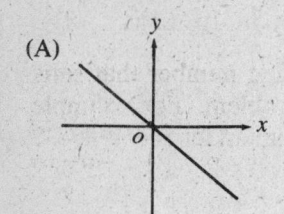

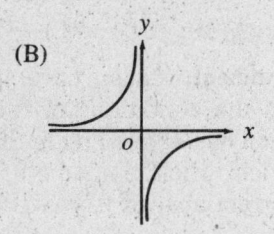

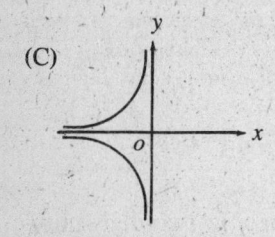

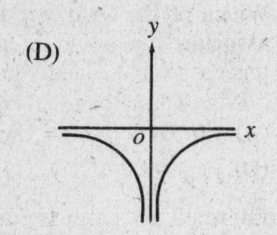

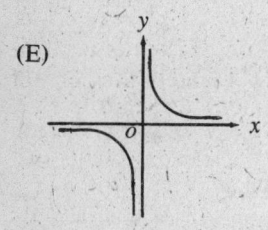

(A) [graph]   (B) [graph]

(C) [graph]   (D) [graph]

(E) [graph]

Comment: For any value of $x \neq 0$, $y = \dfrac{1}{x}$ and $y = \dfrac{1}{-x}$ have the same absolute values for y, but they are opposite in sign. Therefore, $y = f(-x)$ is the mirror image of $y = f(x)$ reflected in the x-axis. Since the graph of $f(x) = \dfrac{1}{x}$ looks like the sketch below, the graph of $y = f(-x)$ must look like (B).

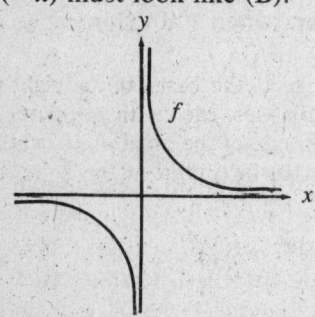

[graph of f]

8. If the line $y = k$ is tangent to the circle $(x-2)^2 + y^2 = 9$, then $k =$

(A) -1 or 4 (B) -3 or 3 (C) -4 or 1
(D) -6 or 6 (E) -9 or 9

Comment: $y = k$ is a horizontal line, tangent to a circle whose center is at $(2,0)$ and with radius 3. Therefore, the line is either 3 units above the center or 3 units below the center. Thus, $k = \pm 3$. Correct answer (B).

9. In the figure, if $\text{Arcsin } x = \text{Arccos } x$, then $k =$

(A) x (B) x^2 (C) 1 (D) $1-x$ (E) $\dfrac{1}{x}$

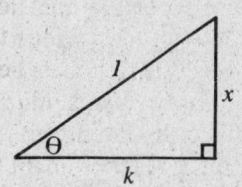

[right triangle with hypotenuse 1, angle θ, legs k and x]

Note: Figure is not drawn to scale

Comment: A good technique is to simplify notation. Let $A = \arcsin x$ and let $B = \arccos x$. Then the question really states that $\sin A = \cos B$ since both equal x. If the cofunctions of two angles are equal, the angles are complementary. But the question also tells us that they are equal. Therefore, $A = B = 45°$; the triangle shown is isosceles, and $x = k$. Correct answer (A).

10. $(-i)^n$ is a negative real number if $n =$

(A) 21   (B) 22   (C) 23   (D) 24   (E) 25

Comment: Remember $i^4 = 1$. n must be even for $(-i)^n$ to be a real number. This eliminates answers (A), (C), (E). If n is divisible by 4, $(-i)^n = (-1)^n(i)^n = (-1)^n(i)^{4k} = (1)(i^4)^k = (1)(1)^k = (1)(1) = 1$, which eliminates (D). Correct answer (B).

11. $\dfrac{(n-1)!}{n!} + \dfrac{(n+1)!}{n!} =$

(A) $\dfrac{n-1}{n}$ (B) $\dfrac{n^2+1}{n}$ (C) $\dfrac{n^2-1}{n}$ (D) $\dfrac{n+1}{n}$

(E) $\dfrac{n^2+n+1}{n}$

Comment: If the method of solution is not obvious, try expanding the factorials a bit to see if you find a clue to finding a solution.

$$\dfrac{(n-1)!}{n!} + \dfrac{(n+1)!}{n!}$$
$$= \dfrac{(n-1)(n-2)!}{n(n-1)(n-2)!} + \dfrac{(n+1)(n!)}{n!}$$
$$= \dfrac{1}{n} + \dfrac{n+1}{1}$$
$$= \dfrac{n^2+n+1}{n}.$$

Correct answer (E).

12. What is the range of the function defined by $f(x) = \frac{1}{x} + 2$?

    (A) All real numbers
    (B) All real numbers except $-\frac{1}{2}$
    (C) All real numbers except 0
    (D) All real numbers except 2
    (E) All real numbers between 2 and 3

Comment: It often helps to reduce the problem to a simpler and more manageable one. $f(x) = \frac{1}{x}$ is simpler than $f(x) = \frac{1}{x} + 2$ but related to it. The graph of $y = \frac{1}{x}$ looks like the one below. Its range is all values of y except zero. The graph of $f(x)$ is the same but it is shifted up 2 units. Therefore, the range of $f(x)$ is all real numbers except 2. Correct answer (D).

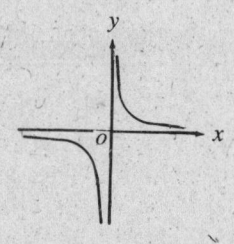

13. How many different sets of two parallel edges are there in a cube?

    (A) 6 (B) 8 (C) 12 (D) 18 (E) 24

Comment: Draw a sketch of a cube and try to find a pattern. Take one edge, a, and find all edges parallel to it. There are 3 (b,c,d). Now take b and find all edges parallel to it except a. There are 2 (c,d). Now take c and find all edges parallel to it except a and b. There is 1 (d). Thus, for this set of edges, there are 6 sets of parallel edges. There are two other sets of edges, (E,F,G,H) and (1,2,3,4). Therefore, there is a total of 18 sets of two parallel edges. Correct answer (D).

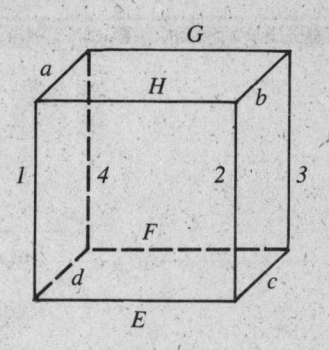

14. The "spread" of a point (x,y) in the rectangular coordinate plane is defined as $|x| + |y|$. Which of the following points has the same spread as $\left(\frac{3}{2}, -\frac{1}{2}\right)$?

    (A) $(-1, 0)$ (B) $\left(0, \frac{1}{2}\right)$ (C) $\left(\frac{1}{2}, \frac{1}{2}\right)$
    (D) $(1, -1)$ (E) $(2, 1)$

Comment: Often problems are presented where a new operation (such as "spread") is defined. Usually this new operation is just a combination of simple, familiar operations. In this case, the "spread" is just the sum of absolute values. $\left|\frac{3}{2}\right| + \left|-\frac{1}{2}\right| = \frac{3}{2} + \frac{1}{2} = 2$. By inspection the correct answer is (D).

15. The least positive integer N for which each of $\frac{N}{2}, \frac{N}{3}, \frac{N}{4}, \frac{N}{5}, \frac{N}{6}, \frac{N}{7}, \frac{N}{8}$, and $\frac{N}{9}$ is an integer is

    (A) $9 \cdot 8 \cdot 7 \cdot 6 \cdot 5 \cdot 4 \cdot 3 \cdot 2$
    (B) $9 \cdot 8 \cdot 6 \cdot 5 \cdot 4 \cdot 3 \cdot 2$
    (C) $9 \cdot 8 \cdot 7 \cdot 6 \cdot 5$
    (D) $9 \cdot 8 \cdot 7 \cdot 5$
    (E) $9 \cdot 8 \cdot 7$

Comment: Since all the denominators must divide into N, start with the largest one and work down. 9, 8, 7, and 5 are the needed factors of N. 6, 4, 3, and 2 are not needed because all their factors are already included in 9 and 8. Correct answer (D).

DIAGNOSTIC TEST

SECTION 1

The diagnostic test is designed to help you pinpoint the weak spots in your background. The answer explanations are keyed to sections of the book. In order to make the best use of this diagnostic test, set aside between 1 and $1\frac{1}{2}$ hours so you will be able to do the whole test at one sitting.

Do the problems as if it were a regular testing session. When finished, check your answers with those at the end of the test. For those that you got wrong, note the sections containing the material that you must review. If you do not fully understand how you got some of the correct answers, you should review those sections also.

Directions: For each of the following problems, decide which is the best of the choices given. Then blacken the corresponding space on the answer sheet on page 137. Answers may be found on page 6, and solutions begin on page 7.

Notes: (1) Figures that accompany problems in this test are intended to provide information useful in solving the problems. They are drawn as accurately as possible EXCEPT when it is stated in a specific problem that its figure is not drawn to scale. All figures lie in a plane unless otherwise indicated.

(2) Unless otherwise specified, the domain of a function f is assumed to be the set of all real numbers x for which $f(x)$ is a real number.

1. A linear function, f, has a slope of -2. $f(1) = 2$ and $f(2) = q$. Find q.

   [A] 0 [B] 4 [C] $\frac{3}{2}$ [D] $\frac{5}{2}$ [E] 3

2. $\dfrac{\sin 120° \cdot \cos \frac{2\pi}{3}}{\tan 315°} =$

   [A] $\dfrac{\sqrt{3}}{2}$ [B] $-\dfrac{\sqrt{3}}{4}$ [C] $\dfrac{\sqrt{6}}{4}$

   [D] $-\dfrac{\sqrt{3}}{2}$ [E] $\dfrac{\sqrt{3}}{4}$

3. A function is said to be even if $f(x) = f(-x)$. Which of the following is not an even function?

   [A] $y = |x|$ [B] $y = \sec x$ [C] $y = \log x^2$
   [D] $y = x^2 + \sin x$ [E] $y = 3x^4 - 2x^2 + 17$

4. What is the radius of a sphere, with center at the origin, that passes through the point (2,3,6)?

   [A] 6 [B] 7 [C] 11 [D] 12 [E] 36

5. If $f(x) = \frac{1}{2}x^2 - 8$ is defined when $-4 \le x \le 4$, the maximum value of the graph of $|f(x)|$ is

   [A] -8 [B] 0 [C] 8 [D] 4 [E] 2

6. $\displaystyle\sum_{j=1}^{5} 2\left(\dfrac{3}{2}\right)^{j-1} =$

   [A] $26\dfrac{3}{8}$ [B] $26\dfrac{5}{8}$ [C] $26\dfrac{1}{2}$

   [D] $26\dfrac{1}{8}$ [E] $26\dfrac{7}{8}$

7. If $f(x) = \dfrac{k}{x}$ for all non-zero real numbers, for what value of k does $f(f(x)) = x$?

   [A] 1 only
   [B] 0 only
   [C] all real numbers
   [D] all real numbers except 0
   [E] no real numbers

8. $\dfrac{\log 9}{\log 3} =$

   [A] 3 [B] 6 [C] 2
   [D] $\log 6$ [E] none of these

9. If $P(x) = 3x^3 - 4x^2 - 2x + 2 = (x-1)Q(x) + R(x)$, then $R(x) =$

   [A] -3 [B] -1 [C] 0 [D] 1 [E] 5

10. $F(x) = \begin{cases} \dfrac{3x^2 - 3}{x - 1}, & \text{when } x \ne 1 \\ k, & \text{when } x = 1 \end{cases}$

    For what value(s) of k will F be a continuous function?

    [A] 1 [B] 2 [C] 3 [D] 6
    [E] no value of k

11. If $\tan \theta = \dfrac{2}{3}$, then $\sin \theta =$

    [A] $\dfrac{2\sqrt{13}}{13}$ [B] $\pm\dfrac{2\sqrt{13}}{13}$ [C] $\dfrac{3\sqrt{13}}{13}$

    [D] $\pm\dfrac{2}{5}$ [E] $\dfrac{2\sqrt{5}}{5}$

12. If a circle has a central angle of $67\frac{1}{2}°$ which intercepts an arc of length 24π, how long is the radius?

[A] 8 [B] 64 [C] 16π [D] 8π [E] 16

13. If $f(x,y) = 2x^2 - y^2$ and $g(x) = 2^x$, the value of $g(f(1,2))$

[A] 1 [B] 4 [C] $\frac{1}{4}$ [D] -4 [E] 0

14. What is the amplitude of the graph of the function $y = \cos^4 x - \sin^4 x$?

[A] $\frac{1}{2}$ [B] $\frac{\sqrt{2}}{2}$ [C] 1

[D] $1 + \frac{\sqrt{2}}{2}$ [E] 2

15. Which of the following could be the equation of the graph in the figure?
  I. $y = \sin 4x$
  II. $y = \cos\left(4x - \frac{\pi}{2}\right)$
  III. $y = -\sin(4x + \pi)$

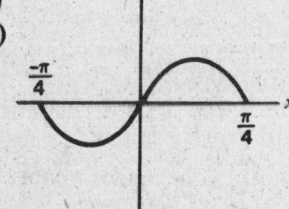

[A] I only [B] I & II only
[C] II & III only [D] II only
[E] I, II, & III

16. If $2 \cdot \sin^2 x - 3 = 3 \cdot \cos x$ and $90° < x < 270°$, the number of values that satisfy the equation is

[A] 0 [B] 1 [C] 2 [D] 3 [E] 4

17. If $A = \text{Arctan}\left(\frac{3}{4}\right)$ and $A + B = 315°$, then $B =$

[A] $\text{Arctan}\frac{3}{5}$ [B] $\text{Arctan}\frac{4}{5}$

[C] $\text{Arctan}\left(-\frac{3}{5}\right)$ [D] $\text{Arctan}\left(-\frac{4}{5}\right)$

[E] $\text{Arctan}(-7)$

18. The units digit of 1567^{93} is

[A] 1 [B] 3 [C] 7 [D] 9
[E] none of these

19. The area of a triangle with sides 3, 5, and 7 is

[A] 7.5 [B] $\frac{15\sqrt{3}}{4}$ [C] 3.75

[D] $\frac{15\sqrt{3}}{2}$ [E] $\frac{3\sqrt{10}}{4}$

20. In the figure at right, S is the set of points in the shaded region. Which of the following represents the set T consisting of all points $(2x-y, y)$, where (x,y) is a point in S?

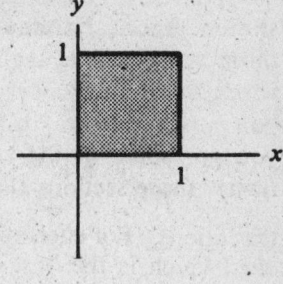

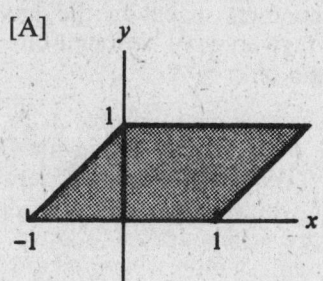

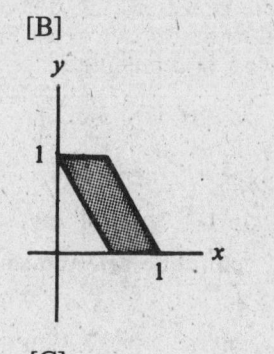

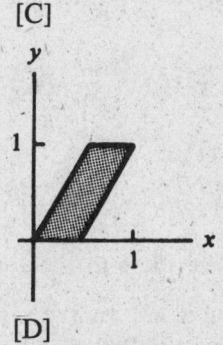

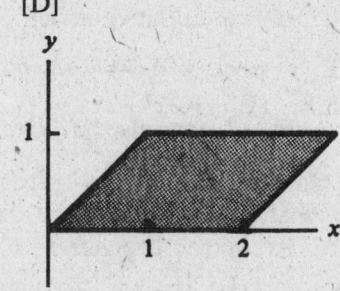

[A]

[B]

[C]

[D]

[E]

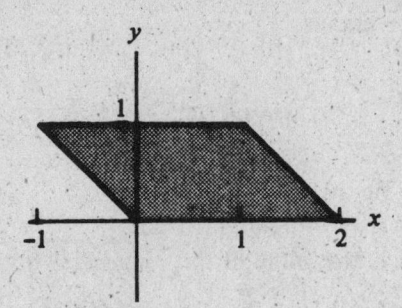

[Graph showing parallelogram shape from -1 to 2 on x-axis, up to y=1]

21. If the graph below represents the function $f(x)$, which of the following could represent the equation of the inverse of f?

[A] $x = y^2 - 8y - 1$ [B] $x = y^2 + 11$
[C] $x = (y-4)^2 - 3$ [D] $x = (y+4)^2 - 3$
[E] $x = (y+4)^2 + 3$

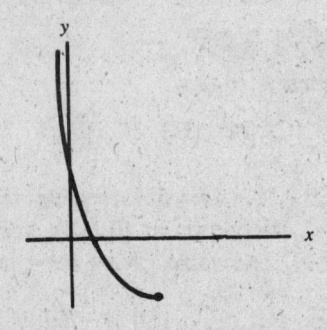

[Graph showing decreasing curve in first quadrant]

22. If $\log_2 m = x$ and $\log_2 n = y$, then $mn =$

[A] 2^{x+y} [B] 2^{xy} [C] 4^{xy}
[D] 4^{x+y} [E] cannot be determined

23. The distance between the point (2,4) and the line $3x - 4y = 8$ is

[A] $\dfrac{9\sqrt{5}}{5}$ [B] $\dfrac{22}{5}$ [C] $\dfrac{18}{5}$

[D] 2   [E] $\dfrac{11\sqrt{5}}{5}$

24. How many integers are there in the solution set of $|x - 2| < 5$?

[A] 11   [B] 0   [C] an infinite number
[D] 9   [E] 7

25. If $f(x) = \sqrt{x^2}$, then $f(x)$ can also be expressed as

[A] x [B] $-x$ [C] $\pm x$ [D] $|x|$
[E] cannot be determined because x is unknown

26. If $f(x) = i$, where i is an integer such that $i \leq x < i+1$, and $g(x) = f(x) - |f(x)|$, what is the maximum value of $g(x)$?

[A] 0   [B] 1   [C] -1 [D] 2   [E] i

27. The graph of $(x^2 - 1)y = x^2 - 4$ has

[A] 1 horizontal and 1 vertical asymptote
[B] 2 vertical but no horizontal asymptotes
[C] 1 horizontal and 2 vertical asymptotes
[D] 2 horizontal and 2 vertical asymptotes
[E] neither a horizontal nor a vertical asymptote

28. $\displaystyle\lim_{x \to \infty} \left(\dfrac{3x^2 + 4x - 5}{6x^2 + 3x + 1} \right) =$

[A] $\dfrac{1}{2}$ [B] 1   [C] -5 [D] $\dfrac{1}{5}$
[E] undefined

29. The graph of the parametric equations

$$\begin{cases} x = r^2 + 4 \\ y = (r+2)^2 \end{cases}$$

is a portion of a

[A] line   [B] parabola   [C] circle
[D] hyperbola   [E] ellipse

30. The polar coordinates of a point P are $(2, 240°)$. The Cartesian (rectangular) coordinates of P are

[A] $(-1, -\sqrt{3})$ [B] $(-1, \sqrt{3})$
[C] $(-\sqrt{3}, -1)$ [D] $(-\sqrt{3}, 1)$
[E] none of the above

31. The figure below most closely resembles the graph whose equation is

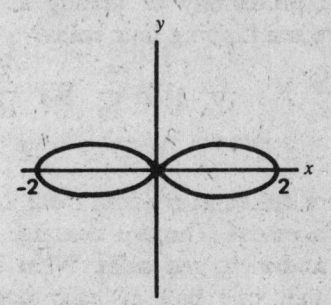

[Graph showing lemniscate-like figure with lobes extending to -2 and 2 on x-axis]

[A] $r = 2 \cdot \cos 2\theta + 2$
[B] $r = 4 \cdot \cos \theta$
[C] $r^2 = 4 \cdot \cos^2 2\theta$
[D] $r^2 = 4 \cdot \cos 2\theta$
[E] $r = \sin \theta + 4$

32 The height of a cone is equal to the radius of its base. The radius of a sphere is equal to the radius of the base of the cone. The ratio of the volume of the *cone* to the volume of the *sphere* is

[A] $\frac{1}{3}$ [B] $\frac{1}{4}$ [C] $\frac{1}{12}$ [D] $\frac{1}{1}$ [E] $\frac{4}{3}$

33 In how many different ways may the seven letters in the word *MINIMUM* be arranged, if all of the letters are used each time?

[A] 7 [B] 42 [C] 420
[D] 840 [E] 5040

34 How many different three-member committees can be selected from a group of ten people?

[A] $\frac{10!}{7!}$ [B] 3! [C] $\frac{10!}{7!3!}$
[D] $\frac{10!}{3!}$ [E] 10^3

35 What is the value of k if the line $y = 3x + k$ is to be tangent to the parabola $y = 2x^2 - 5x + 1$?

[A] ± 7 [B] -7 only [C] 7 only
[D] 9 only [E] -9 only

36 If $i = \sqrt{-1}$, the fourth term of the expansion of $(1+i)^6$ is

[A] -15 [B] 15 [C] $-20i$
[D] $20i$ [E] $6i$

37 What is the probability of getting at least three heads when flipping four coins?

[A] $\frac{3}{4}$ [B] $\frac{1}{4}$ [C] $\frac{7}{16}$ [D] $\frac{5}{16}$ [E] $\frac{3}{16}$

38 Two balls are drawn at random from an urn without replacement. The urn contains three white balls and eight red balls. What is the probability that one ball of each color is drawn?

[A] $\frac{31}{55}$ [B] $\frac{28}{55}$ [C] $\frac{49}{110}$
[D] $\frac{12}{55}$ [E] $\frac{24}{55}$

39 If $f(x) = ax^2 + bx + c$ and $f(2) = 3$ and $f(3) = 3$, then $\frac{a}{b}$ equals

[A] 5 [B] $-\frac{13}{5}$ [C] $-\frac{1}{5}$
[D] $-\frac{5}{13}$ [E] -5

40 What is the value of the remainder obtained when $6x^4 + 5x^3 - 2x + 8$ is divided by $x - \frac{1}{2}$?

[A] 2 [B] 4 [C] 6 [D] 8 [E] 10

41 How many positive real roots could the equation $x^5 - 5x^4 - 3x^3 + 2x^2 - 3 = 0$ have?

[A] 2 or 0 [B] 3 or 1 [C] 4 or 2 or 0
[D] 0 [E] 5 or 3 or 1

42 If $f_{n+1} = f_{n-1} + 2 \cdot f_n$ for $n = 2, 3, 4, \ldots$, and $f_1 = 1$ and $f_2 = 1$, then f_5 equals

[A] 7 [B] 41 [C] 11 [D] 21 [E] 17

43 The plane $2x + 3y - 4z = 5$ intersects the x-axis at $(a, 0, 0)$, the y-axis at $(0, b, 0)$, and the z-axis at $(0, 0, c)$. The value of $a + b + c$ is

[A] 5 [B] $\frac{35}{12}$ [C] $\frac{65}{12}$ [D] 1 [E] 9

44 In the figure, S is the set of all points in the shaded region. Which of the following represents the set consisting of all points $(2x, y)$, where (x, y) is a point in S?

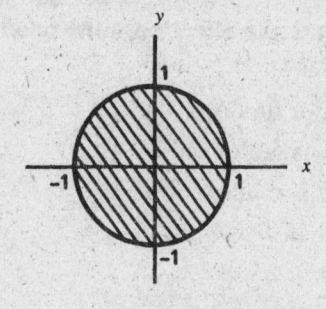

[C]

[D]

[E]

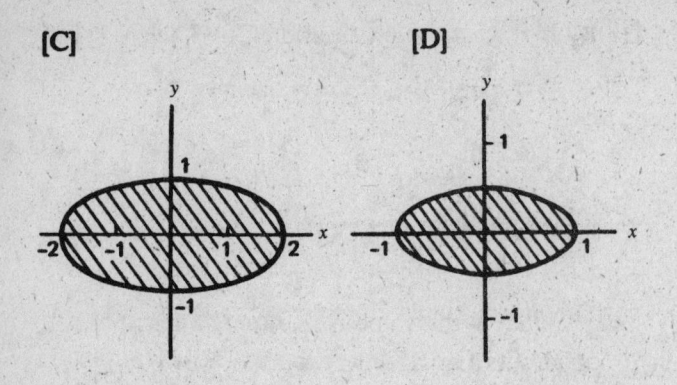

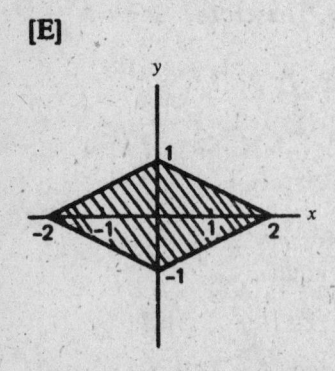

45 If a square prism is inscribed in a right circular cylinder of radius r and height h, the volume inside the cylinder but outside the prism is

[A] $2\pi rh - r^2h$ [B] $(\pi-1)r^2h$
[C] $\pi r^2h - \frac{1}{2}r^2h$ [D] $2\pi rh - \frac{1}{2}r^2h$
[E] $(\pi-2)r^2h$

46 If y varies jointly as x, w, and the square of z, what is the effect on w when x, y, and z are doubled?

[A] w is doubled
[B] w is multiplied by 4
[C] w is multiplied by 8
[D] w is divided by 4
[E] w is divided by 2

47 Given the set of data 1, 1, 2, 2, 2, 3, 3, 4. Which one of the following statements is true?

[A] mean ⩽ median ⩽ mode
[B] median ⩽ mean ⩽ mode
[C] median ⩽ mode ⩽ mean
[D] mode ⩽ mean ⩽ median
[E] The relationship cannot be determined because the median cannot be calculated.

48 Given the statement "All girls play tennis," which of the following negates this statement?

[A] All boys play tennis.
[B] Some girls play tennis.
[C] All boys do not play tennis.
[D] At least one girl doesn't play tennis.
[E] All girls do not play tennis.

49 In a plane, the *homogeneous coordinates* of a point P, whose rectangular coordinates are (x,y), are any three numbers a, b, and c for which $\frac{a}{c} = x$ and $\frac{b}{c} = y$. If the coordinates of P are $(3,4)$ and a, b, and c are integers, then the sum of a, b, and c could be

[A] 2   [B] 5   [C] 8   [D] 11   [E] 14

50 If $[x]$ is defined to represent the greatest integer less than or equal to x, and $f(x) = \left| x - [x] - \frac{1}{2} \right|$, what is the period of $f(x)$?

[A] 1   [B] $\frac{1}{2}$ [C] 2   [D] 4

[E] f is not a periodic function

ANSWER KEY

1. A	18. C	35. B
2. E	19. B	36. C
3. D	20. E	37. D
4. B	21. C	38. E
5. C	22. A	39. C
6. A	23. C	40. D
7. D	24. A	41. B
8. C	25. D	42. E
9. B	26. A	43. B
10. D	27. C	44. C
11. B	28. A	45. E
12. B	29. B	46. D
13. C	30. A	47. C
14. C	31. D	48. D
15. E	32. B	49. C
16. D	33. C	50. A
17. E	34. C	

ANSWER EXPLANATIONS

The following explanations to questions included on the diagnostic test are keyed to the review portions of this book. For example, material covered in question 1 is reviewed in sections 1.2 and 2.2. If you have had trouble answering any of these questions, be sure to check the review topics indicated.

1. **A** $f(1)=2$ means the line goes through the point $(1,2)$. $f(2)=q$ means that the line goes through the point $(2,q)$. $-2=\text{slope}=\dfrac{\Delta y}{\Delta x}=\dfrac{q-2}{2-1}$ implies $-2=\dfrac{q-2}{1}$, and $q=0$. [2.2, 1.2].

2. **E** $\sin 120°=\dfrac{\sqrt{3}}{2}$, $\cos\dfrac{2\pi}{3}=-\dfrac{1}{2}$, $\tan 315°=-1$. [3.3].

3. **D** Since $\sin x \neq \sin(-x)$, [D] is not an even function. [1.4].

4. **B** Since the radius of a sphere is the distance between the center and a point on the surface, $r=\sqrt{(2-0)^2+(3-0)^2+(6-0)^2}=7$. [2.2, 5.5].

5. **C** f is a parabola symmetric about y-axis, which opens up. $f(-4)=0$, $f(0)=-8$, $f(4)=0$ are the extreme values. Therefore, maximum occurs at $|-8|=8$. [2.3, 4.3].

6. **A** The terms are $2+3+\dfrac{9}{2}+\dfrac{27}{4}+\dfrac{81}{8}=26\dfrac{3}{8}$. [5.4].

7. **D** $f(f(x))=\dfrac{k}{f(x)}=k\div\dfrac{k}{x}=k\cdot\dfrac{x}{k}=x$. Since the k's divide out, k can equal any real number except zero (since you cannot divide by zero). [1.2].

8. **C** $\log 9 = \log 3^2 = 2\log 3$. Therefore, $\dfrac{\log 9}{\log 3}=2$. [4.2].

9. **B** The problem implies that when $P(x)$ is divided by $(x-1)$, the remainder is $R(x)$. The Remainder Theorem states that $R(x)=P(1)=-1$. [2.4].

10. **D** $\dfrac{3x^2-3}{x-1}=\dfrac{3(x+1)(x-1)}{(x-1)}=3(x+1)$. Since $\lim_{x\to 1} 3(x+1)=6$, k must equal 6 when $x=1$. [4.5, 2.4].

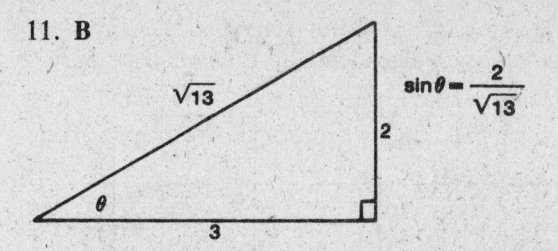

11. **B**

$\sin\theta=\dfrac{2}{\sqrt{13}}$

Therefore, $\sin\theta=\pm\dfrac{2\sqrt{13}}{13}$ since θ could be in quadrants I or III where $\tan\theta$ is positive. [3.1].

12. **B** $s=r\cdot\theta^R$, $67\dfrac{1}{2}°=\dfrac{3\pi}{8}$. Therefore, $24\pi=r\cdot\dfrac{3\pi}{8}$. $r=64$. [3.2].

13. **C** $f(1,2)=2(1^2)-2^2=-2$. $g(f(1,2))=g(-2)=2^{-2}=\dfrac{1}{4}$. [1.2, 1.5, 4.2].

14. **C** $\cos^4 x - \sin^4 x = (\cos^2 x+\sin^2 x)(\cos^2 x-\sin^2 x) = 1\cdot(\cos 2x)$. Amplitude $=1$. [3.4].

15. **E** All three have a period of $\dfrac{\pi}{2}$. I. It is a normal sine curve. II. Phase shift $\dfrac{\pi}{8}$. Cosine curve fits. III. Phase shift $\dfrac{\pi}{4}$. Sine curve shifted and reflected about x-axis. Sine curve fits. Therefore, all three are equations of the graph. [3.4].

16. **D** $2(1-\cos^2 x)-3=3\cos x$. $2\cos^2 x+3\cos x+1=0$. $\cos x=-\dfrac{1}{2}$ or -1. Therefore, $x=120°, 240°, 180°$. [3.1].

17. **E** $\tan(A+B)=\tan 315°$. So $\dfrac{\tan A+\tan B}{1-\tan A\cdot\tan B}=-1$, and $\tan A=\dfrac{3}{4}$. Substituting, $\tan B=-7$. Therefore $\text{Arctan}(-7)=B$. [3.6].

18. **C** When 1567 is multiplied by itself successively, the units digit takes on values of 7,9,3,1,7,9,3,1,... Every fourth multiplication ends in 1. Therefore 1567^{92} ends in 1. [5.9].

19. **B** By law of cosines, cosine of largest angle $=-\dfrac{1}{2}$. Therefore, largest angle (opposite side 7) $=120°$. Area $=\dfrac{1}{2}ab\sin C$. Area $=\dfrac{15\sqrt{3}}{4}$. [3.7].

20. **E** Setting up the following table of a few representative values leads to this answer. [5.5].

x	0	0	0	$\frac{1}{2}$	$\frac{1}{2}$	$\frac{1}{2}$	1	1	1
y	0	$\frac{1}{2}$	1	0	$\frac{1}{2}$	1	0	$\frac{1}{2}$	1
$2x-y$	0	$-\frac{1}{2}$	-1	1	$\frac{1}{2}$	0	2	$\frac{3}{2}$	1

21. **C** If the graph is folded about the line $y=x$ to get the inverse, the graph is one half of a parabola opening to the right with vertex at a point with negative x-coordinate and positive y-coordinate. (A) has a negative y-coordinate. (C) is the only possible answer with vertex at $(-3,4)$. [1.3].

22. **A** Add the two equations: $\log_2 m + \log_2 n = x+y$, which becomes $\log_2 mn = x+y$. Basic property of logs: $2^{x+y} = mn$. [4.2].

23. **C** Distance $= \frac{|3\cdot 2 - 4\cdot 4 - 8|}{\sqrt{3^2 + (-4)^2}} = \frac{|-18|}{5} = \frac{18}{5}$. [2.2].

24. **A** x is less than or equal to 5 units from 2. Therefore, $-3 \leq x \leq 7$. [2.5, 4.3].

25. **D** $\sqrt{x^2}$ indicates the need for the *positive* square root of x^2. Therefore, $\sqrt{x^2} = x$ if $x \geq 0$ and $\sqrt{x^2} = -x$ if $x < 0$. This is just the definition of absolute value so $\sqrt{x^2} = |x|$ is the only answer for all values of x. [4.3].

26. **A** If $i \geq 0$, $f(x) = |f(x)|$ and therefore $g(x) = 0$. If $i < 0$, $|f(x)| = -f(x)$ and therefore $g(x) = 2i$, which is negative. Therefore, maximum $= 0$. [4.4].

27. **C** $y = \frac{x^2 - 4}{x^2 - 1}$. Asymptotes when $x = \pm 1$. As $x \to \infty$, $y \to 1$. Therefore, 2 vertical and 1 horizontal asymptotes. [4.5]

28. **A** Divide numerator and denominator through by x^2. As $x \to \infty$, the fraction $\to \frac{3}{6} = \frac{1}{2}$. [4.5]

29. **B** $x \geq 4$ and $y \geq 0$. $y = r^2 + 4 + 4r$. Substitute x for $r^2 + 4$ and $\sqrt{x-4}$ for r. $y = x + 4\sqrt{x-4}$. $(y-x)^2 = 16(x-4)$. $x^2 - 2xy + y^2 - 16x + 64 = 0$. $B^2 - 4AC = 4 - 4\cdot 1\cdot 1 = 0$. Therefore, a parabola. [4.6, 4.1].

30. **A** From the figure $(-1, -\sqrt{3})$. [4.7]

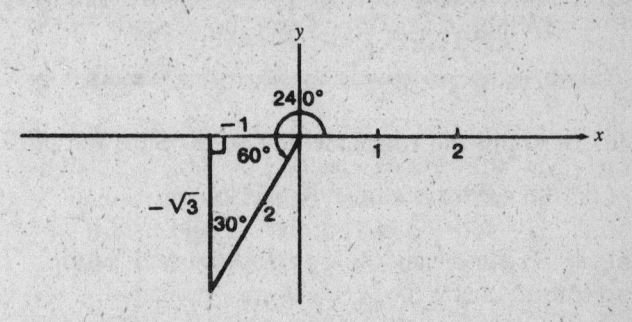

31. **D** By letting $\theta = 0$, [A], [B], and [E] can be ruled out because they do not allow the graph to cross the x-axis at ± 2. Plotting a few points rules out [C]. [4.7]

32. **B** Since $r = h$ in the cone,
$\frac{\text{Volume of cone}}{\text{Volume of sphere}} = \frac{\frac{1}{3}\pi r^2 h}{\frac{4}{3}\pi r^3} = \frac{\frac{1}{3}\pi r^3}{\frac{4}{3}\pi r^3} = \frac{1}{4}$. [5.5].

33. **C** Permutation with repetitions is $\frac{7!}{3!2!} = 420$. [5.1].

34. **C** Combination is $\binom{10}{3} = \frac{10!}{7!3!}$. [5.1].

35. **B** Setting the two equations equal, $3x + k = 2x^2 - 5x + 1$, which is the same as $2x^2 - 8x + (1-k) = 0$. If they are to be tangent there is only one point of intersection and one solution to this equation. Therefore, the discriminant must equal zero. $b^2 - 4ac = (-8)^2 - 4\cdot 2\cdot(1-k) = 0$. Therefore, $k = -7$. [2.3].

36. **C** Fourth term is $\binom{6}{3} \cdot 1^3 \cdot i^3 = 20i^3 = -20i$. [5.2].

37. **D** There are 16 outcomes in the sample space. $\binom{4}{3} = 4$ ways to get 3 heads. $\binom{4}{4} = 1$ way to get 4 heads. [5.3].

38. **E** $\binom{8}{1} = 8$ ways to draw the red ball. $\binom{3}{1} = 3$ ways to draw the white ball. $\binom{11}{2} = 55$ ways to draw any two balls from the urn. Probability of 1 white and 1 red $= \frac{\binom{8}{1}\binom{3}{1}}{\binom{11}{2}} = \frac{24}{55}$. [5.3].

39. **C** Substituting: $4a+2b+c=3$ and $9a+3b+c=3$. Subtract the first from the second. $5a+b=0$. $5a=-b$. Divide by $5b$. $\frac{a}{b}=\frac{-1}{5}$. [2.3].

40. **D** Apply the remainder theorem. Substituting $\frac{1}{2}$ for x gives a remainder of 8. [2.4].

41. **B** By Descartes' Rule of Signs, 3 or 1. [2.4].

42. **E** $f_3=3$, $f_4=7$, $f_5=17$. [5.4].

43. **B** Substituting the points into the equation gives $a=\frac{5}{2}$, $b=\frac{5}{3}$, and $z=-\frac{5}{4}$. [5.5].

44. **C** The y values remain the same but the x values are doubled so the circle is stretched along the x-axis. [5.5].

45. **E** Volume of cylinder $=\pi r^2h$. Volume of prism $=$(area of base)h. Area of base is a square whose side is $r\sqrt{2}$. Volume of prism$=2r^2h$. [5.5].

46. **D** $\frac{y}{xwz^2}=K$. The doubling of y and x cancel each other out. The doubling of z is squared so w has to be divided by 4. [5.6].

47. **C.** mode $=2$, median $=\frac{2+2}{2}=2$, mean $=\frac{2+6+6+4}{8}=\frac{18}{8}=2.25$.
    Thus, median $\leq$ mode $\leq$ mean [5.8]

48. **D** "All G are T" is negated by "Some G are not T." [5.7].

49. **C** $\frac{a}{c}=3$ and $\frac{b}{c}=4$. Therefore $a=3c$ and $b=4c$. If $c=1$, then $a=3$ and $b=4$. [5.9].

50. **A** Plotting a few points, the graph looks like this. Therefore, the period is 1. [4.4, 4.3].

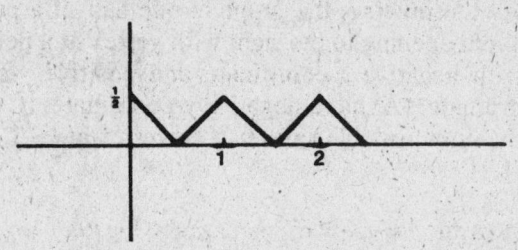

REVIEW OF MAJOR TOPICS

SECTION 2

INTRODUCTION TO FUNCTIONS

CHAPTER 1

1. DEFINITION

A *relation* is a set of ordered pairs. A *function* is a relation such that for each first element there is one and only one second element. The set of numbers which make up all first elements of the ordered pairs is called the *domain* of the function, and the resulting set of second elements is called the *range* of the function.

EXAMPLE 1: $\{(1,2),(3,4),(5,6),(6,1),(2,2)\}$
This is a function because every ordered pair has a different first element. Domain = $\{1,2,3,5,6\}$ Range = $\{1,2,4,6\}$

EXAMPLE 2: $f(x)=3x+2$
This is a function because for each value substituted for x there is one and only one value for $f(x)$. Domain = {all real numbers} Range = {all real numbers}

EXAMPLE 3: $\{(1,2),(3,2),(1,4)\}$
This is a relation but *not* a function because when the first element is 1 the second element could be either 2 or 4. Domain = $\{1,3\}$ Range = $\{2,4\}$

EXAMPLE 4: $\{(x,y): y^2 = x\}$
(This should be read, "The set of all ordered pairs (x,y) such that $y^2 = x$.") This is a relation but *not* a function because for each non-negative number that is substituted for x there are 2 values for y. For example, $x=4$, $y=+2$ or $y=-2$. Domain = {all non-negative real numbers} Range = {all real numbers}

EXERCISES

1. If $\{(3,2),(4,2),(3,1),(7,1),(2,3)\}$ is to be a function, which one of the following must be removed from the set?

   [A] (3,2)   [B] (4,2)
   [C] (2,3)   [D] (7,1)
   [E] none of above

2. $f(x) = 3x^2 + 4$, $g(x) = 2$, and $h = \{(1,1),(2,1),(3,1)\}$

   [A] f is the only function
   [B] h is the only function
   [C] f and g are the only functions
   [D] g and h are the only functions
   [E] f, g, and h are all functions

2. FUNCTION NOTATION

$f=\{(x,y):y=x^2\}$ and $f(x)=x^2$ both name the same function. f is the rule that pairs any number with its square. Thus, $f(x)=x^2$, $f(a)=a^2$, $f(z)=z^2$ all name the same function. The symbol $f(2)$ is the value of the function f when $x=2$. Thus, $f(2)=4$.

If f and g name two functions, the following rules apply:

$$(f+g)(x) = f(x) + g(x)$$
$$(f \cdot g)(x) = f(x) \cdot g(x)$$
$$\frac{f}{g}(x) = \frac{f(x)}{g(x)}$$

if and only if $g(x) \neq 0$

$$(f \circ g)(x) = f(x) \circ g(x) = f[g(x)]$$

(this is called the *composition* of functions)

EXAMPLE: If $f(x)=3x-2$ and $g(x)=x^2-4$, then

$$(f+g)(x) = f(x) + g(x) = (3x-2) + (x^2-4)$$
$$= x^2 + 3x - 6$$

$$(f-g)(x) = f(x) - g(x) = (3x-2) - (x^2-4)$$
$$= -x^2 + 3x + 2$$

$$(f \cdot g)(x) = f(x) \cdot g(x) = (3x-2)(x^2-4)$$
$$= 3x^3 - 2x^2 - 12x + 8$$

$$\frac{f}{g}(x) = \frac{f(x)}{g(x)} = \frac{3x-2}{x^2-4} \text{ and } x \neq \pm 2$$

$$(f \circ g)(x) = f(x) \circ g(x) = f[g(x)] = 3(g(x)) - 2$$
$$= 3(x^2-4) - 2 = 3x^2 - 14$$

$$(g \circ f)(x) = g(x) \circ f(x) = g(f(x)) = (f(x))^2 - 4$$
$$= (3x-2)^2 - 4 = 9x^2 - 12x$$

(notice that the composition of functions is not commutative)

EXERCISES

[3] What value(s) must be excluded from the domain of $f = \left\{ (x,y) : y = \dfrac{x+2}{x-2} \right\}$?

[A] 2 [B] −2 [C] 0
[D] 2 and −2 [E] no value

[1] If $f(x)=3x^2-2x+4$, $f(-2)=$
[A] −2 [B] 20 [C] −4
[D] 12 [E] −12

[2] If $f(x)=4x-5$ and $g(x)=3^x$, then $f(g(2))=$
[A] 27 [B] 9 [C] 3 [D] 31
[E] none of these

[3] If $f(g(x))=4x^2-8x$ and $f(x)=x^2-4$, then $g(x)=$
[A] $2x-2$ [B] x [C] $4x$
[D] $4-x$ [E] x^2

[4] What values must be excluded from the domain of $\dfrac{f}{g}(x)$ if $f(x)=3x^2-4x+1$ and $g(x)=3x^2-3$?
[A] no values [B] 0 [C] 3
[D] 1 [E] both ± 1

[5] If $g(x)=3x+2$ and $g(f(x))=x$, then $f(2)=$
[A] 2 [B] 6 [C] 0 [D] 8 [E] 1

[6] If $p(x)=4x-6$ and $p(a)=0$, then $a=$
[A] −6 [B] 2 [C] $\dfrac{3}{2}$
[D] $\dfrac{2}{3}$ [E] $-\dfrac{3}{2}$

3. INVERSE FUNCTIONS

The *inverse* of a function f, denoted by f^{-1}, is a relation which has the property that $f(x) \circ f^{-1}(x) = f^{-1}(x) \circ f(x) = x$. f^{-1} is not necessarily a function.

EXAMPLE 1: $f(x)=3x+2$. Find the inverse.

$$f^{-1}(x) = \frac{x-2}{3}$$

To verify this:

$$f(x) \circ f^{-1}(x) = f(f^{-1}(x))$$
$$= f\left(\frac{x-2}{3}\right) = 3\left(\frac{x-2}{3}\right) + 2 = x$$

and

$$f^{-1}(x) \circ f(x) = f^{-1}(f(x)) = f^{-1}(3x+2)$$
$$= \frac{(3x+2)-2}{3} = x.$$

In this case $f^{-1}(x)$ is a function.

EXAMPLE 2: $f = \{(1,2),(2,3),(3,2)\}$ **Find the inverse.**
$$f^{-1} = \{(2,1),(3,2),(2,3)\}$$
To verify this, check $f \circ f^{-1}$ and $f^{-1} \circ f$ term by term.

$f \circ f^{-1} = f(f^{-1}(x))$; when $x = 2, f(f^{-1}(2)) = f(1) = 2$
when $x = 3, f(f^{-1}(3)) = f(2) = 3$
when $x = 2, f(f^{-1}(2)) = f(3) = 2$

Thus, for each x, $f(f^{-1}(x)) = x$.

$f^{-1} \circ f = f^{-1} \circ f(x)$; when $x = 1, f^{-1}(f(1)) = f^{-1}(2) = 1$
when $x = 2, f^{-1}(f(2)) = f^{-1}(3) = 2$
when $x = 3, f^{-1}(f(3)) = f^{-1}(2) = 3$

Thus, for each x, $f^{-1}(f(x)) = x$. In this case f^{-1} is *not* a function.

If the point with coordinates (a,b) belongs to a function f, then the point with coordinates (b,a) belongs to the inverse of f. Because this is true of a function and its inverse, the graph of the inverse is the reflection of the graph of f about the line $y = x$.

EXAMPLE 3:
f^{-1} is *not* a function

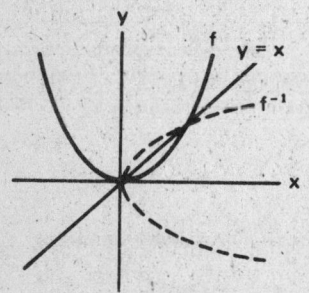

EXAMPLE 4:
f^{-1} is a function

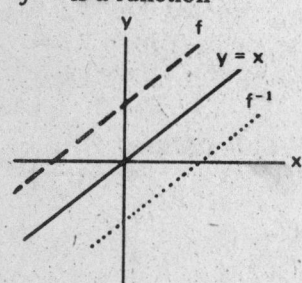

As can be seen from the above examples, if the graph of f is given, the graph of f^{-1} is the image obtained by folding the graph of f about the line $y = x$. Algebraically the equation of the inverse of f can be found by interchanging the variables.

EXAMPLE 5: $f = \{(x,y): y = 3x + 2\}$. Find f^{-1}.
In order to find f^{-1} interchange the x and y and solve for y: $x = 3y + 2$, which becomes $y = \dfrac{x-2}{3}$.
Thus, $f^{-1} = \{(x,y): y = \dfrac{x-2}{3}\}$

EXAMPLE 6: $f = \{(x,y): y = x^2\}$. Find f^{-1}.
Interchange x and y: $x = y^2$
Solve for $y = y = \pm \sqrt{x}$
Thus, $f^{-1} = \{(x,y): y = \pm \sqrt{x}\}$, which is *not* a function.

The inverse of any function f can always be made a function by limiting the domain of f. In Example 6 above the domain of f could be limited to all non-negative numbers or all non-positive numbers. In this way f^{-1} would become either $y = +\sqrt{x}$ or $y = -\sqrt{x}$, both of which are functions.

EXAMPLE 7: $f = \{(x,y): y = x^2 \text{ and } x \geqslant 0\}$. Find f^{-1}.
$f^{-1} = \{(x,y): x = y^2 \text{ and } y \geqslant 0\}$, which could also be written
$$f^{-1} = \{(x,y): y = +\sqrt{x}\}$$
f^{-1} is a function

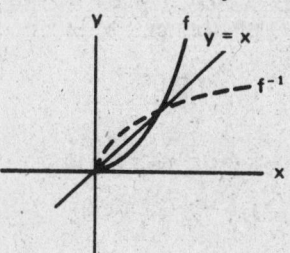

EXAMPLE 8: $f = \{(x,y): y = x^2 \text{ and } x \leqslant 0\}$. Find f^{-1}.
$$f^{-1} = \{(x,y): x = y^2 \text{ and } y \leqslant 0\},$$
which could also be written
$$f^{-1} = \{(x,y): y = -\sqrt{x}\}.$$
f^{-1} is a function

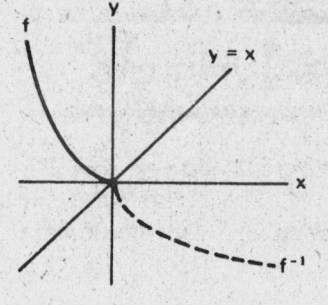

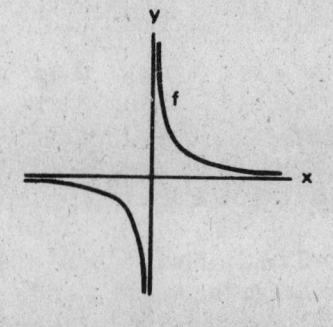

EXERCISES

[1] If $f(x)=2x-3$, the inverse of f, f^{-1}, could be represented by

[A] $f^{-1}(x)=3x-2$
[B] $f^{-1}(x)=\dfrac{1}{2x-3}$
[C] $f^{-1}(x)=\dfrac{x-2}{3}$
[D] $f^{-1}(x)=\dfrac{x+2}{3}$
[E] $f^{-1}(x)=\dfrac{x+3}{2}$

[2] If $f(x)=x$, the inverse of f, f^{-1}, could be represented by

[A] $f^{-1}(x)=x$ [B] $f^{-1}(x)=1$
[C] $f^{-1}(x)=\dfrac{1}{x}$ [D] $f^{-1}(x)=y$
[E] f^{-1} does not exist

[3] The inverse of $f=\{(1,2),(2,3),(3,4),(4,1),(5,2)\}$ would be a function if the domain of f is limited to

[A] $\{1,3,5\}$ [B] $\{1,2,3,4\}$ [C] $\{1,5\}$
[D] $\{1,2,4,5\}$ [E] $\{1,2,3,4,5\}$

[4] Which of the following could represent the equation of the inverse of the graph in the figure?

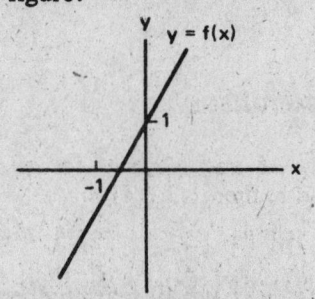

[A] $y=-2x+1$ [B] $y=2x+1$
[C] $y=\dfrac{1}{2}x+1$ [D] $y=\dfrac{1}{2}x-1$
[E] $y=\dfrac{1}{2}x-\dfrac{1}{2}$

4. ODD AND EVEN FUNCTIONS

A function (or a relation) is said to be *even* if it is symmetric about the y-axis. This occurs whenever $f(x)=f(-x)$ for all real numbers x in the domain of f.

EXAMPLE 1: $f(x)=x^2$ and $f(-x)=(-x)^2=x^2$

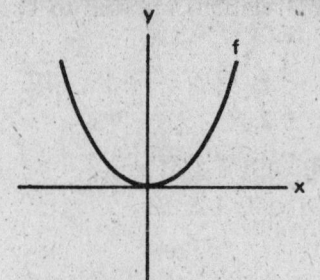

EXAMPLE 2: $f(x)=|x|$ and $f(-x)=|-x|=|-1\cdot x|=|-1|\cdot|x|=|x|$.

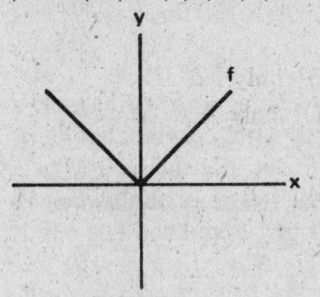

A function (or a relation) is said to be *odd* if it is symmetric about the origin. (For example, if the graph looks the same after it has been rotated 180° about the origin.) This occurs whenever $f(x)=-f(-x)$ for all real numbers in the domain of f.

EXAMPLE 3: $f(x)=x^3$ and $f(-x)=(-x)^3=-x^3$. Therefore, $-f(-x)=x^3=f(x)$.

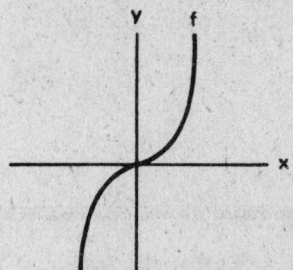

EXAMPLE 4: $f(x)=\dfrac{1}{x}$ and $f(-x)=\dfrac{1}{-x}=-\left(\dfrac{1}{x}\right)$. Therefore, $-f(-x)=\dfrac{1}{x}=f(x)$.

EXERCISES

1 Which of the following relations are said to be *even*?
  I. $y = 2$
  II. $y = x$
  III. $x^2 + y^2 = 1$

  [A] only I [B] only I & II
  [C] only II & III [D] only I & III
  [E] I, II, & III

2 Which of the following relations are said to be *odd*?
  I. $y = 2$
  II. $y = x$
  III. $x^2 + y^2 = 1$

  [A] only II [B] only I & II
  [C] only I & III [D] only II & III
  [E] I, II, & III

3 Which of the following relations are said to be both *odd* and *even*?
  I. $x^2 + y^2 = 1$
  II. $x^2 - y^2 = 0$
  III. $x + y = 0$

  [A] III only [B] I & II only
  [C] I & III only [D] II & III only
  [E] I, II, & III

4 Which of the following functions is neither *odd* nor *even*?
  [A] $\{(1,2), (4,7), (-1,2), (0,4), (-4,7)\}$
  [B] $\{(1,2), (4,7), (-1,-2), (0,0), (-4,-7)\}$
  [C] $\{(x,y) : y = x^3 - 1\}$
  [D] $\{(x,y) : y = x^2 - 1\}$
  [E] $f(x) = -x$

5. MULTI-VARIABLE FUNCTIONS

At times it is necessary or convenient to express a function or relation in terms of more than one variable. For instance, the area of a triangle is expressed in terms of its base, b, and its altitude, h. Thus, in function notation, $A(b,h) = \frac{1}{2} bh$, where b and h are the independent variables. The formula for the surface area of a rectangular solid expresses the surface area in terms of three independent variables, l, w, and h. $S(l,w,h) = 2lw + 2lh + 2wh$.

EXAMPLE 1: If $f(x,y,z) = (x+y)(y+z)$, what does $f(1,-2,3) = ?$
Substituting 1 for x, -2 for y, and 3 for z,
$$f(1,-2,3) = (1-2)(-2+3) = (-1)(1) = -1.$$

EXAMPLE 2: If $f(x,y) = 2x + 3y - 4$, what does $f(2x, -3y+1) = ?$
Substituting $2x$ for x and $-3y + 1$ for y,
$$f(2x, -3y+1) = 2(2x) + 3(-3y+1) - 4$$
$$= 4x - 9y - 1.$$

EXAMPLE 3: If $f(x,y) = x^2 + y^2 + y$ and $g(z) = z - 2$, what does $f(2, g(1)) = ?$
Substituting 2 for x and $g(1) = 1 - 2 = -1$ for y,
$$f(2, g(1)) = 2^2 + (g(1))^2 + g(1)$$
$$= 4 + (-1)^2 + (-1) = 4.$$

EXERCISES

1 If $f(x,y) = 3x + 2y - 8$ and $g(z) = z^2$ for all real numbers x, y, and z, then $f(3, g(4)) =$

  [A] 9 [B] 33 [C] 81 [D] 7 [E] 5

2 If $f(x,y,z) = x^2 + y^2 - 2z$ for all real numbers x, y, and z, then $f(-1, 1, -1) =$

  [A] 0 [B] 2 [C] 4 [D] -2 [E] 1

3 If $f(x,y) = 3x + 2y$ and $g(x,y) = x^2 - y^2$ for all real numbers x and y, then $f(g(1,2), 3) =$

  [A] 40 [B] -3 [C] 9 [D] 6 [E] 3

POLYNOMIAL FUNCTIONS

CHAPTER 2

1. DEFINITION

A *polynomial* is an algebraic expression of the form

$$a_nx^n + a_{n-1}x^{n-1} + \cdots + a_1x + a_0$$

where x is a variable, n is a non-negative integer, and the coefficients $a_n, a_{n-1}, \ldots, a_1, a_0$ are complex numbers. If the coefficients are all real numbers, it is called a real polynomial. If the coefficients are all rational numbers, it is called a rational polynomial.

2. LINEAR FUNCTIONS

Linear functions are polynomials with the largest exponent a 1. The graph is always a straight line. Although the general form of the equation is $Ax + By + C = 0$ where A, B, and C are constants, the most useful form occurs when it is solved for y. This is known as the *slope-intercept* form and is written $y = mx + b$. The slope of the line is represented by m and is defined to be the ratio of $\dfrac{y_1 - y_2}{x_1 - x_2}$, where (x_1, y_1) and (x_2, y_2) are any two points on the line. The y-intercept is b (the point where the graph crosses the y-axis).

Parallel lines have the same slope. The slopes of two perpendicular lines are negative reciprocals of one another.

EXAMPLE 1: The equation of line l_1 is $y = 2x + 3$ and the equation of line l_2 is $y = 2x - 5$.
These lines are parallel because the slope of each line is 2, and the y-intercepts are different.

EXAMPLE 2: The equation of line l_1 is $y = \dfrac{5}{2}x - 4$ and the equation of line l_2 is $y = -\dfrac{2}{5}x + 9$.
These lines are perpendicular because the slope of l_2, $-\dfrac{2}{5}$, is the negative reciprocal of the slope of l_1, $\dfrac{5}{2}$.

The distance between two points P and Q whose coordinates are (x_1, y_1) and (x_2, y_2) is given by the formula:

$$\text{Distance} = \sqrt{(x_1 - x_2)^2 + (y_1 - y_2)^2}$$

and the midpoint, M, of the segment $\overline{PQ}$ has coordinates $\left(\dfrac{x_1 + x_2}{2}, \dfrac{y_1 + y_2}{2}\right)$

2. Linear Functions

EXAMPLE 3: Given the point $(2,-3)$ and the point $(-5,4)$, find the length of $\overline{PQ}$ and the coordinates of the midpoint, M.

$$PQ = \sqrt{(2-(-5))^2+(-3-4)^2} = \sqrt{(7)^2+(-7)^2}$$
$$= \sqrt{98} = 7\sqrt{2}\,.$$
$$M = \left(\frac{2+(-5)}{2},\frac{-3+4}{2}\right) = \left(\frac{-3}{2},\frac{1}{2}\right)$$

The perpendicular distance between a line $Ax+By+C=0$ and a point $P(x_1,y_1)$ not on the line is given by the formula:

$$\text{Distance} = \frac{|Ax_1+By_1+C|}{\sqrt{A^2+B^2}}$$

The angle, θ, between two lines, ℓ_1 and ℓ_2, can be found by using the formula:

$$\text{Tan}\,\theta = \frac{m_1-m_2}{1+m_1m_2}$$

where m_1 is the slope of ℓ_1, and m_2 is the slope of ℓ_2. If $\tan\theta > 0$, θ is the acute angle formed by the two lines. If $\tan\theta < 0$, θ is the obtuse angle formed by the two lines.

EXAMPLE 4: Find the distance between the line, $3x+4y=5$ and the origin.

$$d = \frac{|3\cdot 0+4\cdot 0-5|}{\sqrt{9+16}} = \frac{5}{\sqrt{25}} = 1.$$

EXAMPLE 5: Find the tangent of the angle formed by the lines $2x+3y+5=0$ and $3x-5y+8=0$.
Putting these two equations in the slope-intercept form, the slope of $2x+3y+5=0$ is found to be $-\frac{2}{3}$, and the slope of $3x-5y+8=0$ is found to be $\frac{3}{5}$.

$$\text{Tan}\,\theta = \frac{-\frac{2}{3}-\frac{3}{5}}{1+\left(-\frac{2}{3}\right)\left(\frac{3}{5}\right)} = \frac{\left(-\frac{2}{3}-\frac{3}{5}\right)\cdot 15}{\left(1-\frac{2}{5}\right)\cdot 15}$$
$$= \frac{-2\cdot 5-3\cdot 3}{15-2\cdot 3} = \frac{-10-9}{15-6} = \frac{-19}{9}.$$

Therefore, if θ is acute, $\tan\theta = \frac{19}{9}$, and if θ is obtuse, $\tan\theta = \frac{-19}{9}$.

EXERCISES

1 The slope of the line through the points $A(3,-2)$ and $B(-2,-3)$ is

[A] -5 [B] $-\frac{1}{5}$ [C] $\frac{1}{5}$
[D] 1 [E] 5

2 The slope of the line $8x+12y+5=0$ is

[A] 2 [B] $\frac{2}{3}$ [C] 3
[D] $-\frac{3}{2}$ [E] $-\frac{2}{3}$

3 The slope of the line perpendicular to the line $3x-5y+8=0$ is

[A] $\frac{3}{5}$ [B] $\frac{5}{3}$ [C] $-\frac{3}{5}$
[D] $-\frac{5}{3}$ [E] 3

4 The y-intercept of the line through the two points whose coordinates are $(5,-2)$ and $(1,3)$ is

[A] $\frac{5}{4}$ [B] $-\frac{5}{4}$ [C] 17
[D] $\frac{17}{4}$ [E] 7

5 The equation of the perpendicular bisector of the segment joining the points whose coordinates are $(1,4)$ and $(-2,3)$ is

[A] $3x-2y+5=0$
[B] $x-3y+2=0$
[C] $3x+y-2=0$
[D] $x-3y+11=0$
[E] $x+3y-10=0$

6 The length of the segment joining the points with coordinates $(-2,4)$ and $(3,-5)$ is

[A] $2\sqrt{2}$ [B] $\sqrt{106}$ [C] $\sqrt{14}$
[D] 10 [E] none of these

7 The slope of the line parallel to the line whose equation is $2x+3y=8$ is

[A] $\frac{2}{3}$
[B] $-\frac{2}{3}$
[C] -2
[D] $-\frac{3}{2}$
[E] $\frac{3}{2}$

8 If the point $P(m,2m)$ is 5 units from the line $12x+5y=1$, m could equal

[A] $\frac{43}{11}$ [B] -3 [C] $-\frac{65}{22}$
[D] 5 [E] 3

[9] If θ is the angle between the lines, $2x-3y+4=0$ and $2x-y-3=0$, $\tan\theta$ could equal

[A] $-\dfrac{8}{7}$ [B] -8 [C] 4

[D] $\dfrac{4}{7}$ [E] $-\dfrac{2}{3}$

3. QUADRATIC FUNCTIONS

Quadratic functions are polynomials with the largest exponent being a 2. The graph is always a parabola. The general form of the equation is $y = ax^2 + bx + c$. If $a > 0$, the parabola opens up and has a minimum value. If $a < 0$, the parabola opens down and has a maximum value. The x-coordinate of the vertex of the parabola is equal to $-\dfrac{b}{2a}$, and the axis of symmetry is the vertical line whose equation is $x = -\dfrac{b}{2a}$. In order to find the minimum (or maximum) value of the function, substitute $-\dfrac{b}{2a}$ for x to determine y. Thus, in the general case the coordinates of the vertex are $\left(-\dfrac{b}{2a}, c - \dfrac{b^2}{4a}\right)$ and the minimum (or maximum) value of the function is $c - \dfrac{b^2}{4a}$. Unless specifically limited the domain of a quadratic function is all real numbers, and the range is all values of y greater than or equal to the minimum value (or all values of y less than or equal to the maximum value) of the function.

EXAMPLE 1: Determine the coordinates of the vertex and the equation of the axis of symmetry of $y = 3x^2 + 2x - 5$. Does the quadratic function have a minimum or maximum value? What is it?
Axis of symmetry is

$$x = -\dfrac{b}{2a} = -\dfrac{2}{2\cdot 3} = -\dfrac{1}{3}$$

y-coordinate of the vertex is

$$y = 3\left(-\dfrac{1}{3}\right)^2 + 2\left(-\dfrac{1}{3}\right) - 5 = \dfrac{1}{3} - \dfrac{2}{3} - 5 = -5\dfrac{1}{3}$$

Vertex is, therefore, at $\left(-\dfrac{1}{3}, -5\dfrac{1}{3}\right)$. The function has a minimum value because $a = 3 > 0$. The minimum value is $-5\dfrac{1}{3}$. The graph of $y = 3x^2 + 2x - 5$ is shown below.

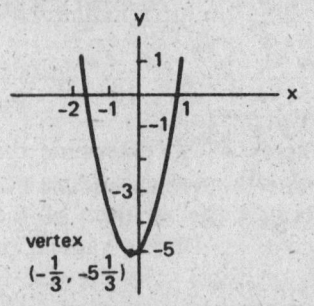

The points where the graph crosses the x-axis are called the *zeros* of the function and occur when $y = 0$. To find the zeros of $y = 3x^2 + 2x - 5$, solve the quadratic equation $3x^2 + 2x - 5 = 0$. The roots of this equation will be the zeros of the polynomial. Often the quickest way to do this is to factor the polynomial. However, if it is not immediately obvious how to do the factoring, substitute the coefficients into the general quadratic formula.

Every quadratic equation can be changed into the form $ax^2 + bx + c = 0$ (if it is not already in that form), which can be solved by completing the square. The solutions are $x = \dfrac{-b \pm \sqrt{b^2 - 4ac}}{2a}$, the *General Quadratic Formula*. In the case of $3x^2 + 2x - 5 = 0$, factoring is the best method to use. $3x^2 + 2x - 5 = (3x + 5)(x - 1) = 0$. Thus, $3x + 5 = 0$ or $x - 1 = 0$, which leads to $x = -\dfrac{5}{3}$ or 1. The zeros of the polynomials are $-\dfrac{5}{3}$ and 1.

EXAMPLE 2: Find the zeros of $y = 2x^2 + 3x - 4$.
Solve the equation $2x^2 + 3x - 4 = 0$. This does not factor easily. Using the general quadratic formula where $a = 2$, $b = 3$, and $c = -4$, $x = \dfrac{-3 \pm \sqrt{9 + 32}}{4} = \dfrac{-3 \pm \sqrt{41}}{4}$. Thus, the zeros are $\dfrac{-3 + \sqrt{41}}{4}$ and $\dfrac{-3 - \sqrt{41}}{4}$.

It is interesting to note that the sum of the two zeros, $\dfrac{-b + \sqrt{b^2 - 4ac}}{2a}$ and $\dfrac{-b - \sqrt{b^2 - 4ac}}{2a}$, equals $-\dfrac{b}{a}$, and their product equals $\dfrac{c}{a}$. This information can be used to check whether the correct zeros have been found. In Example 2 above, the sum and product of the zeros can be determined by inspection from the equation: Sum $= -\dfrac{3}{2}$ and Product = $\frac{-4}{2}$ = −2. Adding the zeros $\frac{-3+\sqrt{41}}{4}$ and $\frac{-3-\sqrt{41}}{4}$ gives $\frac{-6}{4} = -\frac{3}{2}$. Multiplying the zeros $\frac{-3+\sqrt{41}}{4}$ and $\frac{-3-\sqrt{41}}{4}$ gives $\frac{9-41}{16} = \frac{-32}{16}$ = −2. Thus, the zeros are correct.

At times it is only necessary to determine the *nature* of the roots of a quadratic equation and not the roots themselves. Because the b^2-4ac of the general quadratic formula is under the radical, it determines the nature of the roots. The b^2-4ac is called the *discriminant* of a quadratic equation.

(i) If $b^2-4ac=0$, the two roots become $\frac{-b+0}{2a}$ and $\frac{-b-0}{2a}$, which are the same, and the graph of the function is tangent to the x-axis.

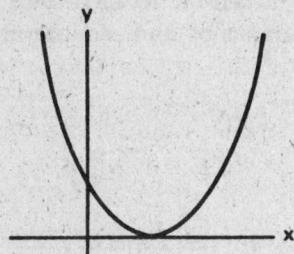

(ii) If $b^2-4ac<0$ there is a negative number under the radical, which gives two complex numbers (of the form $p+qi$ and $p-qi$ where $i=\sqrt{-1}$) as roots, and the graph of the function does not intersect the x-axis.

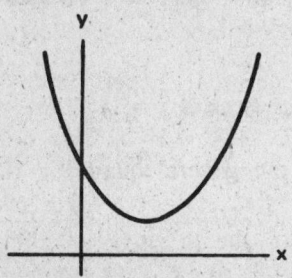

(iii) If $b^2-4ac>0$ there is a positive number under the radical, which gives two different real roots, and the graph of the function intersects the x-axis in two points.

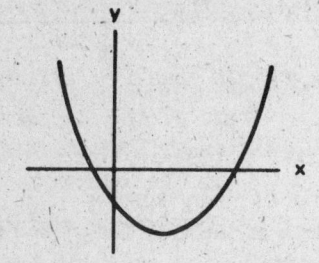

EXERCISES

1. The coordinates of the vertex of the parabola whose equation is $y=2x^2+4x-5$ are

   [A] (2,11)   [B] (−1,−7)   [C] (1,1)
   [D] (−2,−5) [E] (−4,11)

2. The range of the function $f=\{(x,y):y=5-4x-x^2\}$ is

   [A] $\{y:y\leq 0\}$ [B] $\{y:y\geq -9\}$
   [C] $\{y:y\leq 9\}$ [D] $\{y:y\geq 0\}$
   [E] $\{y:y\leq 1\}$

3. The equation of the axis of symmetry of the function $y=2x^2+3x-6$ is

   [A] $x=\frac{1}{3}$ [B] $x=-\frac{3}{4}$ [C] $x=-\frac{1}{3}$
   [D] $x=\frac{3}{4}$ [E] $x=-\frac{3}{2}$

4. Find the zeros of $y=2x^2+x-6$.

   [A] 3 & 2   [B] −3 & 2   [C] $\frac{1}{2}$ & $\frac{3}{2}$
   [D] $-\frac{3}{2}$ & 1   [E] $\frac{3}{2}$ & −2

5. The sum of the zeros of $y=3x^2-6x-4$ is

   [A] $\frac{4}{3}$ [B] 2   [C] 6   [D] −2   [E] $-\frac{4}{3}$

6. $x^2+2x+3=0$ has

   [A] 2 real rational roots
   [B] 2 real irrational roots
   [C] 2 equal real roots
   [D] 2 equal rational roots
   [E] 2 complex conjugate roots

4. HIGHER DEGREE POLYNOMIAL FUNCTIONS

Polynomial functions of degree greater than two (largest exponent greater than 2) are usually treated together since there are no simple formulas (like the general quadratic formula) that aid in finding zeros. Facts about the graphs of polynomial functions:

(1) They are always continuous curves. (The graph can be drawn without removing the pencil from the paper.)

(2) If the largest exponent is an even number, both ends of the graph leave the coordinate system either at the top or at the bottom.

EXAMPLES:

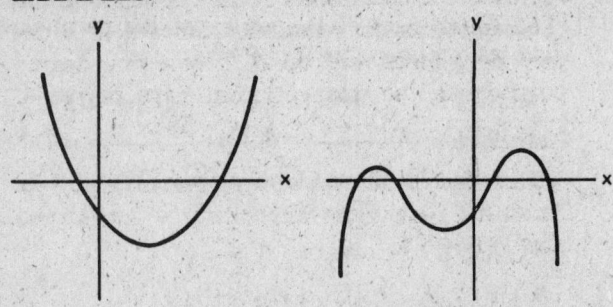

(3) If the largest exponent is an odd number, the ends of the graph leave the coordinate system at opposite ends.

EXAMPLES:

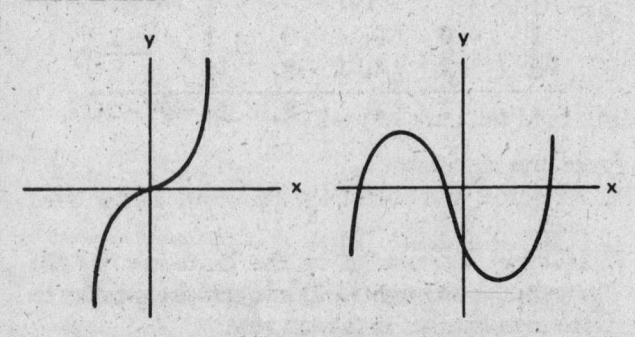

(4) If all exponents are even numbers, the polynomial is an *even function* and is symmetric about the y-axis.

EXAMPLE: $y = 3x^4 + 2x^2 - 8$

(5) If all exponents are odd numbers and there is no constant term, the polynomial is an *odd function* and is symmetric about the origin of the coordinate system.

EXAMPLE: $y = 4x^5 + 2x^3 - 3x$

In order to find the zeros (if possible) of a higher degree polynomial function, $a_n x^n + a_{n-1} x^{n-1} + \cdots + a_1 x + a_0$, set $y = 0$ and attempt to solve the resulting polynomial equation.

Facts useful in solving polynomial equations:

(1) Remainder Theorem—If a polynomial $P(x)$ is divided by $x - r$ (where r is any constant), then the remainder is $P(r)$.
(2) Factor Theorem—r is a zero of the polynomial $P(x)$ if and only if $x - r$ is a divisor of $P(x)$.
(3) Every polynomial of degree n has exactly n zeros.
(4) Rational Zero (Root) Theorem—If $\frac{p}{q}$ is a rational zero (reduced to lowest terms) of a polynomial $P(x)$ with integral coefficients, then p is a factor of a_0 (the constant term) and q is a factor of a_n (the leading coefficient).
(5) If $P(x)$ is a polynomial with rational coefficients, then irrational zeros occur as conjugate pairs. (E.g., if $p + \sqrt{q}$ is a zero, then $p - \sqrt{q}$ is also a zero.)
(6) If $P(x)$ is a polynomial with real coefficients, then complex zeros occur as conjugate pairs. (E.g., if $p + qi$ is a zero, then $p - qi$ is also a zero.)
(7) Descartes' Rule of Signs—The number of positive real zeros of a polynomial $P(x)$ either is equal to the number of variations of the sign between terms or is less than that number by an even integer. The number of negative real zeros of $P(x)$ either is equal to the number of variations of the sign between the terms of $P(-x)$ or is less than that number by an even integer.

EXAMPLE: $P(x) = 18x^4 - 2x^3 + 7x^2 + 8x - 5$

Three sign changes indicate there will be either 3 or 1 positive zeros of $P(x)$.

$P(-x) = 18x^4 + 2x^3 + 7x^2 - 8x - 5$

One sign change indicates there will be exactly 1 negative zero of $P(x)$.

(8) Relation between zeros and coefficients—In a polynomial $P(x)$ with zeros $z_1, z_2, z_3, \ldots, z_n$,

$-\dfrac{a_{n-1}}{a_n}$ = sum of the zeros = $z_1 + z_2 + z_3 + \cdots + z_n$

$\dfrac{a_{n-2}}{a_n}$ = sum of the products of the zeros taken 2 at a time = $z_1 z_2 + z_1 z_3 + \cdots + z_1 z_n + z_2 z_3 + z_2 z_4 + \cdots + z_2 z_n + \cdots z_{n-1} z_n$

$-\dfrac{a_{n-3}}{a_n}$ = sum of the products of the zeros taken 3 at a time = $z_1 z_2 z_3 + z_1 z_2 z_4 + \cdots + z_1 z_2 z_n + z_2 z_3 z_4 + \cdots + z_2 z_3 z_n + \cdots + z_{n-2} z_{n-1} z_n$

$(-1)^k \cdot \dfrac{a_0}{a_n}$ = product of all the zeros = $z_1 z_2 z_3 \ldots z_n$ where k is the number of zeros in each product

EXAMPLE: $P(x) = 18x^4 - 2x^3 + 7x^2 + 8x - 5$ has four zeros.

$z_1 + z_2 + z_3 + z_4 = -\frac{-2}{18} = \frac{1}{9}$

$z_1z_2 + z_1z_3 + z_1z_4 + z_2z_3 + z_2z_4 + z_3z_4 = \frac{7}{18}$

$z_1z_2z_3 + z_1z_2z_4 + z_1z_3z_4 + z_2z_3z_4 = -\frac{8}{18} = -\frac{4}{9}$

$z_1z_2z_3z_4 = \frac{-5}{18}$

(9) Synthetic Division—Synthetic division greatly decreases the amount of work necessary when dividing a polynomial by a divisor of the form $x - r$. The easiest way to learn how to use synthetic division is through examples.

EXAMPLE 1: Divide $P(x) = 3x^5 - 4x^4 - 15x^2 - 88x - 12$ by $x - 3$.

Write just the coefficients of $P(x)$, inserting a zero for any missing term. Put the value of r to the right of the row of coefficients.

$$\begin{array}{rrrrrr|r} 3 & -4 & 0 & -15 & -88 & -12 & \underline{3} \\ \downarrow & 9 & 15 & 45 & 90 & 6 & \\ \hline 3 & 5 & 15 & 30 & 2 & -6 = \text{remainder} = P(3) \end{array}$$

Procedure to follow:
1. Bring the first coefficient (3) down to the third row.
2. Multiply the number in the third row (3) by the number on the right (3) and place the product under the -4.
3. Add the -4 and the 9, putting the sum in the third row.
4. Multiply the number in the third row (5) by the number on the right (3) and place the product under the 0.
5. Add the 0 and the 15, putting the sum in the third row.
6. Multiply the number in the third row (15) by the number on the right (3) and place the product under the -15.
7. Add the -15 and the 45, putting the sum in the third row.
8. Multiply the number in the third row (30) by the number on the right (3) and place the product under the -88.
9. Add the -88 and the 90, putting the sum in the third row.
10. Multiply the number in the third row (2) by the number on the right (3) and place the product under the -12.
11. Add the -12 and the 6, putting the sum in the third row. The numbers on the bottom row represent the coefficients of the quotient and the -6 represents the remainder. Since a fifth degree polynomial was divided by a first degree polynomial, the quotient must have degree 4. Therefore, $\frac{3x^5 - 4x^4 - 15x^2 - 88x - 12}{x - 3} = 3x^4 + 5x^3 + 15x^2 + 30x + 2$ with a remainder of -6. From the Remainder Theorem it is determined that $P(3) = -6$.

EXAMPLE 2: Divide $P(x) = x^4 + 8$ by $x + 2$.

Since the divisor should be in the form $x - r$, $x + 2$ is changed to $x - (-2)$. In the synthetic division format zeros must be inserted as coefficients for the x^3, x^2, and x terms that are missing.

$$\begin{array}{rrrrr|r} 1 & 0 & 0 & 0 & 8 & \underline{-2} \\ \downarrow & -2 & 4 & -8 & 16 & \\ \hline 1 & -2 & 4 & -8 & 24 = P(-2) \end{array}$$

Procedure to follow:
1. Bring the first coefficient (1) down to the third row.
2. Multiply the number in the third row by the number on the right (-2) and add the product to the next number in the top row.
3. Repeat step #2 until the third row is filled.

The resulting quotient will be $x^3 - 2x^2 + 4x - 8$ with a remainder of 24. Thus, from the Remainder Theorem $P(-2) = 24$.

(10) Upper and lower bounds on zeros—If $P(x)$ is divided by $x - r$ where r is a positive number, and if the numbers in the third row of the synthetic division are all positive or zero, there is no zero of the polynomial greater than r. Thus, r is an upper bound of all the zeros.

If $P(x)$ is divided by $x - r$ where r is a negative number, and if all the numbers in the third row of the synthetic division are alternately positive and negative (or zero), there is no zero of the polynomial which is less than r. Thus r is a lower bound of all the zeros.

The following are a series of examples using one or more of these ten facts.

EXAMPLE 1: What is the remainder when $3x^3 + 2x^2 - 5x - 8$ is divided by $x + 2$?

Method 1: If $P(x) = 3x^3 + 2x^2 - 5x - 8$, the remainder is given to be $P(-2)$ and $P(-2) = 3(-2)^3 + 2(-2)^2 - 5(-2) - 8 = -24 + 8 + 10 - 8 = -14$. Therefore, the remainder $= -14$.

Method 2: Using synthetic division,

$$\begin{array}{rrrr|r} 3 & 2 & -5 & -8 & \underline{-2} \\ \downarrow & -6 & 8 & -6 & \\ \hline 3 & -4 & 3 & -14 = P(-2) \end{array}$$

Therefore, the remainder $= -14$.

EXAMPLE 2: Is $x-99$ a factor of $P(x) = x^4 - 100x^3 + 97x^2 + 200x - 197$?

If $x-99$ is a factor, the remainder upon division by $x-99$ will be zero. It would be very tedious to determine whether $P(99) = 0$ directly (as in Method 1 above), but synthetic division is quite easy.

$$\begin{array}{rrrrr|r} 1 & -100 & 97 & 200 & -197 & \underline{99} \\ \downarrow & 99 & -99 & -198 & 198 & \\ \hline 1 & -1 & -2 & 2 & 1 = P(99) \end{array}$$

Therefore, $x-99$ is not a factor of $P(x)$.

EXAMPLE 3: If $3+2i$, 2, and $2-3i$ are all zeros of $P(x) = 3x^5 - 36x^4 + 2x^3 - 8x^2 + 9x - 338$, what are the other zeros?

Since $P(x)$ must have five zeros because it is a fifth degree polynomial, and since complex zeros come in conjugate pairs, the two remaining zeros must be $3-2i$ and $2+3i$.

EXAMPLE 4: What are all the possible rational zeros of $P(x) = 3x^3 + 2x^2 + 4x - 6$?

Note that this question does not ask what the zeros are, only which rational numbers might be zeros. Using the Rational Root Theorem, the rational roots, $\frac{p}{q}$, are such that p is a factor of 6 and q is a factor of 3. Thus, $p \in \{\pm 1, \pm 2, \pm 3, \pm 6\}$ and $q \in \{\pm 1, \pm 3\}$. Forming all possible fractions

$$\frac{p}{q} \in \left\{ \pm 1, \pm 2, \pm 3, \pm 6, \pm \frac{1}{3}, \pm \frac{2}{3} \right\}.$$

Thus, these 12 numbers are the only possible rational numbers that could be zeros of $P(x)$. It could turn out that none of them actually is a zero, meaning that the three zeros are irrational and/or complex numbers.

EXAMPLE 5: How many positive real zeros and how many negative real zeros could you expect to find for the polynomial $P(x) = 3x^3 + 2x^2 + 4x - 6$?

Since there is only one sign change in $P(x)$, there will be exactly one positive real zero.

$$P(-x) = \underbrace{-3x^3 + 2x^2}_{1} \underbrace{- 4x - 6}_{2}$$

Since there are two sign changes in $P(-x)$, there will be either two negative real zeros or no negative real zeros. Since $P(x)$ is a third degree polynomial and has three zeros, there will be either one positive real zero and 2 negative real zeros, or one positive real zero and 2 complex conjugate zeros.

EXAMPLE 6: What is the smallest positive integer which is an upper bound of the zeros of $P(x) = 3x^3 + 2x^2 + 4x - 6$?

Use synthetic division dividing by successive integers:

$$\begin{array}{rrrr|r} 3 & 2 & 4 & -6 & \underline{1} \\ \downarrow & 3 & 5 & 9 & \\ \hline 3 & 5 & 9 & 3 = P(1) \end{array}$$

Since all the numbers in the third row are positive, there are no zeros greater than 1. Therefore, 1 is the smallest positive integer which is an upper bound for the zeros of $P(x)$.

EXAMPLE 7: Find the zeros of $P(x) = x^6 - x^5 - 4x^4 - x^3 + 5x^2 + 8x + 4$.

The only possible rational zeros are $\pm 1, \pm 2, \pm 4$. By Descartes' Rule of Signs there are 2 or 0 positive real zeros, and 4, 2, or 0 negative real zeros. Using synthetic division, divide by $x-1$:

$$\begin{array}{rrrrrrr|r} 1 & -1 & -4 & -1 & 5 & 8 & 4 & \underline{1} \\ \downarrow & 1 & 0 & -4 & -5 & 0 & 8 & \\ \hline 1 & 0 & -4 & -5 & 0 & 8 & 12 = P(1) \end{array}$$

Thus, $x-1$ is not a factor of $P(x)$ and 1 is not a zero. Divide by $x-2$.

$$\begin{array}{rrrrrrr|r} 1 & -1 & -4 & -1 & 5 & 8 & 4 & \underline{2} \\ \downarrow & 2 & 2 & -4 & -10 & -10 & -4 & \\ \hline 1 & 1 & -2 & -5 & -5 & -2 & 0 = P(2) \end{array}$$

Thus, $x-2$ is a factor, and 2 is a zero. $P(x)$ can be factored into $(x-2)Q_1(x)$ where $Q_1(x) = x^5 + x^4 - 2x^3 - 5x^2 - 5x - 2$ using the numbers in the third row of the synthetic division. Since all future factors of $P(x)$ must also be factors of $Q_1(x)$, it is sufficient to search for factors (and zeros) of $Q_1(x)$. Since $Q_1(x)$ was obtained from a polynomial of higher degree it is called a *depressed equation*. Continuing the synthetic division by dividing $Q_1(x)$ by $x-2$,

$$\begin{array}{rrrrrr|r} 1 & 1 & -2 & -5 & -5 & -2 & \underline{2} \\ \downarrow & 2 & 6 & 8 & 6 & 2 & \\ \hline 1 & 3 & 4 & 3 & 1 & 0 = Q_1(2) = P(2) \end{array}$$

we find that both positive zeros of $P(x)$ are 2, and 2 is said to be a *double zero* or a *zero of multiplicity 2*. In factored form $P(x) = (x-2)^2(x^4 + 3x^3 + 4x^2 + 3x$

+1). Descartes' Rule of Signs indicates that there are no more positive zeros. Attempting to find negative zeros, divide $Q_2(x)=x^4+3x^3+4x^2+3x+1$ by $x+1$.

$$\begin{array}{rrrrr|r} 1 & 3 & 4 & 3 & 1 & \underline{-1} \\ \downarrow & -1 & -2 & -2 & -1 & \\ \hline 1 & 2 & 2 & 1 & 0=Q_2(-1)=P(-1) \end{array}$$

Thus, $x+1$ is a factor, and -1 is a zero. Continuing the synthetic division by dividing $Q_3(x)=x^3+2x^2+2x+1$ by $x+1$,

$$\begin{array}{rrrr|r} 1 & 2 & 2 & 1 & \underline{-1} \\ \downarrow & -1 & -1 & -1 & \\ \hline 1 & 1 & 1 & 0=Q_3(-1)=P(-1) \end{array}$$

we find that -1 is also a zero of multiplicity 2. In factored form $P(x)=(x-2)^2(x+1)^2(x^2+x+1)$. Since the final quotient is a second degree polynomial, its zeros can be found by using the General Quadratic Formula. $x^2+x+1=0$ where $a=1$, $b=1$, and $c=1$. Therefore,

$$x = \frac{-1 \pm \sqrt{1-4}}{2} = \frac{-1 \pm i\sqrt{3}}{2}$$

and the entire list of zeros of $P(x)=x^6-x^5-4x^4-x^3+5x^2+8x+4$ is 2, 2, -1, -1, $\dfrac{-1+i\sqrt{3}}{2}$, and $\dfrac{-1-i\sqrt{3}}{2}$.

EXAMPLE 8: Is $2x+1$ a factor of $P(x)=2x^3+3x^2-3x-2$?

Since $2x+1$ is not in the form $x-r$, it appears that synthetic division cannot be used. However, if $2x+1$ is written as $2\left(x+\dfrac{1}{2}\right)$, synthetic division can be used with the factor $\left(x+\dfrac{1}{2}\right)$. $2x+1$ will be a factor if and only if $x+\dfrac{1}{2}$ is a factor.

$$\begin{array}{rrrr|r} 2 & 3 & -3 & -2 & \underline{-\tfrac{1}{2}} \\ \downarrow & -1 & -1 & 2 & \\ \hline 2 & 2 & -4 & 0=P\left(-\tfrac{1}{2}\right) \end{array}$$

Thus, both $2x+1$ and $x+\dfrac{1}{2}$ are factors of $P(x)$. In order to find other factors the third row of the synthetic division must be divided by the 2 that was factored out of the original divisor. In factored form $P(x) = \left(x+\dfrac{1}{2}\right)(2x^2+2x-4) = (2x+1)(x^2+x-2)$. This quadratic factor can itself be factored into $(x+2)(x-1)$. Thus, $P(x)=(2x+1)(x+2)(x-1)$.

EXERCISES

1 The graph of $P(x)=3x^5+5x^3-8x+5$ can cross the x-axis in no more than r points. What is the value of r?

[A] 0 [B] 1 [C] 2 [D] 3 [E] 5

2 Between which two consecutive integers is there a zero of $P(x)=28x^3-11x^2+15x-2$?

[A] -2 & -1 [B] -1 & 0 [C] 0 & 1
[D] 1 & 2 [E] 2 & 3

3 $P(x)=6x^4+x^3-26x^2-4x+8$. If $P(x)$ increases without bound as x increases without bound, then as x decreases without bound, $P(x)$

[A] increases without bound
[B] decreases without bound
[C] approaches zero from above the x-axis
[D] approaches zero from below the x-axis
[E] cannot be determined

4 Which of the following is an odd function?
  I. $f(x)=3x^3+5$
  II. $g(x)=4x^6+2x^4-3x^2$
  III. $h(x)=7x^5-8x^3+12x$

[A] I only [B] II only [C] III only
[D] I & II only [E] I & III only

5 Using the Rational Root Theorem, how many possible rational roots are there for $2x^4+4x^3-6x^2+15x-12=0$?

[A] 8 [B] 4 [C] 16 [D] 12 [E] 6

6 How many positive real zeros would you expect to find for the polynomial function $P(x)=3x^4-2x^2-12$?

[A] 0 [B] 2 or 0 [C] 1
[D] 3 or 1 [E] 4 or 2 or 0

7 If both $x-1$ and $x-2$ are factors of $x^3-3x^2+2x-4b$, then b must be

[A] 0 [B] 1 [C] 2 [D] 3 [E] 4

8 If i is a root of $x^4+2x^3-3x^2+2x-4=0$, what are the real roots?

[A] ± 2 [B] $1,-4$ [C] $-1\pm\sqrt{5}$
[D] $1\pm\sqrt{5}$ [E] $-1,4$

9 How many positive real roots does $x^4+x^3-3x^2-3x=0$ have?

[A] 0 [B] 1 [C] 2 [D] 3 [E] 4

10. The sum of the zeros of $P(x) = 8x^3 - 2x + 3$ is

[A] 2 [B] $-\frac{3}{8}$ [C] $-\frac{1}{4}$
[D] $\frac{3}{8}$ [E] $\frac{1}{4}$

11. If $3x^3 - 9x^2 + Kx - 12$ is divisible by $x - 3$, then it is also divisible by

[A] $3x^2 - x + 4$ [B] $3x^2 - 4$ [C] $3x^2 + 4$
[D] $3x - 4$ [E] $3x + 4$

12. Write the equation of lowest degree with real coefficients, if two of its roots are -1 and $1 + i$.

[A] $x^3 + x^2 + 2 = 0$

[B] $x^3 - x^2 - 2 = 0$

[C] $x^3 - x + 2 = 0$

[D] $x^3 - x^2 + 2 = 0$

[E] none of these

5. INEQUALITIES

Given any algebraic expression $f(x)$, there are exactly three situations that can exist:
 (1) for some values of $x, f(x) < 0$
 (2) for some values of $x, f(x) = 0$
 (3) for some values of $x, f(x) > 0$
If all three of these sets of numbers are indicated on a number line, the set of values that satisfy $f(x) < 0$ is always separated from the set of values that satisfy $f(x) > 0$ by those values of x that satisfy $f(x) = 0$.

EXAMPLE: Find the set of values for x that satisfies $x^2 - 3x - 4 < 0$.
Consider the associated equation:
$$x^2 - 3x - 4 = 0$$
$$(x - 4)(x + 1) = 0$$
$$x - 4 = 0 \text{ or } x + 1 = 0$$
Therefore, $x = 4$ or $x = -1$.

$$-2 \ -1 \ 0 \ 1 \ 2 \ 3 \ 4 \ 5$$

These two points, -1 and 4, separate the set of x-values that satisfy $x^2 - 3x - 4 < 0$ from the set of x-values that satisfy $x^2 - 3x - 4 > 0$. The correct region can be determined by direct substitution of numbers for x from each of the three regions indicated on the number line into the original inequality.
Let $x = -2$: $x^2 - 3x - 4 = 4 + 6 - 4 > 0$.
Let $x = 0$: $x^2 - 3x - 4 = 0 + 0 - 4 < 0$.
Let $x = 5$: $x^2 - 3x - 4 = 25 - 15 - 4 > 0$.
Therefore, the region which contains zero is the only one that satisfies $x^2 - 3x - 4 < 0$, and the solution set is $\{x : -1 < x < 4\}$. The graph of this solution set on a number line is indicated below.

$$-2 \ -1 \ 0 \ 1 \ 2 \ 3 \ 4 \ 5$$

Although this may seem to be a rather roundabout method of solution for inequalities, it does have the advantage of working for all types of inequalities. No special rules for special cases must be remembered.

EXERCISES

1. Which of the following is equivalent to
$$3x^2 - x < 2?$$

[A] $-\frac{3}{2} < x < 1$ [B] $-1 < x < \frac{2}{3}$
[C] $-\frac{2}{3} < x < 1$ [D] $-1 < x < \frac{3}{2}$
[E] $x < -\frac{2}{3}$ or $x > 1$

2. If $3 < x < 7$ and $-12 < y < -6$, what are all possible values of xy?

[A] $-72 < xy < -21$
[B] $-84 < xy < -18$
[C] $-42 < xy < -36$
[D] $-84 < xy < -36$
[E] $-36 < xy < -42$

3. The number of integers which satisfy the inequality $x^2 + 48 < 16x$ is

[A] 0 [B] 4 [C] 7
[D] an infinite number
[E] none of the above

TRIGONOMETRIC FUNCTIONS

CHAPTER 3

1. DEFINITIONS

The general definition of the six trigonometric functions is obtained from an angle placed in standard position on a rectangular coordinate system. When an angle θ is placed so that its vertex is at the origin, its initial side is placed along the positive x-axis, and its terminal side is anywhere on the coordinate system, it is said to be in *standard position*. The angle is given a positive value if it is measured in a counter-clockwise direction from the initial side to the terminal side and a negative value if it is measured in a clockwise direction.

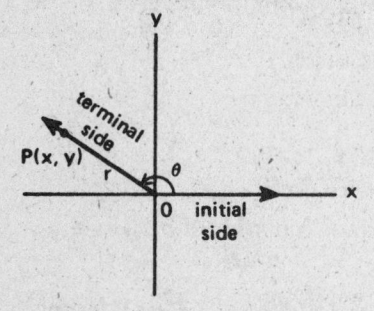

Let $P(x,y)$ be any point on the terminal side of the angle, and let r represent the distance between O and P. The six trigonometric functions are defined to be:

$$\sin\theta = \frac{\text{ordinate of } P}{OP} = \frac{y}{r}$$

$$\cos\theta = \frac{\text{abscissa of } P}{OP} = \frac{x}{r}$$

$$\tan\theta = \frac{\text{ordinate of } P}{\text{abscissa of } P} = \frac{y}{x}$$

$$\csc\theta = \frac{OP}{\text{ordinate of } P} = \frac{r}{y}$$

$$\sec\theta = \frac{OP}{\text{abscissa of } P} = \frac{r}{x}$$

$$\cot\theta = \frac{\text{abscissa of } P}{\text{ordinate of } P} = \frac{x}{y}$$

Sine and cosine, tangent and cotangent, and secant and cosecant are *co-functions*. From these definitions it follows that:

$$\sin\theta \cdot \csc\theta = 1 \qquad \tan\theta = \frac{\sin\theta}{\cos\theta}$$

$$\cos\theta \cdot \sec\theta = 1 \qquad \cot\theta = \frac{\cos\theta}{\sin\theta}$$

$$\tan\theta \cdot \cot\theta = 1$$

The distance OP is always positive and the ordinate and abscissa of P are positive or negative depending upon which quadrant the terminal side of $\angle\theta$ lies in. The signs of the trigonometric functions are indicated in the following table:

Quadrant	I	II	III	IV
Function: $\sin\theta$ & $\csc\theta$	+	+	−	−
$\cos\theta$ & $\sec\theta$	+	−	−	+
$\tan\theta$ & $\cot\theta$	+	−	+	−

Each angle θ whose terminal side lies in quadrants II, III, & IV has associated with it two acute angles called reference angles. $\angle\alpha$ is the acute angle formed by the x-axis and the terminal side. $\angle\beta$ is the acute angle formed by the y-axis and the terminal side.

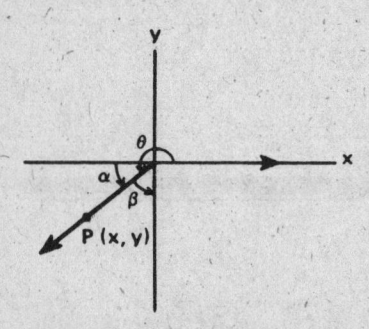

Any function of $\angle\theta = \pm$ the same function of $\angle\alpha$ and any function of $\angle\theta = \pm$ the co-function of $\angle\beta$. The sign is determined by the quadrant in which the terminal side lies.

EXAMPLE 1: Express $\sin 320°$ in terms of $\angle\alpha$ and $\angle\beta$.
$$\alpha = 360° - 320° = 40°$$
$$\beta = 320° - 270° = 50°$$

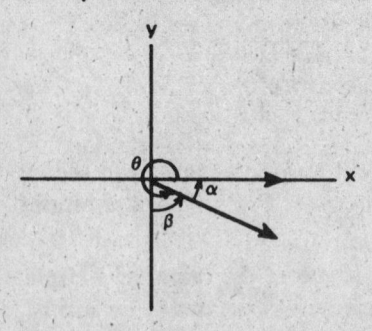

Since the sine is negative in quadrant IV, $\sin 320° = -\sin 40° = -\cos 50°$.

EXAMPLE 2: Express $\cot 200°$ in terms of $\angle\alpha$ and $\angle\beta$.
$$\alpha = 200° - 180° = 20°$$
$$\beta = 270° - 200° = 70°$$

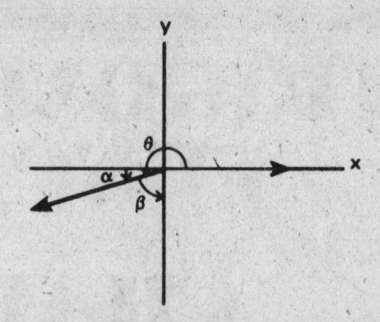

Since the cotangent is positive in quadrant III, $\cot 200° = \cot 20° = \tan 70°$.

EXAMPLE 3: Express $\cos 130°$ in terms of $\angle\alpha$ and $\angle\beta$.
$$\alpha = 180° - 130° = 50°$$
$$\beta = 130° - 90° = 40°$$

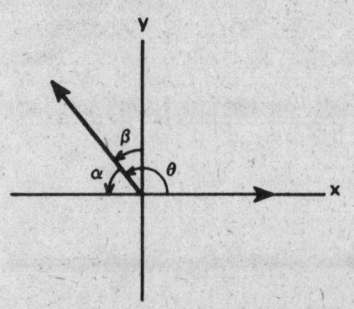

Since the cosine is negative in quadrant II, $\cos 130° = -\cos 50° = -\sin 40°$.

These examples indicate that any function of $\alpha = $ the co-function of β and that α and β are *complementary* angles because their sum is 90°. A very useful property is obtained from this observation: *Co-functions of complementary angles are equal.*

EXAMPLE 4: If both the angles are acute and $\sin(3x + 20°) = \cos(2x - 40°)$, find x.
Since these co-functions are equal, the angles must be complementary.
Therefore, $(3x + 20°) + (2x - 40°) = 90°$
$$5x - 20° = 90°$$
$$x = 22°$$

EXERCISES

1. Express $\cos 320°$ as a function of an angle between 0° and 45°.

   [A] $\cos 40°$ [B] $\sin 40°$ [C] $-\cos 40°$
   [D] $-\sin 40°$ [E] none of the above

2. If the point $P(-5,12)$ lies on the terminal side of $\angle \theta$ in standard position, $\sin\theta =$

[A] $\frac{-5}{12}$ [B] $\frac{12}{13}$ [C] $\frac{-5}{13}$
[D] $\frac{12}{5}$ [E] $-\frac{12}{13}$

3. If $\sec\theta = -\frac{5}{4}$ and $\sin\theta > 0$, then $\tan\theta =$

[A] $\frac{4}{3}$ [B] $\frac{3}{4}$ [C] $-\frac{4}{3}$
[D] $-\frac{3}{4}$ [E] none of the above

4. If x is an angle in the third quadrant and $\tan(x - 30°) = \cot x$, find x.

[A] 210° [B] 240° [C] 225°
[D] 60° [E] none of the above

5. If $90° < \alpha < 180°$ and $270° < \beta < 360°$, then which of the following *cannot* be true?

[A] $\sin\alpha = \sin\beta$ [B] $\tan\alpha = \sin\beta$
[C] $\tan\alpha = \tan\beta$ [D] $\sin\alpha = \cos\beta$
[E] $\sec\alpha = \csc\beta$

2. ARCS AND ANGLES

Although the degree is the chief unit used to measure an angle in elementary mathematics courses, the radian has several advantages in more advanced mathematics courses. Degrees and radians are related by the equation: $\pi^R = 180°$.

EXAMPLE 1: Convert the degrees to radians and the radians to degrees.

[A] 30° [B] 270° [C] $\frac{\pi^R}{4}$
[D] $\frac{17\pi}{3}$ [E] 24

(If no unit of measurement is indicated, radians are assumed.)

[A] $\frac{x^R}{30°} = \frac{\pi^R}{180°}$. Solving for x, $x = \frac{\pi^R}{6}$. Since $\pi^R = 180°$ on the right of the equation, $\frac{\pi^R}{6}$ must equal 30° on the left of the equation.

[B] $\frac{x^R}{270°} = \frac{\pi^R}{180°} \cdot x = \frac{3\pi^R}{2}$

[C] $\frac{\frac{\pi}{4}}{x°} = \frac{\pi^R}{180°} \cdot x = 45°$

[D] $\frac{\frac{17\pi}{3}}{x°} = \frac{\pi^R}{180°} \cdot x = 1020°$

[E] $\frac{24}{x°} = \frac{\pi^R}{180°} \cdot x = \left(\frac{4320}{\pi}\right)°$

In a circle of radius r inches with an arc subtended by a central angle of θ^R two important formulas can be derived. The length of the arc, s, is equal to $r\theta$, and the area of the sector, AOB, is equal to $\frac{1}{2}r^2\theta$.

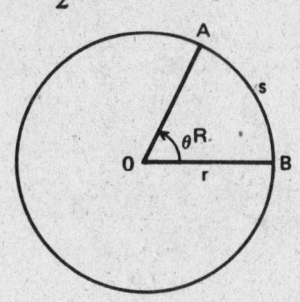

EXAMPLE 2: Find the area of the sector and the length of the arc subtended by a central angle of $\frac{2\pi}{3}$ radians in a circle whose radius is 6 inches.

$s = r\theta$ $\qquad A = \frac{1}{2}r^2\theta$

$s = 6 \cdot \frac{2\pi}{3} = 4\pi$ in. $\quad A = \frac{1}{2} \cdot 36 \cdot \frac{2\pi}{3} = 12\pi$ sq. in.

EXAMPLE 3: In a circle of radius 8 inches, find the area of the sector whose arc length is 6π inches.

$s = r\theta$ $\qquad A = \frac{1}{2}r^2\theta$

$6\pi = 8\theta$ $\qquad A = \frac{1}{2} \cdot 64 \cdot \frac{3\pi}{4} = 24\pi$ sq. in.

$\theta = \frac{3\pi^R}{4}$

EXAMPLE 4: Find the length of the radius of a circle in which a central angle of 60° subtends an arc of length 8π inches.

The 60° angle must be converted to radians:

$\frac{x^R}{60°} = \frac{\pi^R}{180°}$, therefore $x = \frac{\pi}{3}$. $s = r\theta$

$8\pi = r \cdot \frac{\pi}{3}$

$r = 24$ in.

EXERCISES

1 An angle of 30 radians is equal to how many degrees?

[A] $\dfrac{\pi}{6}$ [B] $\dfrac{30}{\pi}$ [C] $\dfrac{5400}{\pi}$

[D] $\dfrac{540}{\pi}$ [E] $\dfrac{\pi}{30}$

2 If a sector of a circle has an arc length of 2π inches and an area of 6π square inches, what is the radius of the circle?

[A] 1 [B] 2 [C] 3 [D] 6 [E] 12

3 If a circle has a circumference of 16 inches, the area of a sector with a central angle of 1.5π radians is

[A] 48 [B] 12 [C] $\dfrac{96}{\pi^2}$

[D] $\dfrac{48}{\pi}$ [E] 24π

4 A central angle of 40° in a circle of radius 1 inch intercepts an arc whose length is s. Find s.

[A] $\dfrac{4\pi}{9}$ [B] 40 [C] $\dfrac{2\pi}{9}$

[D] $\dfrac{\pi}{9}$ [E] π

---

3. SPECIAL ANGLES

Although most of the values of the trigonometric functions are difficult to determine (and are usually obtained from tables), those with angles which are multiples of $\dfrac{\pi}{6}$, $\dfrac{\pi}{4}$, $\dfrac{\pi}{3}$, or $\dfrac{\pi}{2}$ are easy to find. Any angle that is a multiple of $\dfrac{\pi}{2}$ is known as a *quadrantal* angle because, in standard position, its terminal side lies on one of the axes between quadrants. Any function of these angles is easily determined because one of the coordinates of the point P on the terminal side is zero.

	0	$\dfrac{\pi}{2}$	π	$\dfrac{3\pi}{2}$	2π
sine	0	1	0	-1	0
cosine	1	0	-1	0	1
tangent	0	und	0	und	0
cotangent	und	0	und	0	und
secant	1	und	-1	und	1
cosecant	und	1	und	—	und

und means the function is undefined because the definition of the function necessitates division by zero.

Any function of an odd multiple of $\dfrac{\pi}{4}=\pm$ that function of any other odd multiple of $\dfrac{\pi}{4}$. (The sign is determined by the quadrant in which the terminal side of the angle lies.)

Any function of $k\cdot\dfrac{\pi}{6}$ (where k is not a multiple or factor of 6) $=\pm$ that function of any other similar multiple of $\dfrac{\pi}{6}$.

Any function of $k\cdot\dfrac{\pi}{3}$ (where k is not a multiple or factor of 3) $=\pm$ that function of any other similar multiple of $\dfrac{\pi}{3}$.

Because of these three properties it is only necessary to learn the values of the trigonometric functions of $\dfrac{\pi}{4}$, $\dfrac{\pi}{6}$, and $\dfrac{\pi}{3}$, and then to attach a + or − depending on the quadrant in which the terminal side of the angle lies.

	$\dfrac{\pi}{6}$ or 30°	$\dfrac{\pi}{4}$ or 45°	$\dfrac{\pi}{3}$ or 60°
sine	$\dfrac{1}{2}$	$\dfrac{\sqrt{2}}{2}$	$\dfrac{\sqrt{3}}{2}$
cosine	$\dfrac{\sqrt{3}}{2}$	$\dfrac{\sqrt{2}}{2}$	$\dfrac{1}{2}$
tangent	$\dfrac{\sqrt{3}}{3}$	1	$\sqrt{3}$
cotangent	$\sqrt{3}$	1	$\dfrac{\sqrt{3}}{3}$
secant	$\dfrac{2\sqrt{3}}{3}$	$\sqrt{2}$	2
cosecant	2	$\sqrt{2}$	$\dfrac{2\sqrt{3}}{3}$

EXAMPLE: Determine the value of each of the following:

[A] $\cos 0$ [B] $\csc\dfrac{3\pi}{2}$

[C] $\cos\dfrac{7\pi}{6}$ [D] $\tan\dfrac{5\pi}{4}$

[E] $\sec\dfrac{3\pi}{4}$ [F] $\sin 300°$

[G] $\cos(-390°)$ [H] $\cot 60°$

[I] $\tan(-45°)$

Procedures are as follows:

[A] From the first table, $\cos 0 = 1$.

[B] $\frac{3\pi}{2}$ is a quadrantal angle with its terminal side along the negative y-axis. Therefore, $P(0, -r)$ where $OP = r$. $\csc \frac{3\pi}{2} = \frac{r}{-r} = -1$. (Or, from the first table, $\csc \frac{3\pi}{2} = -1$.)

[C] $\frac{7\pi}{6}$ is a multiple of $\frac{\pi}{6}$ so $\cos \frac{7\pi}{6} = \pm \cos \frac{\pi}{6} = \pm \frac{\sqrt{3}}{2}$. Since the terminal side of $\frac{7\pi}{6}$ is in the third quadrant, $\cos \frac{7\pi}{6} = -\frac{\sqrt{3}}{2}$.

[D] $\frac{5\pi}{4}$ is a multiple of $\frac{\pi}{4}$, and its terminal side lies in the third quadrant. Therefore, $\tan \frac{5\pi}{4}$ is positive, and $\tan \frac{5\pi}{4} = \tan \frac{\pi}{4} = 1$.

[E] $\frac{3\pi}{4}$ is a multiple of $\frac{\pi}{4}$, and its terminal side lies in the second quadrant. Therefore, $\sec \frac{3\pi}{4}$ is negative, and $\sec \frac{3\pi}{4} = -\sec \frac{\pi}{4} = -\sqrt{2}$.

[F] $300° = \frac{5\pi}{3}$, which is a multiple of $\frac{\pi}{3}$, and its terminal side lies in the fourth quadrant. Therefore, $\sin 300°$ is negative, and $\sin 300° = \sin \frac{5\pi}{3} = -\sin \frac{\pi}{3} = -\frac{\sqrt{3}}{2}$.

[G] $-390° = -\frac{13\pi}{6}$, which is a multiple of $\frac{\pi}{6}$, and its terminal side lies in the fourth quadrant. Therefore, $\cos(-390°)$ is positive, and $\cos(-390°) = \cos(-\frac{13\pi}{6}) = \cos \frac{\pi}{6} = \frac{\sqrt{3}}{2}$.

[H] $60° = \frac{\pi}{3}$. Therefore, $\cot 60° = \cot \frac{\pi}{3} = \frac{\sqrt{3}}{3}$.

[I] $-45° = -\frac{\pi}{4}$, which is a multiple of $\frac{\pi}{4}$, and its terminal side lies in the fourth quadrant. Therefore, $\tan(-45°)$ is negative, and $\tan(-45°) = \tan(-\frac{\pi}{4}) = -\tan \frac{\pi}{4} = -1$.

EXERCISES

1. $\operatorname{Tan}(-60°)$ equals
   [A] $-\tan 30°$ [B] $\cot 30°$ [C] $-\tan 60°$
   [D] $-\cot 60°$ [E] $\tan 60°$

2. $\operatorname{Tan}(-135°) + \cot 315°$ equals
   [A] 1 [B] 2 [C] -1
   [D] -2 [E] 0

3. $\operatorname{Cos} \pi - \sin 570° - \csc(-\frac{\pi}{2}) + \sec 0°$ equals
   [A] 1 [B] $1\frac{1}{2}$ [C] $-1\frac{1}{2}$
   [D] 0 [E] $-\frac{1}{2}$

4. $\operatorname{Sec} \frac{11\pi}{6} \cdot \tan \frac{2\pi}{3} \cdot \sin \frac{7\pi}{4}$ equals
   [A] -1 [B] $\sqrt{2}$ [C] $\frac{\sqrt{6}}{3}$
   [D] $-\sqrt{2}$ [E] $-\frac{\sqrt{6}}{3}$

5. $\operatorname{Sin} 300°$ equals
   [A] $\cos 60°$ [B] $\sin 120°$ [C] $\cos 240°$
   [D] $\sin 240°$ [E] $\cos 120°$

4. GRAPHS

Since the values of all the trigonometric functions repeat themselves at regular intervals, and, for some number $p, f(x) = f(x + p)$ for all numbers x, they are called *periodic functions*. The smallest positive value of p for which this property holds is called the *period* of the function.

The sine, cosine, secant, and cosecant have a period of 2π, and tangent and cotangent have a period of π. The graphs of the six trigonometric functions are shown below, demonstrating that the tangent and cotangent repeat on intervals of length π, and the others repeat on intervals of length 2π.

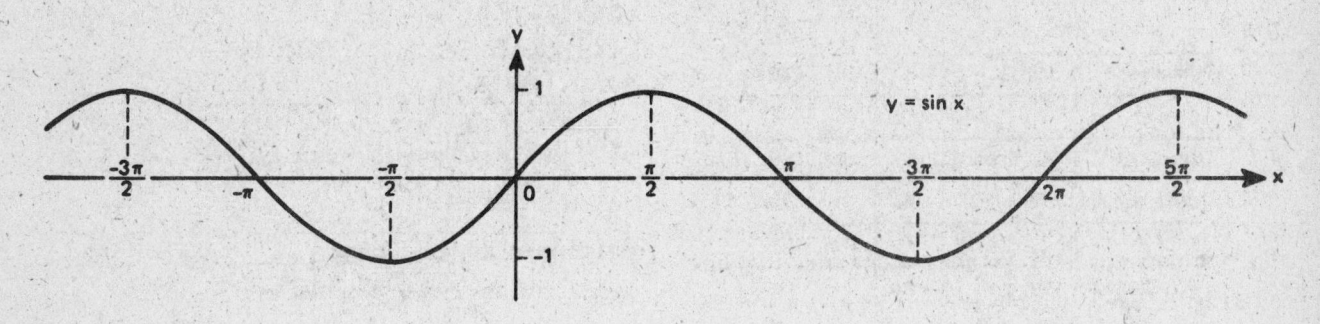

$y = \sin x$

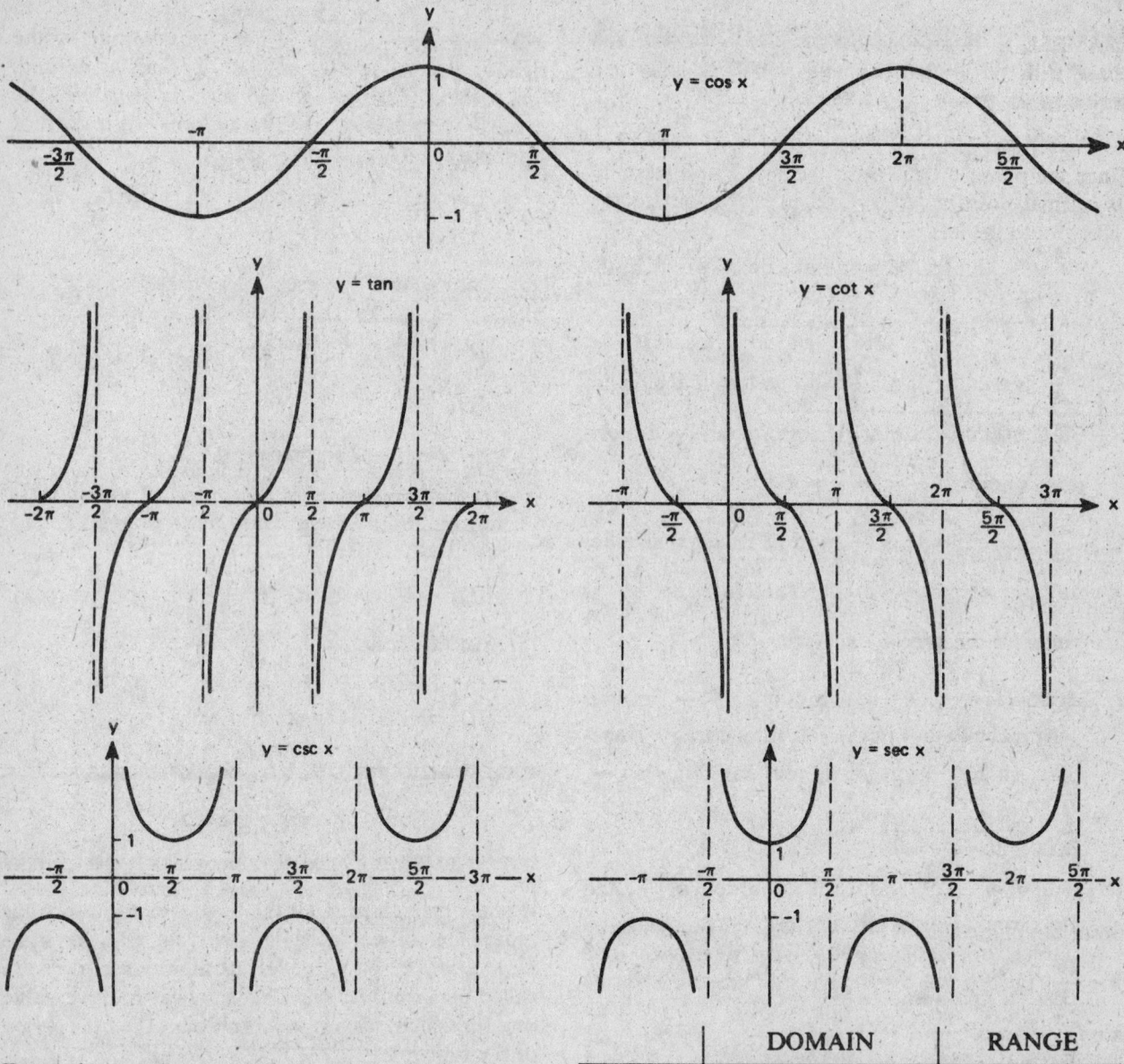

From the graphs or from the table of values it is possible to determine the domain and range of each of the six trigonometric functions.

The general form of a trigonometric function, f, is given to be $y = A \cdot f(Bx + C)$ where $|A|$ is the amplitude, $\dfrac{\text{normal period of } f}{B}$ is the period of the graph, and $-\dfrac{C}{B}$ is the phase shift. Although the amplitude is associated with only the sine and cosine, the others can be associated with any of the trigonometric functions; they usually are associated with the sine and cosine graphs. The *amplitude* is one-half the vertical distance between the lowest and highest point of the graph. The *phase shift* is the distance to the right or left from its normal position that the graph is changed. The *frequency* is equal to $\dfrac{1}{\text{period}}$

	DOMAIN	RANGE
sine	all real numbers	$-1 \leqslant \sin x \leqslant 1$
cosine	all real numbers	$-1 \leqslant \cos x \leqslant 1$
tangent	all real numbers except odd multiples of $\dfrac{\pi}{2}$	all real numbers
cotangent	all real numbers except all multiples of π	all real numbers
secant	all real numbers except odd multiples of $\dfrac{\pi}{2}$	$\sec x \leqslant -1$ or $\sec x \geqslant 1$
cosecant	all real numbers except all multiples of π	$\csc x \leqslant -1$ or $\csc x \geqslant 1$

4. Graphs

EXAMPLE 1: Determine the amplitude, period, and phase shift of $y = 2\sin 2x$ and sketch at least one period of the graph.

Amplitude = 2. Period = $\frac{2\pi}{2} = \pi$. Phase shift = 0. Since the phase shift is zero, the sine graph starts at its normal position, (0,0), and is drawn out to the right and to the left.

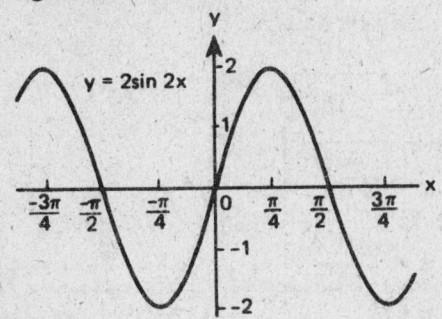

EXAMPLE 2: Determine the amplitude, period, and phase shift of $y = \frac{1}{2}\cos\left(\frac{1}{2}x - \frac{\pi}{3}\right)$ and sketch at least one period of the graph.

Amplitude = $\frac{1}{2}$.

Period = $\frac{2\pi}{\frac{1}{2}} = 4\pi$.

Phase shift = $\frac{\frac{\pi}{3}}{\frac{1}{2}} = \frac{2\pi}{3}$.

Since the phase shift is $\frac{2\pi}{3}$ the cosine graph starts at $x = \frac{2\pi}{3}$ instead of $x = 0$ and one period ends at $x = \frac{2\pi}{3} + 4\pi$ or $\frac{14\pi}{3}$.

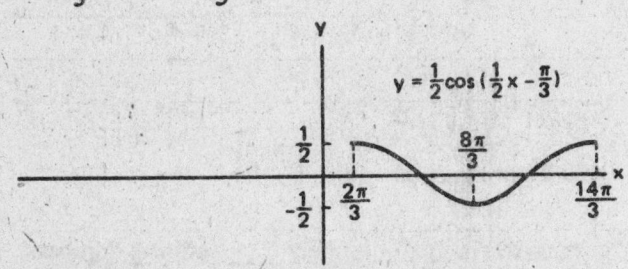

EXAMPLE 3: Determine the amplitude, period, and phase shift of $y = -2\sin(\pi x + 3\pi)$ and sketch at least one period of the graph.

Amplitude = 2.

Period = $\frac{2\pi}{\pi} = 2$.

Phase shift = $-\frac{3\pi}{\pi} = -3$.

Since the phase shift is -3, the sine graph starts at $x = -3$ instead of $x = 0$, and one period ends at $-3 + 2$ or $x = -1$. The graph can continue to the right and to the left for as many periods as desired. Since the coefficient of the sine is negative, the graph starts down as x increases from -3 instead of up as the normal sine graph does.

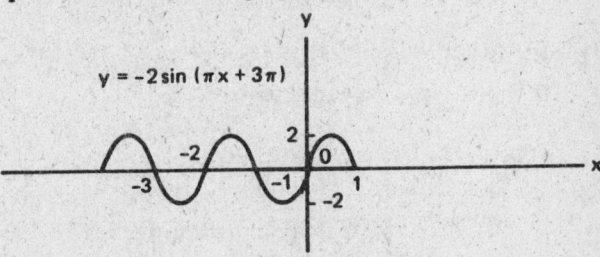

EXERCISES

1. In the figure, part of the graph of $y = \sin 2x$ is shown. What are the coordinates of point P?

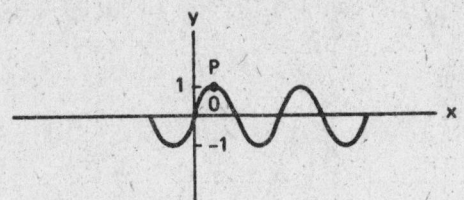

    [A] $\left(\frac{\pi}{2}, 1\right)$ [B] $(\pi, 1)$
    [C] $\left(\frac{\pi}{4}, 1\right)$ [D] $\left(\frac{\pi}{2}, 2\right)$
    [E] $(\pi, 2)$

2. The figure could be a portion of the graph whose equation is

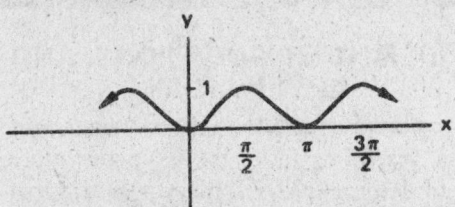

    [A] $y - 1 = \sin x \cdot \cos x$
    [B] $y \sec x = 1$
    [C] $2y + 1 = \sin 2x$
    [D] $2y + 1 = \cos 2x$
    [E] $1 - 2y = \cos 2x$

3. As θ increases from $\frac{\pi}{4}$ to $\frac{5\pi}{4}$ the value of $4\cos\frac{1}{2}\theta$.

    [A] increases then decreases
    [B] decreases then increases
    [C] decreases throughout
    [D] increases throughout
    [E] decreases, increases, then decreases again

4. The function $f(x) = \sqrt{3}\cos x + \sin x$ has an amplitude of

[A] $\sqrt{3}$ [B] $\sqrt{3}+1$ [C] 2
[D] $2\sqrt{3}$ [E] $\dfrac{\sqrt{3}+1}{2}$

5. For what value of P is the period of the function $y = \dfrac{1}{3}\cos Px$ equal to $\dfrac{2\pi}{3}$?

[A] $\dfrac{1}{3}$ [B] 2 [C] 3 [D] 6 [E] $\dfrac{2}{3}$

6. If $0 < x < \dfrac{\pi}{2}$, what is the maximum value of the function $f(x) = \sin\dfrac{1}{3}x$?

[A] 0 [B] 1 [C] $\dfrac{1}{3}$ [D] $\dfrac{1}{2}$ [E] $\dfrac{\sqrt{3}}{2}$

7. If the graph in the figure has an equation of the form $y = \sin(Mx + N)$, what is the value of N?

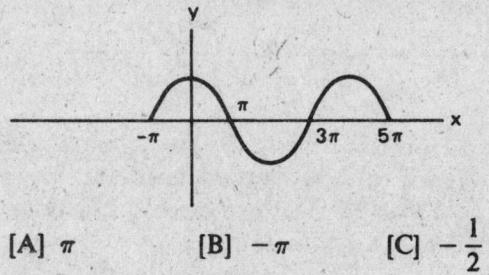

[A] π [B] $-\pi$ [C] $-\dfrac{1}{2}$
[D] $\dfrac{\pi}{2}$ [E] -1

5. IDENTITIES, EQUATIONS, AND INEQUALITIES

Many of the problems involving trigonometry depend upon several formulas that can be used with any angle as long as the value of the function is not undefined. A list of the most used formulas follows:

1. $\sin^2 x + \cos^2 x = 1$
2. $\tan^2 x + 1 = \sec^2 x$ } Pythagorean Identities
3. $\cot^2 x + 1 = \csc^2 x$

4. $\sin(A+B) = \sin A \cdot \cos B + \cos A \cdot \sin B$
5. $\sin(A-B) = \sin A \cdot \cos B - \cos A \cdot \sin B$
6. $\cos(A+B) = \cos A \cdot \cos B - \sin A \cdot \sin B$
7. $\cos(A-B) = \cos A \cdot \cos B + \sin A \cdot \sin B$ } Sum and Difference Formulas
8. $\tan(A+B) = \dfrac{\tan A + \tan B}{1 - \tan A \cdot \tan B}$
9. $\tan(A-B) = \dfrac{\tan A - \tan B}{1 + \tan A \cdot \tan B}$

10. $\sin 2A = 2\sin A \cdot \cos A$
11. $\cos 2A = \cos^2 A - \sin^2 A$
12. $\quad = 2\cos^2 A - 1$
13. $\quad = 1 - 2\sin^2 A$
14. $\tan 2A = \dfrac{2\tan A}{1 - \tan^2 A}$ } Double Angle Formulas

A few formulas that are rarely used on the Level II examination but may be helpful:

15. $\sin\dfrac{1}{2}A = \pm\sqrt{\dfrac{1-\cos A}{2}}$
16. $\cos\dfrac{1}{2}A = \pm\sqrt{\dfrac{1+\cos A}{2}}$
17. $\tan\dfrac{1}{2}A = \pm\sqrt{\dfrac{1-\cos A}{1+\cos A}}$ } Half Angle Formulas
18. $\quad = \dfrac{1-\cos A}{\sin A}$
19. $\quad = \dfrac{\sin A}{1+\cos A}$

The correct sign for formulas 15, 16, and 17 is determined by the quadrant in which the angle $\dfrac{1}{2}A$ lies.

EXAMPLE 1: Simplify $\dfrac{\csc A}{\cot A + \tan A}$ and express the answer in terms of a single trigonometric function.

If none of the formulas seems to be helpful, change all the functions to sines and cosines. From the basic definitions of the trigonometric functions, $\csc A = \dfrac{1}{\sin A}$, $\cot A = \dfrac{\cos A}{\sin A}$, and $\tan A = \dfrac{\sin A}{\cos A}$. Therefore,

$$\dfrac{\csc A}{\cot A + \tan A} = \dfrac{\dfrac{1}{\sin A}}{\dfrac{\cos A}{\sin A} + \dfrac{\sin A}{\cos A}}.$$

Multiplying the numerator and the denominator of the complex fraction by the LCD of the three "little" denominators, $\sin A \cdot \cos A$, the fraction becomes

$$\dfrac{\dfrac{1}{\sin A} \cdot \sin A \cdot \cos A}{\left(\dfrac{\cos A}{\sin A} + \dfrac{\sin A}{\cos A}\right) \cdot \sin A \cdot \cos A} = \dfrac{\cos A}{\cos^2 A + \sin^2 A} = \dfrac{\cos A}{1}$$

by formula 1. Therefore, $\dfrac{\csc A}{\cot A + \tan A} = \cos A$.

EXAMPLE 2: Express $\sin 4x$ in terms of $\sin x$ and $\cos x$.

$\sin 4x = \sin 2(2x)$. Let $A = 2x$ and use formula 10. $\sin 4x = \sin 2A = 2\sin A \cdot \cos A = 2\sin 2x \cdot \cos 2x$. Using

5. Identities, Equations, and Inequalities

formulas 10 and 11,
$$\sin 4x = 2(2\sin x \cdot \cos x) \cdot (\cos^2 x - \sin^2 x)$$
$$= 4\sin x \cdot \cos^3 x - 4\sin^3 x \cdot \cos x$$

EXAMPLE 3: If $\tan A = \dfrac{5}{12}$ and $\sin B = \dfrac{3}{5}$, where A and B are acute angles, find the value of $\cos(A+B)$. From the basic definition of the trigonometric functions and the Pythagorean Theorem, $\sin A = \dfrac{5}{13}$, $\cos A = \dfrac{12}{13}$, and $\cos B = \dfrac{4}{5}$.

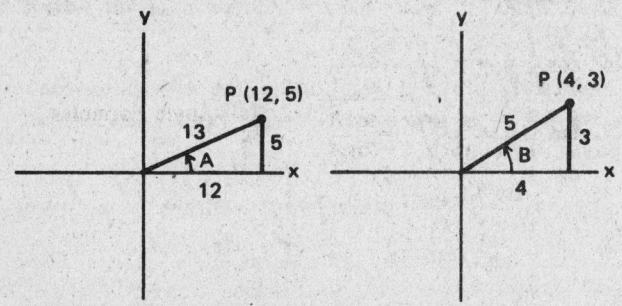

Therefore, using formula 6,
$$\cos(A+B) = \cos A \cdot \cos B - \sin A \cdot \sin B$$
$$= \frac{12}{13} \cdot \frac{4}{5} - \frac{5}{13} \cdot \frac{3}{5}$$
$$= \frac{48}{65} - \frac{15}{65}$$
$$= \frac{33}{65}$$

EXAMPLE 4: If A is an angle in the third quadrant, B is an angle in the second quadrant, $\tan A = \dfrac{3}{4}$, and $\tan B = -\dfrac{1}{2}$, in which quadrant does the angle $(A+B)$ lie?

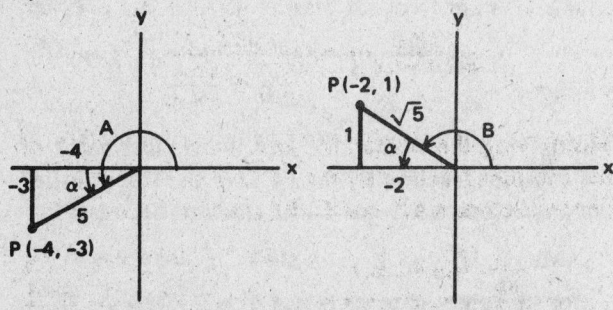

$180° < A < 270°$ and $90° < B < 180°$. Therefore, $270° < A+B < 450°$, which means that the angle $(A+B)$ must lie in quadrants I or IV. Since the sine is positive in quadrant I and negative in quadrant IV, the value of $\sin(A+B)$ will determine the correct quadrant. From the basic definition of the trigonometric functions and the Pythagorean Theorem, $\sin A = -\dfrac{3}{5}$, $\cos A = -\dfrac{4}{5}$, $\sin B = \dfrac{\sqrt{5}}{5}$, and $\cos B = \dfrac{-2\sqrt{5}}{5}$. From formula 4,

$$\sin(A+B) = \sin A \cdot \cos B + \cos A \cdot \sin B$$
$$= \frac{-3}{5} \cdot \frac{-2\sqrt{5}}{5} + \frac{-4}{5} \cdot \frac{\sqrt{5}}{5}$$
$$= \frac{6\sqrt{5}}{25} - \frac{4\sqrt{5}}{25}$$
$$= \frac{2\sqrt{5}}{25} > 0$$

Therefore, the angle $(A+B)$ lies in quadrant I.

EXAMPLE 5: Find the value of $\cos 2\theta - \sin(90° + \theta)$ if $\tan\theta = -\dfrac{3}{4}$ and $\sin\theta$ is positive.

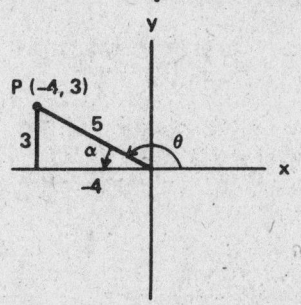

Since $\tan\theta < 0$ and $\sin\theta > 0$, the terminal side of angle θ must lie in the second quadrant. From the basic definition of the trigonometric functions and the Pythagorean Theorem
$$\sin\theta = \frac{3}{5} \text{ and } \cos\theta = \frac{-4}{5}.$$
From formula 11,
$$\cos 2\theta = \cos^2\theta - \sin^2\theta = \frac{16}{25} - \frac{9}{25} = \frac{7}{25}.$$
From formula 4,
$$\sin(90° + \theta) = \sin 90° \cdot \cos\theta + \cos 90° \cdot \sin\theta$$
$$= 1 \cdot \frac{-4}{5} + 0 \cdot \frac{3}{5} = -\frac{4}{5}$$
Therefore,
$$\cos 2\theta - \sin(90° + \theta) = \frac{7}{25} - \left(\frac{-4}{5}\right) = \frac{7}{25} + \frac{4}{5} = \frac{27}{25}$$

EXAMPLE 6: If $\sin\theta = a$, find the value of $\sin 2\theta$ in terms of a.

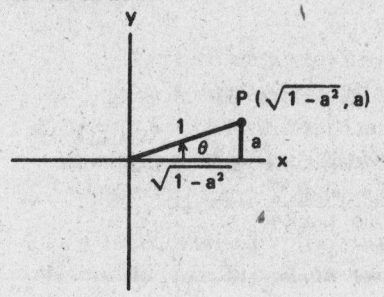

Choose θ to be an angle in any quadrant. (It can be chosen in the first quadrant for convenience and without any loss in generality.) From the basic definition of the trigonometric functions and the Pythagorean Theorem, the coordinates of P are $(\sqrt{1-a^2}, a)$. From formula 10, $\sin 2\theta = 2\sin\theta \cdot \cos\theta = 2 \cdot a \cdot \sqrt{1-a^2}$.

EXAMPLE 7: If $\cos 23° = z$, find the value of $\cos 46°$. Since $46 = 2(23)$, formula 12 can be used: $\cos 2A = 2\cos^2 A - 1$. $\cos 46° = \cos 2(23°) = 2\cos^2 23° - 1 = 2(\cos 23°)^2 - 1 = 2z^2 - 1$.

EXAMPLE 8: Solve $\sin 2x = 3\sin x$ for x where $0 < x < 2\pi$.
Use formula 10 to convert each term into functions of x only.
$2\sin x \cdot \cos x = 3\sin x$
$2\sin x \cdot \cos x - 3\sin x = 0$
Use the distributive property to factor the $\sin x$ out of each term:
$\sin x(2\cos x - 3) = 0$
Therefore, $\sin x = 0$ or $2\cos x - 3 = 0$. The second equation implies that $\cos x = \frac{3}{2}$, which is impossible because the range of $\text{cosine} = \{y : -1 \leq y \leq 1\}$. Therefore $\sin x = 0$ will contribute the only roots of the original equation. $\sin x = 0$ when $x = 0, \pi$.

EXAMPLE 9: Solve $\sin x + \cos 2x = 4\sin^2 x - 1$ for all values of x such that $0 \leq x < 2\pi$.
Use formula 13 to convert every term to $\sin x$. The equation becomes $2\sin x + 1 - 2\sin^2 x = 2\sin^2 x - 1$

$4\sin^2 x - 2\sin x - 2 = 0$
$2\sin^2 x - \sin x - 1 = 0$
$(2\sin x + 1)(\sin x - 1) = 0$
$2\sin x + 1 = 0$ or $\sin x - 1 = 0$
$\sin x = -\frac{1}{2}$ or $\sin x = 1$

Since sine is negative in quadrants III and IV and since $\sin\frac{\pi}{6} = \frac{1}{2}$, $\sin x = -\frac{1}{2}$ when $x = \frac{7\pi}{6}$ and $\frac{11\pi}{6}$. $\sin x = 1$ when $x = \frac{\pi}{2}$. Therefore, the solution set $= \left\{\frac{\pi}{2}, \frac{7\pi}{6}, \frac{11\pi}{6}\right\}$.

EXAMPLE 10: Solve $\tan 3x$ for all values of x such that $0° < x < 360°$.
Tan $3x$ is positive when the angle $3x$ is in quadrants I and III. Therefore, $3x = \frac{\pi}{4} = 45°$ or $3x = \frac{5\pi}{4} = 225°$. However, in order to find all values of x such that $0° < x < 360°$, two angles coterminal with $45°$ and $225°$ must also be found for values of $3x$. $3x = 45°, 45° + 360°, 45° + 720°$ or $3x = 225°, 225° + 360°, 225° + 720°$. Therefore, if $3x = 45°, 225°, 405°, 585°, 765°,$ or $945°$, then $x = 15°, 75°, 135°, 195°, 255°,$ or $315°$.

In this type of problem make sure that enough coterminal angles are chosen in order to determine all values of x. If n is the coefficient of x (in this problem it is 3), $n-1$ angles coterminal to the principal values are necessary.

EXAMPLE 11: Solve $\sin x = \sqrt{3}\cos x$ for all values of x such that $0 \leq x < 2\pi$.

Method 1: Since a solution is not obtained when $\cos x = 0$, it is possible to divide both sides of the equation by $\cos x$.

$\frac{\sin x}{\cos x} = \sqrt{3}$, which becomes $\tan x = \sqrt{3}$

$\tan x = \sqrt{3}$ when $x = \frac{\pi}{3}$ or $\frac{4\pi}{3}$

Method 2: Square both sides of the equation and use formula 1.

$\sin^2 x = 3\cos^2 x$
$\sin^2 x = 3(1 - \sin^2 x) = 3 - 3\sin^2 x$
$4\sin^2 x = 3$
$\sin^2 x = \frac{3}{4}$
$\sin x = \pm\frac{\sqrt{3}}{2}$
Therefore, $x = \frac{\pi}{3}, \frac{2\pi}{3}, \frac{4\pi}{3}, \frac{5\pi}{3}$

This method has introduced extra roots into the solution set, as the squaring of both sides of any equation is apt to do. If these four roots are checked in the original equation, only $\frac{\pi}{3}$ and $\frac{4\pi}{3}$ would be solutions.

Thus, the solution set obtained by either method is $\left\{\frac{\pi}{3}, \frac{4\pi}{3}\right\}$.

EXAMPLE 12: Solve $\sin x \cdot \sin 40° - \cos x \cdot \cos 40° = \frac{1}{2}$ for all values of x such that $0° < x < 360°$.

When this equation is multiplied through by -1, the left side is similar to the right side of formula 6. From formula 6, $\cos(x + 40°) = \cos x \cdot \cos 40° - \sin x \cdot \sin 40°$. Thus, this equation becomes $\cos(x + 40°) = -\frac{1}{2}$. Since $\cos\frac{2\pi}{3} = \cos\frac{4\pi}{3} = -\frac{1}{2}$ and $\frac{2\pi}{3} = 120°$ and $\frac{4\pi}{3} = 240°$, $x + 40° = 120°$ or $x + 40° = 240°$. Solving for x, the solution set is $\{80°, 200°\}$.

5. Identities, Equations, and Inequalities

EXAMPLE 13: Solve $\sqrt{3} \cos x - \sin x = 2$ for all values of x such that $0 < x < 2\pi$.

Method 1: Divide the equation through by 2, obtaining $\frac{\sqrt{3}}{2} \cos x - \frac{1}{2} \sin x = 1$. Since $\sin \frac{\pi}{3} = \frac{\sqrt{3}}{2}$ and $\cos \frac{\pi}{3} = \frac{1}{2}$, the equation becomes $\sin \frac{\pi}{3} \cdot \cos x - \cos \frac{\pi}{3} \cdot \sin x = 1$. Using formula 5, the equation becomes $\sin\left(\frac{\pi}{3} - x\right) = 1$. Therefore, $\frac{\pi}{3} - x = \frac{\pi}{2}$ and $x = \frac{\pi}{3} - \frac{\pi}{2} = -\frac{\pi}{6}$, which is not in the range $0 < x < 2\pi$. Add 2π to this angle in order to find a coterminal angle within this range: $-\frac{\pi}{6} + 2\pi = \frac{11\pi}{6}$.

Therefore, the solution set has only one member: $\left\{\frac{11\pi}{6}\right\}$.

Method 2: Add $\sin x$ to both sides of the equation and then square each side, obtaining:

$$(\sqrt{3} \cos x)^2 = (2 + \sin x)^2$$
$$3\cos^2 x = 4 + 4\sin x + \sin^2 x$$

Using a form of formula 1 ($\cos^2 x = 1 - \sin^2 x$) and setting everything equal to zero:

$$3(1 - \sin^2 x) = 4 + 4\sin x + \sin^2 x$$
$$3 - 3\sin^2 x = 4 + 4\sin x + \sin^2 x$$
$$4\sin^2 x + 4\sin x + 1 = 0$$
$$(2\sin x + 1)^2 = 0$$
$$\sin x = -\frac{1}{2}$$

Therefore, $x = \frac{7\pi}{6}$ or $\frac{11\pi}{6}$. Since $\frac{7\pi}{6}$ will not check in the original equation, the solution set $= \left\{\frac{11\pi}{6}\right\}$.

EXAMPLE 14: For what values of x between 0 and 2π are $\sin x < \cos x$?

Sketch the graphs of $y = \sin x$ and $y = \cos x$ on the same set of axes between 0 and 2π. Observe where the graph of $y = \sin x$ is below the graph of $y = \cos x$ (the sections indicated with xxxxx).

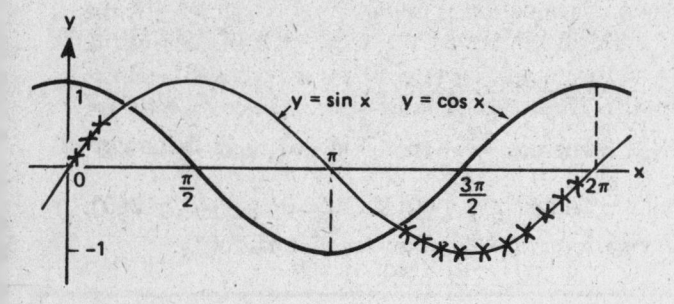

To find the points of intersection of the two graphs solve the equation

$$\sin x = \cos x$$
$$\frac{\sin x}{\cos x} = 1$$
$$\tan x = 1$$
$$x = \frac{\pi}{4} \text{ and } \frac{3\pi}{4}$$

Therefore, the set of x values that solve the inequality is $0 < x < \frac{\pi}{4}$ or $\frac{3\pi}{4} < x < 2\pi$.

EXERCISES

1. Solve $\cos 2x + \cos x + 1 = 0$ when $0 < x < 2\pi$. The solution set is

   [A] $\left\{\frac{\pi}{3}, \frac{\pi}{2}, \frac{3\pi}{2}, \frac{5\pi}{3}\right\}$

   [B] $\left\{\frac{2\pi}{3}, \frac{4\pi}{3}\right\}$

   [C] $\left\{\frac{\pi}{2}, \frac{3\pi}{2}\right\}$

   [D] $\left\{\frac{\pi}{2}, \frac{2\pi}{3}, \frac{4\pi}{3}, \frac{3\pi}{2}\right\}$

   [E] $\left\{\frac{\pi}{3}, \frac{5\pi}{3}\right\}$

2. Which of the following is the solution set of $\sin\theta \cdot \cos\theta = \frac{1}{4}$ when $0° < \theta < 360°$?

   [A] $\{30°, 150°\}$ [B] $\{30°, 150°, 210°, 330°\}$
   [C] $\{15°, 75°\}$ [D] $\{15°, 75°, 195°, 225°\}$
   [E] $\{60°, 300°\}$

3. How many positive values of $x \leq 2\pi$ make $\tan 4x = \sqrt{3}$?

   [A] 0 [B] 1 [C] 2 [D] 4 [E] 8

4. If $\cos\theta = \frac{3}{5}$, $\tan 2\theta$ equals

   [A] $\frac{24}{7}$ [B] $-\frac{24}{7}$ [C] $\frac{24}{5}$

   [D] $-\frac{24}{5}$ [E] $\pm\frac{24}{7}$

5. Solve the equation $\cos 2\theta = -\sin\theta$ for all positive values of $\theta < 2\pi$. θ equals

   [A] $\frac{\pi}{2}, \frac{7\pi}{6}, \frac{11\pi}{6}$ [B] $\frac{\pi}{2}, \frac{7\pi}{6}, \frac{3\pi}{2}, \frac{11\pi}{6}$

   [C] $\frac{\pi}{6}, \frac{\pi}{2}, \frac{5\pi}{6}$ [D] $\frac{\pi}{6}, \frac{\pi}{2}, \frac{5\pi}{6}, \frac{3\pi}{2}$

[6] If $\sin 37° = .6$, then $\sin 74°$ equals
[A] .96   [B] .84   [C] .76
[D] $\dfrac{\sqrt{10}}{10}$ [E] $\dfrac{3\sqrt{10}}{10}$

[7] $\text{Cot}(A+B)$ equals
[A] $\dfrac{\cot A \cdot \cot B + 1}{\cot A - \cot B}$
[B] $\dfrac{\cot A \cdot \cot B - 1}{\cot A + \cot B}$
[C] $\dfrac{\tan A - \tan B}{1 - \tan A \cdot \tan B}$
[D] $\dfrac{1 + \tan A \cdot \tan B}{\tan A - \tan B}$
[E] $\dfrac{\cot A \cdot \cot B + 1}{\cot B - \cot A}$

[8] $\dfrac{\tan 180° + \tan 70°}{1 - \tan 140° \cdot \tan 70°}$ equals
[A] $-\sqrt{3}$ [B] $\dfrac{\sqrt{3}}{3}$ [C] $\dfrac{\sqrt{3}}{1-\sqrt{3}}$
[D] $\sqrt{3}$ [E] $\dfrac{3-\sqrt{3}}{3}$

[9] Which of the following is equal to $\cos^4 40° - \sin^4 40°$?
[A] $\cos 80°$ [B] $\sin 80°$ [C] 0
[D] 1   [E] none of the above

[10] If $\sin A = \dfrac{2}{3}$ and $\cos A < 0$, find the value of $\tan 2A$.
[A] $\dfrac{4\sqrt{5}}{5}$ [B] $-4\sqrt{5}$ [C] $-\dfrac{4\sqrt{5}}{9}$
[D] $\dfrac{4\sqrt{5}}{9}$ [E] $4\sqrt{5}$

[11] What values of x satisfy the inequality $\sin 2x < \sin x$ such that $0 \le x \le 2\pi$?
[A] $\dfrac{\pi}{3} < x < \dfrac{5\pi}{3}$
[B] $0 < x < \dfrac{\pi}{3}$ or $\dfrac{5\pi}{3} < x < 2\pi$
[C] $\dfrac{2\pi}{3} < x < 2\pi$
[D] $\dfrac{\pi}{3} < x < \pi$ or $\dfrac{5\pi}{3} < x < 2\pi$
[E] $0 < x < \dfrac{\pi}{3}$ or $\pi < x < \dfrac{5\pi}{3}$

6. INVERSE FUNCTIONS

If the graph of any trigonometric function, $f(x)$, is reflected about the line $y = x$ (see chapter I, section 3), the graph of the inverse of that trigonometric function, $f^{-1}(x)$, results. In every case the resulting graph is *not* the graph of a function. In order to obtain a function, the range of the inverse relation must be severely limited. The particular range of each inverse trigonometric function is accepted by convention. The range of each inverse function is given as follows.

$-\dfrac{\pi}{2} \le \text{Sin}^{-1}x$ or $\text{Arcsin}\,x \le \dfrac{\pi}{2}$

$0 \le \text{Cos}^{-1}x$ or $\text{Arccos}\,x \le \pi$

$-\dfrac{\pi}{2} < \text{Tan}^{-1}x$ or $\text{Arctan}\,x < \dfrac{\pi}{2}$

$0 < \text{Cot}^{-1}x$ or $\text{Arccot}\,x < \pi$

$0 \le \text{Sec}^{-1}x$ or $\text{Arcsec}\,x \le \pi$ and $\text{Sec}^{-1}x \ne \dfrac{\pi}{2}$

$-\dfrac{\pi}{2} \le \text{Csc}^{-1}x$ or $\text{Arccsc}\,x \le \dfrac{\pi}{2}$ and $\text{Csc}^{-1}x \ne 0$

The last three inverse functions are rarely used. The inverse trigonometric functions are used to represent angles which cannot be expressed any other way without the help of tables or a calculator. It is important to understand that each inverse trigonometric function represents an angle.

EXAMPLE 1: Express each of the following in simpler form:

[A] $\text{Sin}^{-1}\dfrac{\sqrt{3}}{2}$ [B] $\text{Arccos}(-\dfrac{\sqrt{2}}{2})$
[C] $\text{Arctan}(-1)$ [D] $\text{Arccot}\sqrt{3}$
[E] $\text{Sec}^{-1}\sqrt{2}$ [F] $\text{Csc}^{-1}(-2)$

Referring to the ranges of the inverse trigonometric functions:

[A] $\text{Sin}^{-1}\dfrac{\sqrt{3}}{2} = \theta$. Find the principal value of θ (that angle which is in the range of $\text{Sin}^{-1}x$) such that $\text{Sin}\,\theta = \dfrac{\sqrt{3}}{2}$. Since $\sin\theta > 0$, θ must be in quadrant I, $\theta = \dfrac{\pi}{3}$.

[B] $\text{Arccos}(-\dfrac{\sqrt{2}}{2}) = \theta$. Find the principal value of θ such that $\cos\theta = -\dfrac{\sqrt{2}}{2}$. Since $\cos\theta < 0$, θ must be in quadrant II, $\theta = \dfrac{3\pi}{4}$.

[C] Arctan$(-1) = \theta$. Find the principal value of θ such that $\tan \theta = -1$. Since $\tan \theta < 0$, θ must be in the fourth quadrant and $-\frac{\pi}{2} < \text{Arctan } x < \frac{\pi}{2}$, $\theta = -\frac{\pi}{4}$.

[D] Arccot $\sqrt{3} = \theta$. Find the principal value of θ such that $\cot \theta = \sqrt{3}$. Since $\cot \theta > 0$, θ must be in the first quadrant, $\theta = \frac{\pi}{6}$.

[E] $\text{Sec}^{-1}\sqrt{2} = \theta$. Find the principal value of θ such that $\sec \theta = \sqrt{2}$. Since $\sec \theta > 0$, θ must be in the first quadrant, $\theta = \frac{\pi}{4}$.

[F] $\text{Csc}^{-1}(-2) = \theta$. Find the principal value of θ such that $\csc \theta = -2$. Since $\csc \theta < 0$, θ must be in the fourth quadrant and $-\frac{\pi}{2} < \text{Csc}^{-1} x < \frac{\pi}{2}$, $\theta = -\frac{\pi}{6}$.

EXAMPLE 2: Evaluate $\sin\left(\text{Arccos } \frac{3}{5}\right)$.

Let Arccos $\frac{3}{5} = \theta$. The problem now becomes "evaluate $\sin \theta$ when $\cos \theta = \frac{3}{5}$ and θ is in the first quadrant." From the basic definition of the trigonometric functions and the Pythagorean Theorem, $\sin \theta = \frac{4}{5}$.

Therefore, $\sin(\text{Arccos } \frac{3}{5}) = \frac{4}{5}$.

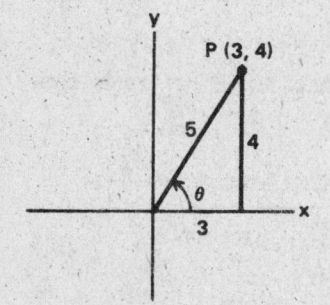

EXAMPLE 3: Evaluate $\cos\left(\text{Arcsin}\left(-\frac{3}{5}\right) + \text{Arccos } \frac{5}{13}\right)$.

Let $\text{Arcsin}\left(-\frac{3}{5}\right) = \alpha$ and Arccos $\frac{5}{13} = \beta$. The problem now becomes "evaluate $\cos(\alpha + \beta)$ when $\sin \alpha = -\frac{3}{5}$ and α is in the fourth quadrant, and $\cos \beta = \frac{5}{13}$ and β is in the first quadrant." From the basic definition of sine and cosine and the Pythagorean Theorem, $\cos \alpha = \frac{4}{5}$ and $\sin \beta = \frac{12}{13}$.

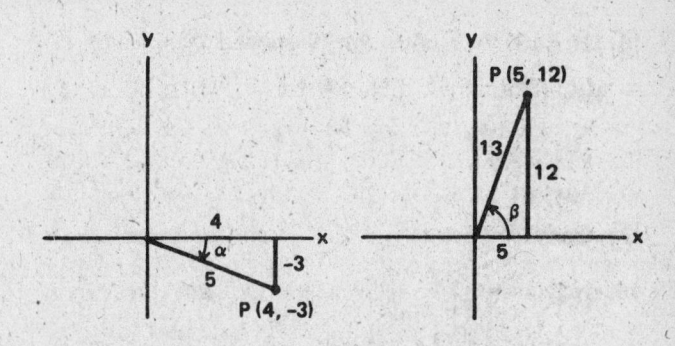

From formula 6,

$$\cos(\alpha + \beta) = \cos \alpha \cdot \cos \beta - \sin \alpha \cdot \sin \beta$$
$$= \frac{4}{5} \cdot \frac{5}{13} - \left(-\frac{3}{5}\right) \cdot \frac{12}{13}$$
$$= \frac{20}{65} + \frac{36}{65} = \frac{56}{65}$$

Therefore, $\cos\left(\text{Arcsin}\left(-\frac{3}{5}\right) + \text{Arccos } \frac{5}{13}\right) = \frac{56}{65}$.

EXAMPLE 4: Evaluate $\sin\left(2 \text{Arctan}\left(-\frac{8}{15}\right)\right)$.

Let $\theta = \text{Arctan}\left(-\frac{8}{15}\right)$. The problem now becomes "evaluate $\sin 2\theta$ when $\tan \theta = -\frac{8}{15}$ and θ is in the fourth quadrant." From the basic definition of the trigonometric functions and the Pythagorean Theorem, $\sin \theta = -\frac{8}{17}$ and $\cos \theta = \frac{15}{17}$.

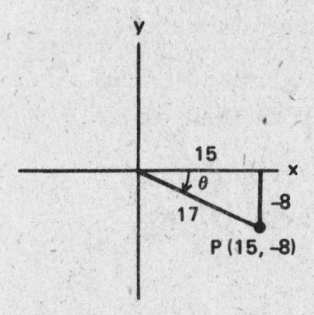

From formula 10,

$$\sin 2\theta = 2 \sin \theta \cdot \cos \theta = 2\left(-\frac{8}{17}\right) \cdot \left(\frac{15}{17}\right) = -\frac{240}{289}.$$

Therefore, $\sin\left(2 \text{Arctan}\left(-\frac{8}{15}\right)\right) = -\frac{240}{289}$.

EXAMPLE 5: Solve $3\sin^2 \theta + 10 \sin \theta - 8 = 0$ for all values of θ such that $0 \leq \theta < 2\pi$. Express angles in terms of the inverse trigonometric functions.

$3\sin^2 \theta + 10 \sin \theta - 8 = 0$ factors into $(3\sin \theta - 2)(\sin \theta + 4) = 0$, which leads to $3\sin \theta - 2 = 0$ or $\sin \theta + 4 = 0$. $\sin \theta = \frac{2}{3}$ or $\sin \theta = -4$. Since -4 is not in the range of the sine function, the second factor gives no solutions. Since an angle θ whose sine is $\frac{2}{3}$ is not readily known, θ must be expressed in terms of Arcsin. Since $\sin\theta > 0$, it is necessary to find angles in the first and second quadrants. In the first quadrant, $\theta = \text{Arcsin}\,\frac{2}{3}$. The second quadrant angle which has the same reference angles as Arcsin $\frac{2}{3}$ is $\pi - \text{Arcsin}\,\frac{2}{3}$. Therefore, the solution set = $\{\text{Arcsin}\,\frac{2}{3}, \pi - \text{Arcsin}\,\frac{2}{3}\}$.

EXERCISES

1. Solve for x: $\text{Arccos}(2x^2 - 2x) = \frac{2\pi}{3}$.

   [A] $\pm \frac{1}{2}$ [B] $\frac{1}{2}$ [C] $-\frac{1}{2}$
   [D] $\frac{\pi}{3}$ [E] 0

2. $\text{Arcsin}(\sin\frac{7\pi}{6}) =$

   [A] $-\frac{1}{2}$ [B] $\frac{\pi}{6}$ [C] $\frac{7\pi}{6}$
   [D] $-\frac{\pi}{6}$ [E] $\frac{1}{2}$

3. Which of the following is (are) true?
   I. $\text{Arcsin}\,1 + \text{Arcsin}(-1) = 0$
   II. $\text{Cos}^{-1}(1) + \text{Cos}^{-1}(-1) = 0$
   III. $\text{Arccos}\,x = \text{Arccos}(-x)$ for all values of x in the domain of Arccos.

   [A] I only   [B] II only
   [C] III only   [D] I and II only
   [E] II and III only

4. Express in terms of an inverse trigonometric function the angle that a diagonal of a cube makes with the base of the cube.

   [A] $\text{Arcsin}\,\frac{\sqrt{2}}{2}$ [B] $\text{Arcsin}\,\frac{\sqrt{3}}{3}$
   [C] $\text{Arccos}\,\frac{\sqrt{6}}{2}$ [D] $\text{Arccos}\,\frac{\sqrt{3}}{3}$
   [E] $\text{Arctan}\,\sqrt{2}$

5. $\tan\left(\text{Arcsin}\left(-\frac{3}{5}\right)\right)$ equals

   [A] $\frac{3}{4}$ [B] $-\frac{3}{4}$ [C] $\frac{4}{3}$
   [D] $-\frac{4}{3}$ [E] $-\frac{4}{5}$

6. Using only principal values, $\text{Arcsin}\left(-\frac{5}{13}\right) + \text{Arccos}\left(-\frac{3}{5}\right)$ represents an angle lying in which quadrant?

   [A] I   [B] II   [C] III   [D] IV   [E] I or II

7. Which of the following is not defined?

   [A] $\text{Arcsin}\,\frac{1}{9}$ [B] $\text{Arccos}\left(-\frac{4}{3}\right)$
   [C] $\text{Arctan}\,\frac{11}{12}$ [D] $\text{Arccot}(-4)$
   [E] $\text{Arcsec}\,3\pi$

8. Which of the following is a solution of $\cos 3x = \frac{1}{2}$?

   [A] $60°$ [B] $\frac{5\pi}{3}$ [C] $\text{Arccos}\,\frac{1}{6}$
   [D] $\text{Arccos}\,\frac{\sqrt{3}}{2}$ [E] $\frac{1}{3}\text{Arccos}\,\frac{1}{2}$

7. TRIANGLES

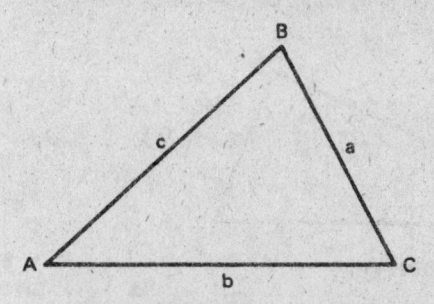

The final topic in trigonometry concerns the relationship between the angles and sides of a triangle which is *not* a right triangle. Depending upon which of the sides and angles of the triangle are supplied, the following formulas are helpful. In $\triangle ABC$.

Law of Sines: $\frac{\sin A}{a} = \frac{\sin B}{b} = \frac{\sin C}{c}$

used when two sides and the angle opposite one or two angles and one side are given.

Law of Cosines: $a^2 = b^2 + c^2 - 2bc \cdot \cos A$
$b^2 = a^2 + c^2 - 2ac \cdot \cos B$
$c^2 = a^2 + b^2 - 2ab \cdot \cos C$

used when two sides and the included angle or three sides are given.

7. Triangles

Area of a △: Area $= \frac{1}{2} bc \cdot \sin A$

Area $= \frac{1}{2} ac \cdot \sin B$

Area $= \frac{1}{2} ab \cdot \sin C$

used when two sides and the included angle are given.

EXAMPLE 1: Find the number of degrees in the largest angle of a triangle whose sides are 3, 5, and 7.

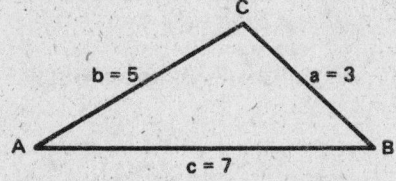

The largest angle is opposite the longest side. Use the law of cosines:

$$c^2 = a^2 + b^2 - 2ab \cdot \cos C$$
$$49 = 9 + 25 - 30 \cdot \cos C$$

Therefore, $\cos C = -\frac{15}{30} = -\frac{1}{2}$.

Since $\cos C < 0$ and $\angle C$ is an angle of a triangle, $90° < \angle C < 180°$.
Therefore, $\angle C = 120°$.

EXAMPLE 2: Find the number of degrees in the other two angles of a △ABC if $c = 75\sqrt{2}$, $b = 150$, and $\angle C = 30°$.

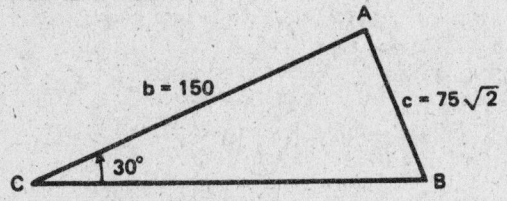

Use the law of sines:

$$\frac{75\sqrt{2}}{\sin 30°} = \frac{150}{\sin B}$$

$$75\sqrt{2} \cdot \sin B = 150 \cdot \sin 30°$$

$$\sin B = \frac{150 \cdot \frac{1}{2}}{75\sqrt{2}} = \frac{75}{75\sqrt{2}} = \frac{1}{\sqrt{2}} = \frac{\sqrt{2}}{2}$$

Therefore, B = 45° or 135°; A = 105° or 15° since there are 180° in the sum of the three angles of a triangle.

EXAMPLE 3: Find the area of the triangle ABC if $a = 180$ inches, $b = 150$ inches, and $\angle C = 30°$.

Area $= \frac{1}{2} ab \cdot \sin C = \frac{1}{2} \cdot 180 \cdot 150 \cdot \sin 30° = \frac{1}{2} \cdot 180 \cdot 150 \cdot \frac{1}{2} = 90 \cdot 75 = 675$ square inches.

If two sides of a triangle and the angle opposite one of those sides is given, it is possible that 2 triangles, 1 triangle, or 0 triangles could be constructed with the data. This is called the *ambiguous case*. If sides a and b and $\angle A$ are given, the length of side b determines the number of triangles that can be constructed.

Case 1: If $\angle A > 90°$ and $a < b$, no triangle can be formed because side a would not reach the base line. If $a > b$, one obtuse triangle can be drawn.

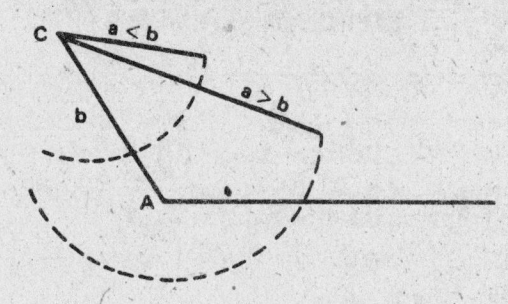

Let the altitude from C to the base line be h. From the basic definition of sine, $\sin A = \frac{h}{b}$, and thus, $h = b \cdot \sin A$.

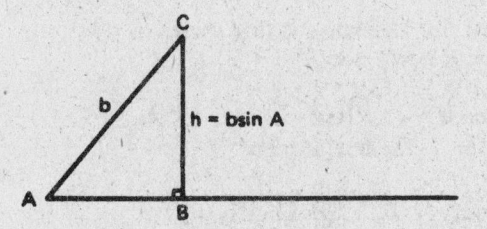

Case 2: If $\angle A < 90°$ and side $a < b \cdot \sin A$, no triangle can be formed. If $a = b \cdot \sin A$, one triangle can be formed. If $a > b$, there also will be only one triangle. If, on the other hand, $b \cdot \sin A < a < b$, there will be two triangles that can be formed.

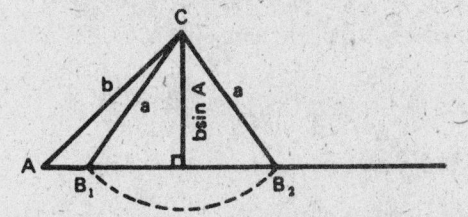

If a compass is opened the length of side a, and a circle is drawn with center at C, the circle will cut the base line at two points, B_1 and B_2. Thus, △AB_1C satisfies the conditions of the problem, as does △AB_2C.

EXAMPLE 1: How many triangles can be formed if $a=24$, $b=31$, and $\angle A = 30°$?

Because $\angle A < 90°$, $b \cdot \sin A = 31 \cdot \sin 30° = 31 \cdot \frac{1}{2} = 15\frac{1}{2}$. Since $b \cdot \sin A < a < b$, there will be two triangles.

EXAMPLE 2: How many triangles can be formed if $a=24$, $b=31$, and $\angle A = 150°$?

Since $\angle A > 90°$ and $a < b$, no triangle can be formed.

EXERCISES

1. In $\triangle ABC$, $\angle A = 30°$, $b = 8$, and $a = 4\sqrt{2}$. $\angle C$ could equal

   [A] 45° [B] 135° [C] 60°
   [D] 15° [E] 90°

2. In $\triangle ABC$, $\angle A = 30°$, $a = 6$, and $c = 8$. Which of the following must be true?

   [A] $0° < \angle C < 90°$
   [B] $90° < \angle C < 180°$
   [C] $45° < \angle C < 135°$
   [D] $0° < \angle C < 45°$ or $90° < \angle C < 135°$
   [E] $0° < \angle C < 45°$ or $135° < \angle C < 180°$

3. The angles of a triangle are in a ratio of $8:3:1$. The ratio of the longest side of the triangle to the next longest side is

   [A] $\sqrt{6}:2$ [B] $8:3$ [C] $\sqrt{3}:1$
   [D] $8:5$ [E] $2\sqrt{2}:\sqrt{3}$

4. The sides of a triangle are in a ratio of $4:5:6$. The cosine of the smallest angle is

   [A] $-\frac{1}{8}$ [B] $\frac{\sqrt{8}}{8}$ [C] $-\frac{9}{16}$
   [D] $\frac{3}{4}$ [E] $-\frac{3}{4}$

5. Find the length of the longer diagonal of a parallelogram if the sides are 6 inches and 8 inches and the smaller angle is 60°.

   [A] $2\sqrt{13}$ [B] $2\sqrt{31}$ [C] $2\sqrt{37}$
   [D] 7 [E] $100 - 48\sqrt{3}$

6. What are all values of side a in the figure such that two triangles can be constructed?

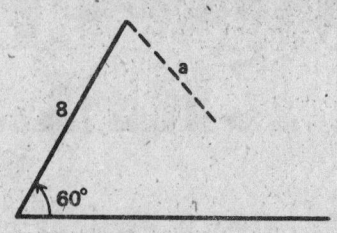

   [A] $a > 4\sqrt{3}$ [B] $a > 8$ [C] $a = 4\sqrt{3}$
   [D] $4\sqrt{3} < a < 8$ [E] $8 < a < 8\sqrt{3}$

7. In $\triangle ABC$, $\angle B = 30°$, $\angle C = 105°$, and $b = 10$. The length of side a equals

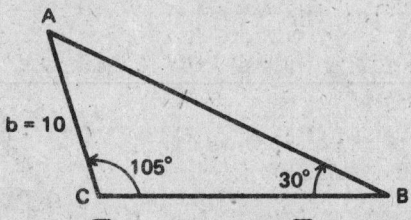

   [A] $5\sqrt{2}$ [B] $10\sqrt{3}$ [C] $5\sqrt{3}$
   [D] 10 [E] $10\sqrt{2}$

8. The area of $\triangle ABC = 24\sqrt{3}$, side $a = 6$, and side $b = 16$. The size of $\angle C$ is

   [A] 30° [B] 30° or 150° [C] 60°
   [D] 60° or 120° [E] none of the above

9. The area of $\triangle ABC = 12\sqrt{3}$, side $a = 6$, and side $b = 8$. Side c equals

   [A] $2\sqrt{37}$ [B] $2\sqrt{13}$
   [C] $2\sqrt{37}$ or $2\sqrt{13}$ [D] 10
   [E] 10 or 12

10. Given the following data, which will form two triangles?

    I. $\angle C = 30°$, $c = 8$, $b = 12$
    II. $\angle B = 45°$, $a = 12\sqrt{2}$, $b = 15\sqrt{2}$
    III. $\angle C = 60°$, $b = 12$, $c = 5\sqrt{3}$

    [A] I only [B] II only [C] III only
    [D] I and II only [E] I and III only

MISCELLANEOUS RELATIONS AND FUNCTIONS

CHAPTER 4

1. CONIC SECTIONS

The general quadratic equation in two variables has the form $Ax^2 + Bxy + Cy^2 + Dx + Ey + F = 0$, where A, B, and C are not all zero. Depending upon the value of the coefficients A, B, or C, the equation represents either a circle, a parabola, a hyperbola, an ellipse, or a degenerate case of one of these (e.g., a point, a line, two parallel lines, two intersecting lines, or no graph at all).

If the graph is not a degenerate case the following indicates which conic section the equation represents:

If $B^2 - 4AC < 0$ and $A = C$, the graph is a circle.
If $B^2 - 4AC < 0$ and $A \neq C$, the graph is an ellipse.
If $B^2 - 4AC = 0$, the graph is a parabola.
If $B^2 - 4AC > 0$, the graph is a hyperbola.

Most of the conic sections encountered on the Level II examination will not have an xy term (i.e., $B = 0$). This term causes the graph to be rotated so that a major axis of symmetry is not parallel to either the x- or y-axis. The general quadratic equation in two variables can be transformed into a more useful form by completing the square separately on the x's and separately on the y's. After completing the square and, for convenience, changing the letters used as constants, the equation becomes:

a) for a circle, $(x-h)^2 + (y-k)^2 = r^2$, where (h,k) are the coordinates of the center and r is the radius of the circle.

b) for an ellipse, if $C > A$, $\dfrac{(x-h)^2}{a^2} + \dfrac{(y-k)^2}{b^2} = 1$, where (h,k) are the coordinates of the center, and the major axis is parallel to the x-axis (Figure a). If $C < A$, $\dfrac{(x-h)^2}{b^2} + \dfrac{(y-k)^2}{a^2} = 1$, where (h,k) are the coordinates of the center, and the major axis is parallel to the y-axis (Figure b).

41

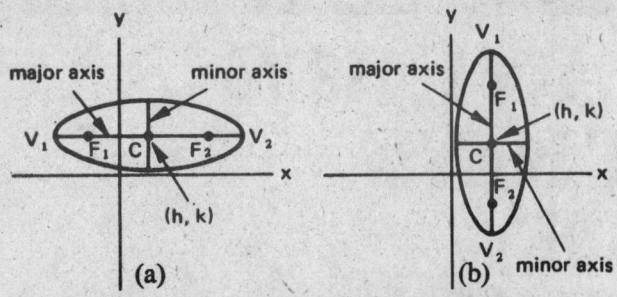

42 Chapter 4 Miscellaneous Relations and Functions

(a) [figure: ellipse with major axis horizontal, vertices V_1, V_2, foci F_1, F_2, center C at (h,k)]

(b) [figure: ellipse with major axis vertical, vertices V_1, V_2, foci F_1, F_2, center C at (h,k)]

In both cases the length of the *major axis* is $2a$, and the length of the *minor axis* is $2b$.

c) for a parabola, if $C=0$, $(x-h)^2 = 4p(y-k)$ (Figure a). If $A=0$, $(y-k)^2 = 4p(x-h)$ (Figure b). In both cases (h,k) are the coordinates of the vertex, and p is the directed distance from the vertex to the focus.

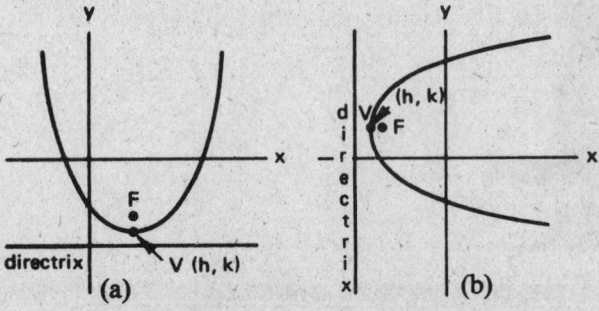

(a) [figure: parabola opening upward, vertex $V(h,k)$, focus F, directrix below]

(b) [figure: parabola opening to the right, vertex $V(h,k)$, focus F, directrix at left]

d) for a hyperbola, $\dfrac{(x-h)^2}{a^2} - \dfrac{(y-k)^2}{b^2} = 1$, where (h,k) are the coordinates of the center, and the graph opens to the side (Figure a). If the graph opens up and down, the equation is $\dfrac{(y-k)^2}{a^2} - \dfrac{(x-h)^2}{b^2} = 1$, where (h,k) are the coordinates of the center (Figure b).

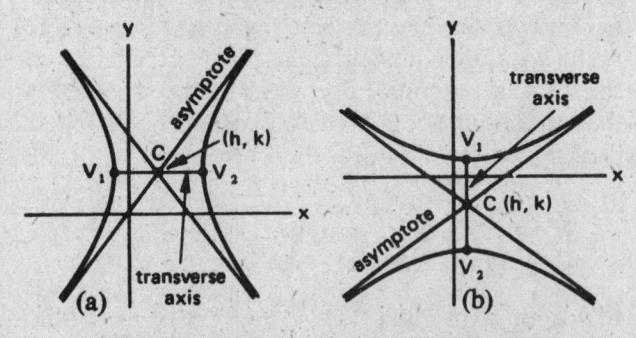

(a) [figure: hyperbola opening sideways with transverse axis horizontal, vertices V_1, V_2, center $C(h,k)$, asymptotes]

(b) [figure: hyperbola opening up/down with transverse axis vertical, vertices V_1, V_2, center $C(h,k)$, asymptotes]

In both cases the length of the *transverse axis* is $2a$, and the length of the *conjugate axis* is $2b$.

An *ellipse* is the set of points in a plane such that the sum of the distances from each point to two fixed points, called *foci*, is a constant equal to $2a$ in the formula above. The distance between the center and a focus is called c and is equal to $\sqrt{a^2 - b^2}$. The *eccentricity* of an ellipse is always less than 1 and is equal to $\dfrac{c}{a}$. The chord through a focus perpendicular to the major axis is called a *latus rectum* and is $\dfrac{2b^2}{a}$ units long.

A *parabola* is the set of points in a plane such that the distance from each point to a fixed point, called the *focus*, is equal to the distance to a fixed line, called the *directrix*. The eccentricity is always equal to one. The chord through the focus perpendicular to the axis of symmetry is called the *latus rectum* and is $4p$ units long.

A *hyperbola* is the set of points in a plane such that the absolute value of the difference of the distances from each point to two fixed points, called *foci*, is a constant equal to $2a$ in the formula above. The distance between the center and a focus is called c and is equal to $\sqrt{a^2 + b^2}$. The eccentricity of a hyperbola is always greater than one and is equal to $\dfrac{c}{a}$. Every hyperbola has associated with it two lines, called *asymptotes*, that intersect at the center. In general, an asymptote is a line which a curve approaches, but never quite touches, as one or both variables get increasingly larger or smaller. If the hyperbola opens to the side, the slopes of the two asymptotes are $\pm \dfrac{b}{a}$. If the hyperbola opens up and down, the slopes of the two asymptotes are $\pm \dfrac{a}{b}$. The chord through a focus perpendicular to the transverse axis is called a *latus rectum* and is equal to $\dfrac{2b^2}{a}$ units.

The equation $xy = k$, where k is a constant, is the equation of a *rectangular hyperbola* whose asymptotes are the x and y-axes. If $k > 0$, the branches of the hyperbola lie in the first and third quadrants. If $k < 0$, the branches lie in the second and fourth quadrants.

EXAMPLE 1: Each of the following is an equation of a conic. State which one and find, if they exist: I the coordinates of the center, II the coordinates of the vertices, III the coordinates of the foci, IV the eccentricity, V the equations of the asymptotes. Sketch the graph.

[A] $9x^2 - 16y^2 - 18x + 96y + 9 = 0$
[B] $4x^2 + 4y^2 - 12x - 20y - 2 = 0$
[C] $4x^2 + y^2 + 24x - 16y = 0$
[D] $y^2 + 6x - 8y + 4 = 0$

In all cases $B = 0$.

[A] $B^2 - 4AC = 0 - 4 \cdot 9 \cdot (-16) > 0$, which means the graph will be a hyperbola. In order to complete the square, group the x-terms and the y-terms. Factor out of each group the coefficient of the quadratic term:

$$9(x^2 - 2x\ ) - 16(y^2 - 6y\ ) + 9 = 0.$$

A perfect trinomial square is obtained within both parentheses by taking $\frac{1}{2}$ of the coefficient of the linear term, squaring it, and adding it to the existing polynomial. A similar amount must be added to the right side of the equation.

$$9(x^2 - 2x + 1) - 16(y^2 - 6y + 9) + 9 = 9 \cdot 1 - 16 \cdot 9$$
$$9(x-1)^2 - 16(y-3)^2 = -144$$

Divide each term by -144 in order to get the equation in the proper form: $\dfrac{(y-3)^2}{9} - \dfrac{(x-1)^2}{16} = 1$

I. Center $(1, 3) \cdot a = 3$, $b = 4$, so $c = \sqrt{9 + 16} = 5$.
II. The vertices are 3 units above and 3 units below the center. Vertices $(1, 6)$ and $(1, 0)$.
III. The foci are 5 units above and 5 units below the center. Foci $(1, 8)$ and $(1, -2)$.
IV. Eccentricity $= \dfrac{c}{a} = \dfrac{5}{3}$.
V. The slopes of the asymptotes are $\pm \dfrac{a}{b} = \pm \dfrac{3}{4}$. The asymptotes pass through the center $(1, 3)$. Therefore, the equations of the asymptotes are $y - 3 = \pm \dfrac{3}{4}(x - 1)$

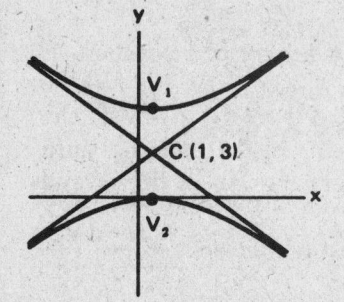

[B] $B^2 - 4AC = 0 - 4 \cdot 4 \cdot 4 < 0$ and $A = C = 4$, which means the graph will be a circle. Divide through by 4, group the x-terms together, group the y-terms together, and complete the square.

$$\left(x^2 - 3x + \frac{9}{4}\right) + \left(y^2 - 5y + \frac{25}{4}\right) - \frac{2}{4} = \frac{9}{4} + \frac{25}{4}$$

$$\left(x - \frac{3}{2}\right)^2 + \left(y - \frac{5}{2}\right)^2 = 9$$

I. Center $\left(\dfrac{3}{2}, \dfrac{5}{2}\right)$. Radius $= 3$. None of the other items is defined.

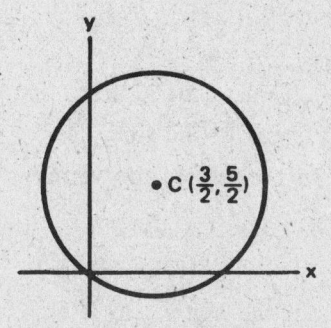

[C] $B^2 - 4AC = 0 - 4 \cdot 4 \cdot 1 < 0$ and $A \neq C$, which means the graph will be an ellipse. Group the x-terms, group the y-terms, and factor out of each group the coefficient of the quadratic term:

$$4(x^2 + 6x\ ) + 1(y^2 - 16y\ ) = 0$$

Complete the square and add similar amounts to both sides of the equation:

$$4(x^2 + 6x + 9) + 1(y^2 - 16y + 64) = 4 \cdot 9 + 1 \cdot 64$$
$$4(x+3)^2 + (y-8)^2 = 100$$

Divide each term by 100 in order to get the equation in the proper form:

$$\frac{(x+3)^2}{25} + \frac{(y-8)^2}{100} = 1$$

I. Center $(-3, 8)$. $a = 10$, $b = 5$, so $c = \sqrt{100 - 25} = \sqrt{75} = 5\sqrt{3}$.
II. Since the major axis is vertical, the vertices are 10 units above and 10 units below the center. Vertices $(-3, 18)$ and $(-3, -2)$.
III. The foci are $5\sqrt{3}$ units above and $5\sqrt{3}$ units below the center. Foci $(-3, 8 + 5\sqrt{3})$ and $(-3, 8 - 5\sqrt{3})$.
IV. Eccentricity $= \dfrac{c}{a} = \dfrac{5\sqrt{3}}{10} = \dfrac{\sqrt{3}}{2}$
V. There are no asymptotes.

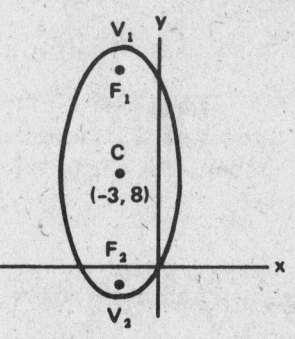

[D] $B^2 - 4AC = 0 - 4 \cdot 0 \cdot 1 = 0$, which means the graph will be a parabola. Group the y-terms and complete the square.

$$(y^2 - 8y + 16) = -6x - 4 + 16$$
$$(y-4)^2 = -6(x-2)$$

I. No center.
II. Vertex (2,4).
III. $4p = -6$, so $p = -\frac{3}{2}$. Since the y-term is squared, the parabola opens to the side. Thus, the focus is $\frac{3}{2}$ units to the left of the vertex.

Focus $(\frac{1}{2}, 4)$.

IV. Eccentricity = 1.
V. No asymptotes.

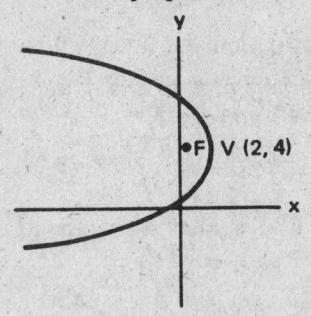

EXAMPLE 2: Find the equation of the hyperbola with center at $(3, -4)$, eccentricity 4, and conjugate axis of length 6.

Two cases are possible: $\frac{(x-3)^2}{a^2} - \frac{(y+4)^2}{b^2} = 1$ or $\frac{(y+4)^2}{a^2} - \frac{(x-3)^2}{b^2} = 1$. Eccentricity $= \frac{c}{a} = 4$. Thus $c = 4a$. Conjugate axis $= 2b = 6$. Thus, $b = 3$. In the case of a hyperbola,

$$c^2 = a^2 + b^2$$
$$(4a)^2 = a^2 + 9$$
$$16a^2 = a^2 + 9$$
$$15a^2 = 9$$
$$a^2 = \frac{9}{15}$$

Therefore, the equations become:

$$\frac{15(x-3)^2}{9} - \frac{(y+4)^2}{9} = 1 \text{ or } \frac{15(y+4)^2}{9} - \frac{(x-3)^2}{9} = 1.$$

The *degenerate cases* occur when one of the variables drops out, or the terms with the variables equal zero or a negative number after completing the square.

EXAMPLE 3: What is the graph of each of the following:

[A] $x^2 + y^2 - 4x + 2y + 5 = 0$
[B] $xy = 0$
[C] $2x^2 - 3y^2 + 8x + 6y + 5 = 0$
[D] $3x^2 + 4y^2 - 6x - 16y + 19 = 0$
[E] $x^2 + y^2 + 5 = 0$

[A] $B^2 - 4AC = 0 - 4 \cdot 1 \cdot 1 < 0$ and $A = C$, which indicates the graph should be a circle. Completing the square on the x's and y's separately gives:

$$(x^2 - 4x + 4) + (y^2 + 2y + 1) + 5 = 0 + 4 + 1$$
$$(x-2)^2 + (y+1)^2 = 0$$

This is a circle with center $(2, -1)$ and radius zero; thus, a degenerate case. The graph is the single point $(2, -1)$.

[B] $xy = 0$ if and only if $x = 0$ or $y = 0$. Thus, the graph is two intersecting lines: the x-axis ($y = 0$) and the y-axis ($x = 0$).

[C] $B^2 - 4AC = 0 - 4 \cdot 2 \cdot (-3) > 0$, which indicates the graph should be a hyperbola. Factoring and completing the square on the x's and y's separately gives:

$$2(x^2 + 4x + 4) - 3(y^2 - 2y + 1) + 5 = 0 + 2 \cdot 4 - 3 \cdot 1$$
$$2(x+2)^2 - 3(y-1)^2 = 0$$

Since the right side of the equation is zero and the left side is the difference between two perfect squares, the left side of the equation can be factored and simplified.

$$[\sqrt{2}(x+2) + \sqrt{3}(y-1)] \cdot [\sqrt{2}(x+2) - \sqrt{3}(y-1)] = 0$$
$$[\sqrt{2}x + 2\sqrt{2} + \sqrt{3}y - \sqrt{3}] = 0$$
or $[\sqrt{2}x + 2\sqrt{2} - \sqrt{3}y + \sqrt{3}] = 0$

$$y = -\frac{\sqrt{2}}{\sqrt{3}}x + \frac{\sqrt{3} - 2\sqrt{2}}{\sqrt{3}}$$

or $y = \frac{\sqrt{2}}{\sqrt{3}}x + \frac{\sqrt{3} + 2\sqrt{2}}{\sqrt{3}}$

Thus, the graph is two lines intersecting at $(-2, 1)$, one with slope $-\frac{\sqrt{2}}{\sqrt{3}}$ and y-intercept $\frac{\sqrt{3} - 2\sqrt{2}}{\sqrt{3}}$ and the other with slope $\frac{\sqrt{2}}{\sqrt{3}}$ and y-intercept $\frac{\sqrt{3} + 2\sqrt{2}}{\sqrt{3}}$.

[D] $B^2 - 4AC = 0 - 4 \cdot 3 \cdot 4 < 0$ and $A \neq C$, which indicates the graph should be an ellipse. Factoring and completing the square on the x's and y's separately gives:

$$3(x^2 - 2x + 1) + 4(y^2 - 4y + 4) + 19 = 0 + 3 \cdot 1 + 4 \cdot 4$$
$$3(x-1)^2 + 4(y-2)^2 = 0$$

Since $3(x-1)^2 \geq 0$ and $4(y-2)^2 \geq 0$, the only point that satisfies the equation is $(1, 2)$, which makes each term on the left side of the equation equal to zero. The graph is the one point $(1, 2)$.

[E] $x^2 + y^2 = -5$. The graph does not exist since $x^2 \geq 0$ and $y^2 \geq 0$, and there is no way that their sum could equal -5.

EXERCISES

1. Which of the following are the coordinates of a focus of $5x^2 + 4y^2 - 20x + 8y + 4 = 0$?

   [A] $(1, -1)$ [B] $(2, -1)$ [C] $(3, -1)$
   [D] $(2, -2)$ [E] $(-2, 1)$

2. If the graphs of $x^2 + y^2 = 4$ and $xy = 1$ are drawn on the same set of axes, how many points will they have in common?

   [A] 0   [B] 2   [C] 4   [D] 1   [E] 3

3. The graph of $(x-2)^2 = 4y$ has

   [A] vertex at $(4, 2)$
   [B] focus at $(2, 0)$
   [C] directrix $y = -1$
   [D] latus rectum 2 units in length
   [E] none of these

4. Which of the following is an asymptote of $3x^2 - 4y^2 - 12 = 0$?

   [A] $y = \frac{4}{3}x$ [B] $y = -\frac{2}{\sqrt{3}}$

   [C] $y = -\frac{3}{4}x$ [D] $y = \frac{\sqrt{3}}{2}x$

   [E] $y = \frac{2\sqrt{3}}{3}x$

5. The graph of $x^2 = (2y+3)^2$ is

   [A] a circle   [B] an ellipse
   [C] a hyperbola   [D] a point
   [E] two intersecting lines

6. The area bounded by the curve $y = \sqrt{4 - x^2}$ and the x-axis is

   [A] 4π [B] 8π [C] 16π
   [D] 2π [E] π

---

2. EXPONENTIAL AND LOGARITHMIC FUNCTIONS

The basic properties of exponents and logarithms and the fact that the exponential function and the logarithmic function are inverses lead to many interesting problems.

The basic exponential properties:

For all positive real numbers x and y, and all real numbers a and b:

$$x^a \cdot x^b = x^{a+b} \qquad x^0 = 1$$
$$\frac{x^a}{x^b} = x^{a-b} \qquad x^{-a} = \frac{1}{x^a}$$
$$[x^a]^b = x^{ab} \qquad x^a \cdot y^a = (xy)^a$$

The basic logarithmic properties:

For all positive real numbers a, b, p, and q, and all real numbers x, where $a \neq 1$ and $b \neq 1$:

$$\log_b(p \cdot q) = \log_b p + \log_b q \qquad \log_b 1 = 0 \qquad b^{\log_b p} = p$$
$$\log_b\left(\frac{p}{q}\right) = \log_b p - \log_b q \qquad \log_b b = 1$$
$$\log_b(p^x) = x \cdot \log_b p \qquad \log_b p = \frac{\log_a p}{\log_a b}$$

The basic property that relates the exponential and logarithmic functions is:

For all real numbers x, and all positive real numbers b and N,

$$\log_b N = x \text{ is equivalent to } b^x = N.$$

EXAMPLE 1: Simplify $x^{n-1} \cdot x^{2n} \cdot (x^{2-n})^2$

This is equal to $x^{n-1} \cdot x^{2n} \cdot x^{4-2n} = x^{n-1+2n+4-2n} = x^{n+3}$.

EXAMPLE 2: Simplify $\dfrac{3^{n-2} \cdot 9^{2-n}}{3^{2-n}}$.

In order to combine exponents using the properties above, the base of each factor must be the same.

$$\frac{3^{n-2} \cdot 9^{2-n}}{3^{2-n}} = \frac{3^{n-2} \cdot (3^2)^{2-n}}{3^{2-n}} = \frac{3^{n-2} \cdot 3^{4-2n}}{3^{2-n}}$$

$$= 3^{n-2+4-2n-(2-n)} = 3^0 = 1.$$

EXAMPLE 3: If $\log_{10} 23 = 1.36$, what does $\log_{10} 2300$ equal?

$\log_{10} 2300 = \log_{10} 23 \cdot 100 = \log_{10} 23 + \log_{10} 100 = 1.36 + \log_{10} 10^2 =$

$1.36 + 2 \cdot \log_{10} 10 = 1.36 + 2 = 3.36.$

EXAMPLE 4: If $\log_{10} 2 = .301$ and $\log_{10} 3 = .477$, find the value of $\log_{10} 15$.

$\log_{10} 15 = \log_{10} 3 \cdot 5 = \log_{10} 3 + \log_{10}\left(\frac{10}{2}\right) = \log_{10} 3 + \log_{10} 10 - \log_{10} 2 =$

$.477 + 1 - .301 = 1.176.$

EXAMPLE 5: Solve for x: $\log(x+5) = \log x + \log 5$. (When no base is indicated, an arbitrary base, b, is understood.)
$\log x + \log 5 = \log 5 \cdot x$
Therefore, $\log(x+5) = \log(5x)$, which is true only when:
$$x + 5 = 5x$$
$$5 = 4x$$
$$x = \frac{5}{4}$$

EXAMPLE 6: Evaluate $\log_{27}\sqrt{54} - \log_{27}\sqrt{6}$
$$\log_{27}\sqrt{54} - \log_{27}\sqrt{6} = \log_{27}\left(\frac{\sqrt{54}}{\sqrt{6}}\right)$$
$$= \log_{27}\sqrt{9} = \log_{27}3 = x.$$

This last equality implies that
$$27^x = 3$$
$$(3^3)^x = 3$$
$$3^{3x} = 3^1$$
Therefore, $3x = 1$ and $x = \frac{1}{3}$.

Thus, $\log_{27}\sqrt{54} - \log_{27}\sqrt{6} = \frac{1}{3}$.

EXAMPLE 7: Evaluate $\log_8 2 - \log_6 216 + \log_{81} 3 - \log_5 (625)^{1/3}$.
Since each term has a different base, they must be evaluated separately.
Let $\log_8 2 = x$. This implies that
$$8^x = 2$$
$$(2^3)^x = 2$$
$$2^{3x} = 2^1$$
Therefore, $3x = 1$ and $x = \frac{1}{3}$.

Let $\log_6 216 = y$. This implies that
$$6^y = 216$$
$$6^y = 6^3$$
Therefore, $y = 3$.

Let $\log_3 81 = z$. This implies that
$$3^z = 81$$
$$3^z = 3^4$$
Therefore, $z = 4$.

$\text{Log}_5(625)^{1/3} = \frac{1}{3} \cdot \log_5 625$. Let $\log_5 625 = w$. This implies that
$$5^w = 625$$
$$5^w = 5^4$$
Therefore, $w = 4$. Thus, $\log_5(625)^{1/3} = \frac{1}{3} \cdot 4 = \frac{4}{3}$. Putting the four parts together, the original expression is equal to $\frac{1}{3} - 3 + 4 - \frac{4}{3} = 0$.

The graphs of all exponential functions $y = b^x$ have roughly the same shape and pass through the point $(0, 1)$. If $b > 1$, the graph increases as x increases and approaches the x-axis as an asymptote as x decreases. The amount of curvature becomes greater as the value of b is made greater. If $0 < b < 1$, the graph increases as x decreases and approaches the x-axis as an asymptote as x increases. The amount of curvature becomes greater as the value of b is made smaller.

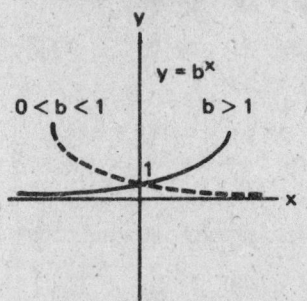

The graphs of all logarithmic functions $y = \log_b x$ have roughly the same shape and pass through the point $(1, 0)$. If $b > 1$, the graph increases as x increases and approaches the y-axis as an asymptote as x approaches zero. The amount of curvature becomes greater as the value of b is made greater. If $0 < b < 1$, the graph decreases as x increases and approaches the y-axis as an asymptote as x approaches zero. The amount of curvature becomes greater as the value of b is made smaller.

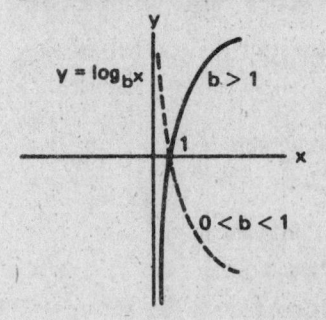

EXERCISES

1. If $x^a \cdot (x^{a+1})^a \cdot (x^a)^{1-a} = x^k$, then k equals
   [A] $2a + 1$ [B] $a + a^2$ [C] $3a$
   [D] $3a + 1$ [E] $a^3 + a$

2. If $\log_8 3 = x \cdot \log_2 3$, then x equals
   [A] 4 [B] $\log_4 3$ [C] 3
   [D] $\frac{1}{3}$ [E] $\log_8 9$

3. If $\log_{10} m = \frac{1}{2}$, then $\log_{10} 10m^2$ equals
   [A] 2.5 [B] 2 [C] 100
   [D] 10.25 [E] 3

4 If $\log_b 5 = a$, $\log_b 2.5 = c$, and $5^x = 2.5$, then x equals

[A] ac [B] $\dfrac{c}{a}$
[C] $a+c$ [D] $c-a$
[E] cannot be determined from the information

5 If $f(x) = \log_2 x$, then $f\left(\dfrac{2}{x}\right) + f(x)$ equals

[A] $\log_2\left(\dfrac{2}{x}\right) + \log_2 x$ [B] 1
[C] $\log_2\left(\dfrac{2+x^2}{x}\right)$ [D] $\log_2(\dfrac{2}{x}) \cdot \log_2 x$
[E] 0

6 If $\log_b(xy) < 0$, then which of the following must be true?

[A] $xy < 0$ [B] $xy < 1$ [C] $xy > 1$
[D] $xy > 0$ [E] none of the above

3. ABSOLUTE VALUE

The absolute value of x (written as $|x|$) is defined as follows:
if $x > 0$, then $|x| = x$
if $x < 0$, then $|x| = -x$ (note that $-x$ is a positive number)
$|x| \geq 0$ for all values of x.

EXAMPLE 1: If $|x-3| = 2$, find x.
Absolute value problems can always be solved by using the definition to restate the problem in two parts.

Part 1: If $x - 3 > 0$, then $|x-3| = x - 3$.
$$\text{Thus, } x - 3 = 2$$
$$x = 5$$

Notice that both conditions, $x - 3 > 0$ and $x = 5$, must be satisfied. They are in this case.

Part 2: If $x - 3 < 0$, then $|x - 3| = -(x-3) = -x + 3$
$$\text{Thus, } -x + 3 = 2$$
$$x = 1$$

Both conditions, $x - 3 < 0$ and $x = 1$, are satisfied. The solution set is $\{1, 5\}$.

EXAMPLE 2: If $|3x + 5| = 2x + 3$, find x.

Part 1: If $3x + 5 > 0$, then $|3x+5| = 3x+5$.
$$\text{Thus, } 3x + 5 = 2x + 3$$
$$x = -2$$

Both conditions, $x > -\dfrac{5}{3}$ and $x = -2$, are not satisfied, so there is no solution supplied by this part.

Part 2: If $3x + 5 < 0$, then $|3x+5| = -(3x+5) = -3x - 5$.
$$\text{Thus, } -3x - 5 = 2x + 3$$
$$-5x = 8$$
$$x = -\dfrac{8}{5}$$

Both conditions, $x < -\dfrac{5}{3}$ and $x = -\dfrac{8}{5}$, are not satisfied, so there is no solution supplied by this part either. Therefore the solution set is the empty set, $\emptyset$.

EXAMPLE 3: Find all values of x such that $|2x+3| > 5$.

Consider the equality $|2x+3| = 5$ which is associated with this inequality. Using the definition of absolute value:

if $2x + 3 > 0$, then $2x + 3 = 5$.
Thus, $x > -\dfrac{3}{2}$ and $x = 1$.
if $2x + 3 < 0$, then $-(2x+3) = 5$.
Thus, $x < -\dfrac{3}{2}$ and $x = -4$.

The solution set of the equality is $\{1, -4\}$. Consider the regions of a number line indicated by these two numbers.

$$\underline{\qquad\bullet\qquad\qquad\qquad\bullet\qquad}$$
$$-5\ -4\ -3\ -2\ -1\ \ 0\ \ 1\ \ 2$$

If $x = -5$, then $|2x+3| = 7 > 5$.
If $x = 0$, then $|2x+3| = 3 < 5$.
If $x = 2$, then $|2x+3| = 7 > 5$.

Therefore, the set of numbers which satisfies the original inequality $|2x+3| > 5$ is $\{x : x < -4 \text{ or } x > 1\}$. The graph of this solution set on a number line is indicated below

$$\longleftarrow\!\!\bullet\qquad\qquad\bullet\!\!\longrightarrow$$
$$-5\ -4\ -3\ -2\ -1\ \ 0\ \ 1\ \ 2$$

Chapter 4 Miscellaneous Relations and Functions

EXAMPLE 4: If the graph of $f(x)$ is shown below, sketch the graph of [A] $|f(x)|$ [B] $f(|x|)$.

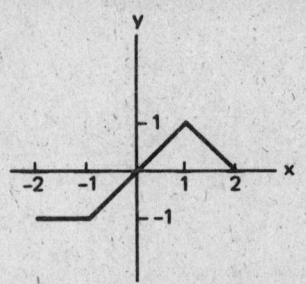

[A] Since $|f(x)| \geq 0$, by the definition of absolute value, the graph cannot have any points below the x-axis. If $f(x) < 0$, then $|f(x)| = -f(x)$. Thus, all points below the x-axis are reflected about the x-axis, and all points above the x-axis remain unchanged.

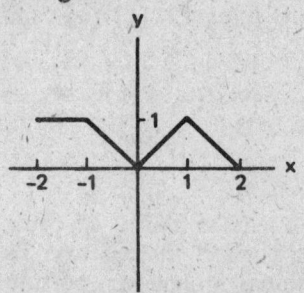

[B] Since the absolute value of x is taken before the function value is found, and since $|x| = -x$ when $x < 0$, any negative values of x will graph the same y-values as the corresponding positive values of x. Thus, the graph to the left of the y-axis will be a reflection of the graph to the right of the y-axis.

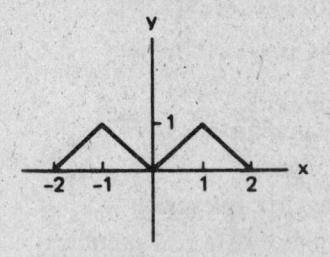

EXAMPLE 5: If $f(x) = |x+1| - 1$, what is the minimum value of $f(x)$?
Since $|x+1| \geq 0$, its smallest value is 0. Therefore, the smallest value of $f(x)$ is $0 - 1 = -1$. The graph of $f(x)$ is indicated below.

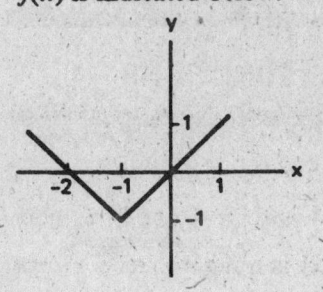

EXERCISES

1 $|2x-1| = 4x+5$ has how many numbers in its solution set?

[A] 0   [B] 1   [C] 2
[D] an infinite number
[E] none of the above

2 Which of the following is equivalent to $1 \leq |x-2| \leq 4$?

[A] $3 \leq x \leq 6$ [B] $x \leq 1$ or $x \geq 3$
[C] $1 \leq x \leq 3$ [D] $x \leq -2$ or $x \geq 6$
[E] $-2 \leq x \leq 1$ or $3 \leq x \leq 6$

3 The area bounded by the relation $|x| + |y| = 2$ is

[A] 8   [B] 1   [C] 2
[D] 4   [E] there is no finite area

4 Given a function, $f(x)$, such that $f(x) = f(|x|)$. Which one of the following could be the graph of $f(x)$?

[A]

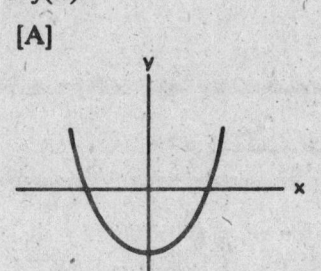

[B]

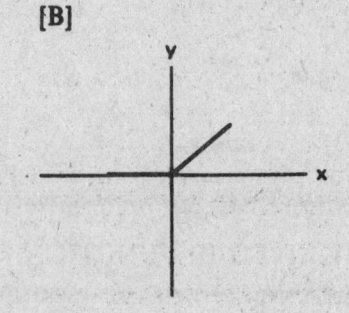

[C]

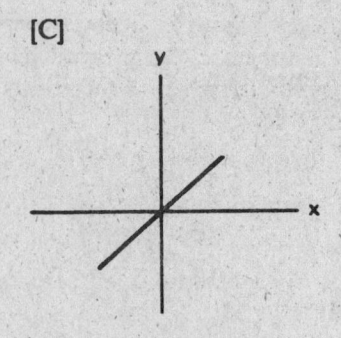

[D]

[E]

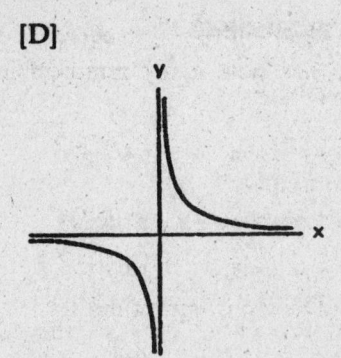

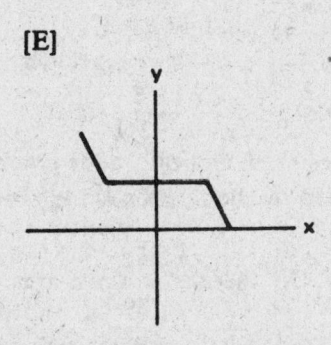

5 The figure shows the graph of which one of the following?

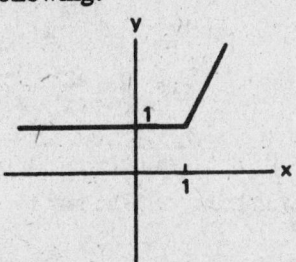

[A] $y=2x-|x|$ [B] $y=|x-1|+x$
[C] $y=|2x-1|$ [D] $y=|x+1|-x$
[E] $y=2|x|-|x|$

4. GREATEST INTEGER FUNCTION

The greatest integer function, denoted by $[x]$, pairs with each real number, x, the greatest integer contained in x. In symbols, where i represents an integer: $f(x)=i$ where $i \leq x < i+1$.

EXAMPLE 1:
[A] $[3.2]=3$ [B] $[1.999]=1$
[C] $[5]=5$ [D] $[-3.12]=-4$
[E] $[-0.123]=-1$.

EXAMPLE 2: Sketch the graph of $f(x)=[x]$.

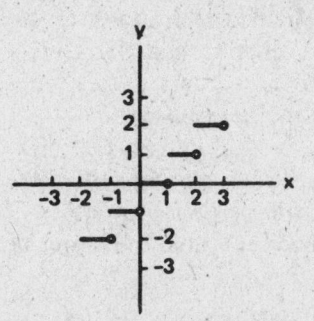

EXAMPLE 3: What is the range of $f(x)=\left[\dfrac{[x]}{x}\right]$?

If x is not an integer, $[x] < x$. Therefore, $\dfrac{[x]}{x}$ represents a decimal between zero and one, and $\left[\dfrac{[x]}{x}\right]=0$. If x is an integer, $[x]=x$, and $\left[\dfrac{[x]}{x}\right]=\left[\dfrac{x}{x}\right]=[1]=1$. Therefore, the range is $\{0,1\}$.

EXERCISES

1 If the postal rate for first class mail is 15 cents for the first ounce or portion thereof and 13 cents for each additional ounce or portion thereof, then the cost in cents of first class postage on a letter weighing N ounces is always

[A] $15+[N-1]\cdot 13$ [B] $[N-15]\cdot 13$
[C] $15+[N]\cdot 13$ [D] $1+[N]\cdot 13$
[E] none of the above

2 If $f(x)=i$ where i is an integer such that
$i \leq x < i+1$, the range of $f(x)$ is

[A] the set of all real numbers
[B] the set of all positive integers
[C] the set of all integers
[D] the set of all negative integers
[E] the set of all non-negative real numbers

3 If $f(x)=[2x]-4x$ with domain $0 \leq x < 2$, then $f(x)$ could also be written as

[A] $2x$ [B] $-x$ [C] $-2x$
[D] x^2-4x [E] none of the above

5. RATIONAL FUNCTIONS

F is a rational function if and only if $F(x)=\dfrac{p(x)}{q(x)}$, where $p(x)$ and $q(x)$ are both polynomial functions and $q(x)$ is not zero. As a general rule, the graphs of rational functions are not continuous. (I.e., they have holes, or sections of the graphs are separated from other sections by asymptotes.) A point of discontinuity occurs at any value of x that would cause $q(x)$ to become zero.

If $p(x)$ and $q(x)$ can be factored so that $F(x)$ can be reduced, removing the factors that caused the discontinuities, the graph will contain only holes. If the factors that caused the discontinuities cannot be removed, asymptotes will occur.

EXAMPLE 1: Sketch the graph of $F(x) = \dfrac{x^2 - 1}{x + 1}$.

There will be a discontinuity at $x = -1$ since this would cause division by zero. Since the fraction $\dfrac{x^2 - 1}{x + 1} = \dfrac{(x-1)(x+1)}{(x+1)} = (x-1)$, the graph of $F(x)$ will be the same as the graph of $y = x - 1$ except for a hole at $x = -1$.

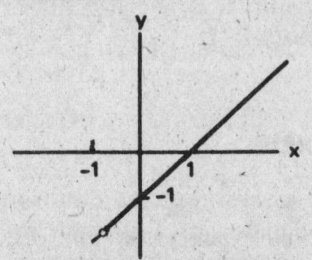

EXAMPLE 2: Sketch the graph of $F(x) = \dfrac{1}{x - 2}$.

Since this fraction will not reduce, and $x = 2$ would cause division by zero, a vertical asymptote will occur when $x = 2$. This is true because as x approaches very close to 2, $F(x)$ gets either extremely large or extremely small. As x gets extremely large or extremely small, $f(x)$ gets closer and closer to zero. This means that a horizontal asymptote occurs when $y = 0$. Plotting a few points indicates that the graph looks like the figure below.

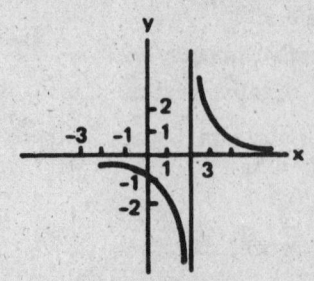

A more compact way of expressing the previous few statements about x approaching 2 or x getting extremely large follows: The statement $\lim_{x \to \infty} f(x) = 0$ is read "the limit of $f(x)$ is zero as x increases without bound (or as x approaches infinity)." $\lim_{x \to 2^-} f(x) = -\infty$ is read "the limit of $f(x)$ decreases without bound (or tends to negative infinity) as x approaches 2 from the negative side."

EXAMPLE 3: What does $\lim_{x \to 1} \dfrac{x^2 - 1}{x + 1}$ equal?

Since $\dfrac{x^2 - 1}{x + 1}$ reduces to $x - 1$,

$$\lim_{x \to 1} \dfrac{x^2 - 1}{x + 1} = \lim_{x \to 1} x - 1 = 0.$$

EXAMPLE 4: What does $\lim_{x \to 2^+} 3x + 5$ equal?

Since "problems" occur only when division by zero appears imminent, this example is extremely easy. As x gets closer and closer to 2, $3x + 5$ seems to be getting closer and closer to 11.
Therefore, $\lim_{x \to 2^+} 3x + 5 = 11$.

EXAMPLE 5: What does $\lim_{x \to 2} \left(\dfrac{3x + 5}{x - 2} \right)$ equal?

The denominator does not factor out so the graph of this rational function will have a vertical asymptote. As x approaches 2 from above (i.e., $2.1, 2.01, 2.001, \ldots$), the numerator and denominator both remain positive so $\dfrac{3x + 5}{x - 2}$ gets larger and larger and approaches positive infinity. As x approaches 2 from below (i.e., $1.9, 1.99, 1.999, \ldots$), the numerator is still positive, but the denominator is negative so $\dfrac{3x + 5}{x - 2}$ gets smaller and smaller and approaches negative infinity. Thus, $\lim_{x \to 2} \dfrac{3x + 5}{x - 2}$ does not exist since $\lim_{x \to 2^-} \dfrac{3x + 5}{x - 2} = -\infty$ and $\lim_{x \to 2^+} \dfrac{3x + 5}{x - 2} = +\infty$, which are not the same.

EXAMPLE 6: If $f(x) = \begin{cases} 3x + 2, & \text{when } x \neq 0 \\ 0, & \text{when } x = 0 \end{cases}$, what does $\lim_{x \to 0} f(x)$ equal?

As x approaches zero, $3x + 2$ approaches 2, in spite of the fact that $f(x) = 0$ when $x = 0$. Therefore, $\lim_{x \to 0} f(x) = 2$.

EXAMPLE 7: What does $\lim_{x \to \infty} \left(\dfrac{3x^2 + 4x + 2}{2x^2 + x - 5} \right)$ equal?

At first glance the answer either seems to be obvious or impossible. The answer could be 1 because if both the numerator and the denominator are increasing without bound, their quotient must approach 1. On the other hand, no matter how large a number is substituted for x, the numerator and the denominator are never the same.

This is, in fact, a rather easy problem to solve. The method most usually used is to divide both the numerator and the denominator through by the variable raised to its highest exponent in the problem. In this case divide through by x^2.

$$\lim_{x\to\infty}\left(\frac{3x^2+4x+2}{2x^2+x-5}\right) = \lim_{x\to\infty}\left[\frac{3+\frac{4}{x}+\frac{2}{x^2}}{2+\frac{1}{x}-\frac{5}{x^2}}\right].$$

Now, as $x\to\infty$, $\frac{4}{x}$, $\frac{2}{x^2}$, $\frac{1}{x}$, and $\frac{5}{x^2}$ each approach zero.

Thus, the entire fraction approaches $\frac{3+0+0}{2+0-0} = \frac{3}{2}$

Therefore, $\lim_{x\to\infty}\left(\frac{3x^2+4x+2}{2x^2+x-5}\right) = \frac{3}{2}$.

EXERCISES

[1] To be continuous at $x=1$, the value of $\frac{x^4-1}{x^3-1}$ must be defined to be equal to

[A] $\frac{4}{3}$ [B] 0 [C] 1 [D] 4 [E] -1

[2] If $f(x) = \begin{cases} \frac{3x^2+2x}{x}, & \text{when } x \neq 0 \\ k, & \text{when } x = 0 \end{cases}$, what must the value of k be equal to in order that $f(x)$ be a continuous function?

[A] 0 [B] 2 [C] $-\frac{2}{3}$ [D] $-\frac{3}{2}$

[E] no value of k will make $f(x)$ a continuous function

[3] $\lim_{x\to 2}\left(\frac{x^3-8}{x^4-16}\right)$ equals

[A] $\frac{3}{8}$ [B] $\frac{1}{2}$ [C] $\frac{4}{7}$

[D] 0 [E] undefined

[4] $\lim_{x\to\infty}\left(\frac{5x^2-2}{3x^2+8}\right)$ equals

[A] ∞ [B] $\frac{3}{11}$ [C] $\frac{5}{3}$

[D] $-\frac{1}{4}$ [E] 0

[5] Which of the following is the equation of an asymptote of $y = \frac{3x^2-2x-1}{9x^2-1}$?

[A] $x = -\frac{1}{3}$ [B] $y = \frac{1}{3}$ [C] $x = 1$

[D] $y = 1$ [E] $y = -\frac{1}{3}$

6. PARAMETRIC EQUATIONS

At times it is convenient to express a relationship between x and y in terms of a third variable. This third variable is called a *parameter* and the equations are called *parametric equations*.

EXAMPLE 1: $\begin{cases} x = 3t+4 \\ y = t-5 \end{cases}$ are parametric equations with a parameter t.

As different values are substituted for the parameter, ordered pairs of points, (x,y), represent points on the graph. At times it is possible to eliminate the parameter and to re-write the equation in familiar x,y form. When eliminating the parameter it is necessary to be aware of the fact that the resulting equation may consist of points not on the graph of the original set of equations.

EXAMPLE 2: $\begin{cases} x = t^2 \\ y = 3t^2+1 \end{cases}$ **Eliminate the parameter and sketch the graph.**

Substituting x for t^2 in the second equation results in $y = 3x+1$, which is the equation of a straight line with a slope of 3 and a y-intercept of 1. However, the original parametric equations indicate that $x \geq 0$ and $y \geq 1$ since t^2 cannot be negative. Thus, the proper way to indicate this set of points without the parameter would be: $y = 3x+1$ and $x \geq 0$. The graph would be the ray indicated in the figure.

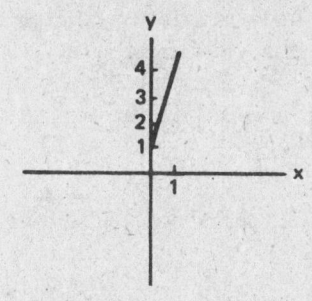

EXAMPLE 3: Sketch the graph of the parametric equations $\begin{cases} x = 4\cos\theta \\ y = 3\sin\theta \end{cases}$.

It is possible to eliminate the parameter, θ, by dividing the first equation by 4 and the second equation by 3, squaring each, and then adding both equations together.

$$\left(\frac{x}{4}\right)^2 = \cos^2\theta \text{ and } \left(\frac{y}{3}\right)^2 = \sin^2\theta$$

$$\frac{x^2}{16} + \frac{y^2}{9} = \cos^2\theta + \sin^2\theta = 1$$

$\frac{x^2}{16} + \frac{y^2}{9} = 1$ is the equation of an ellipse with its center at the origin, $a=4$, and $b=3$. Since $-1 \leq \cos\theta \leq 1$ and $-1 \leq \sin\theta \leq 1$, $-4 \leq x \leq 4$ and $-3 \leq y \leq 3$ from the two parametric equations. In this case the parametric equations do not limit the graph obtained by removing the parameter.

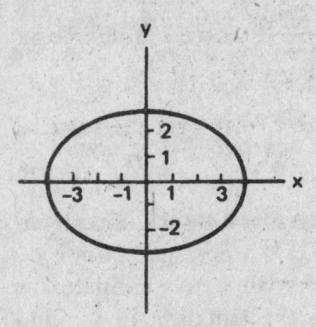

EXAMPLE 4: Sketch the graph of the parametric equations $\begin{cases} x = 2(\theta - \sin\theta) \\ y = 2(1 - \cos\theta) \end{cases}$.

In this case it would be extremely difficult to eliminate the parameter. In order to sketch the graph it will be necessary to choose values of θ, and to determine x and y directly. In order to keep the work to a minimum, choose values of θ that have easily found sines and cosines.

θ	0	$\frac{\pi}{6}$	$\frac{\pi}{3}$	$\frac{\pi}{2}$	$\frac{2\pi}{3}$	$\frac{5\pi}{6}$	π
x	0	.05	.36	1.1	2.5	4.2	6.3
y	0	.27	1	2	3	3.7	4

θ	$\frac{7\pi}{6}$	$\frac{4\pi}{3}$	$\frac{3\pi}{2}$	$\frac{5\pi}{3}$	$\frac{11\pi}{6}$	2π
x	8.3	10.1	11.4	12.2	12.5	12.6
y	3.7	3	2	1	.27	0

The figure shows one period of the graph, which is called a cycloid.

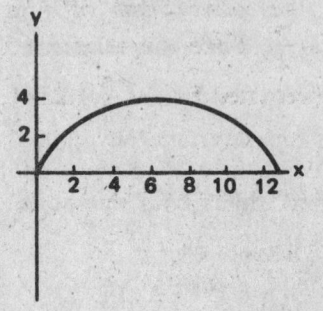

EXERCISES

1 The domain of the function defined by the parametric equations $\begin{cases} x = t^2 + t \\ y = t^2 - t \end{cases}$ is

[A] $\{x : x \geq 0\}$
[B] $\{x : x \geq -\frac{1}{4}\}$
[C] all real numbers
[D] $\{x : x \geq -1\}$
[E] $\{x : x \leq 1\}$

2 The graph of $\begin{cases} x = \sin^2 t \\ y = 2\cos t \end{cases}$ is a

[A] straight line   [B] line segment
[C] parabola   [D] portion of a parabola
[E] semi-circle

3 Which of the following is (are) a pair of parametric equations that represent a circle?

I. $\begin{cases} x = \sin\theta \\ y = \cos\theta \end{cases}$

II. $\begin{cases} x = t \\ y = \sqrt{1 - t^2} \end{cases}$

III. $\begin{cases} x = \sqrt{s} \\ y = \sqrt{1 - s} \end{cases}$

[A] I only   [B] II only   [C] III only
[D] II and III only   [E] I, II, and III

7. POLAR COORDINATES

Although the most common way to represent a point in a plane is in terms of its distance from two perpendicular axes, there are several other ways. One such way is in terms of the distance of the point from the origin and the angle between the positive x-axis and the ray emanating from the origin going through the point.

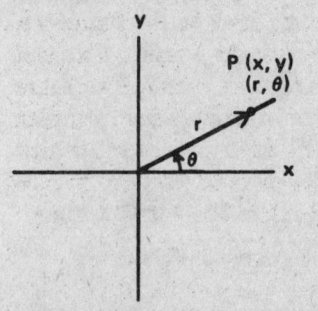

In the figure, the regular rectangular coordinates of P are (x,y), and the *polar coordinates* are (r,θ).

Since $\sin\theta = \frac{y}{r}$ and $\cos\theta = \frac{x}{r}$, there is an easy relationship between rectangular and polar coordinates:

$x = r \cdot \cos\theta$
$y = r \cdot \sin\theta$
$x^2 + y^2 = r^2$, using the Pythagorean Theorem

Unlike rectangular coordinates, each point in the plane can be named by an infinite number of polar coordinates.

EXAMPLE 1: $(2, 30°)$, $(2, 390°)$, $(2, -330°)$, $(-2, 210°)$, $(-2, -150°)$ all name the same point. In general, a point in the plane represented by (r, θ) can also be represented by $(r, \theta + 2\pi n)$ or $(-r, \theta + (2n-1)\pi)$, where n is an integer.

EXAMPLE 2: Express the point P, whose rectangular coordinates are $(3, 3\sqrt{3})$, in terms of polar coordinates.
$r^2 = x^2 + y^2 = 9 + 27 = 36$
$r = 6$
$r \cdot \cos\theta = x$
$\cos\theta = \frac{3}{6} = \frac{1}{2}$
Therefore, $\theta = 60°$ and the coordinates of P are $(6, 60°)$.

EXAMPLE 3: Without sketching, describe the graph of [A] $r = 2$ [B] $r = \frac{1}{\sin\theta}$.

[A] $r^2 = x^2 + y^2$
$r = 2$
Therefore, $x^2 + y^2 = 4$, which is the equation of a circle whose center is at the origin and radius is 2.

[B] $r \cdot \sin\theta = x$
$r = \frac{1}{\sin\theta}$
Therefore, $x = 1$.
Thus, $r = \frac{1}{\sin\theta}$ is the equation of a vertical line one unit to the right of the y-axis.

Since the complex number $a + bi$ can be represented by a point $P(a, b)$ on a coordinate plane, it can also be represented in terms of polar coordinates. Thus, $a = r \cdot \cos\theta$ and $b = r \cdot \sin\theta$. Therefore, $a + bi = r \cdot \cos\theta + i \cdot r \cdot \sin\theta = r(\cos\theta + i \cdot \sin\theta)$. This last statement is often abbreviated as $r \cdot \text{cis}\,\theta$. There are several useful formulas for manipulating complex numbers when they are written in polar form which is sometimes called "trigonometric form." If $z_1 = r_1(\cos\theta_1 + i \cdot \sin\theta_1)$ and $z_2 = r_2(\cos\theta_2 + i \cdot \sin\theta_2)$, then

1) $z_1 \cdot z_2 = r_1 r_2 (\cos(\theta_1 + \theta_2) + i \cdot \sin(\theta_1 + \theta_2))$ or
$z_1 \cdot z_2 = r_1 r_2 \text{cis}(\theta_1 + \theta_2)$

2) $\frac{z_1}{z_2} = \frac{r_1}{r_2}(\cos(\theta_1 - \theta_2) + i \cdot \sin(\theta_1 - \theta_2))$ or
$\frac{z_1}{z_2} = \frac{r_1}{r_2}\text{cis}(\theta_1 - \theta_2)$

DeMoivre's Theorem states that

3) $(z_1)^n = (r_1)^n(\cos n\theta_1 + i \cdot \sin n\theta_1)$ or
$(z_1)^n = (r_1)^n \text{cis}\, n\theta_1$, where n is any positive integer.

The periodicity of the trigonometric functions allows the use of DeMoivre's Theorem to find the nth roots of any complex number.

4) $(z_1)^{1/n} = (r_1)^{1/n}\left(\cos\frac{\theta_1 + 2\pi k}{n} + i \cdot \sin\frac{\theta_1 + 2\pi k}{n}\right)$ or
$(z_1)^{1/n} = (r_1)^{1/n}\text{cis}\frac{\theta_1 + 2\pi k}{n}$, where k is an integer taking on values from 0 up to $n - 1$.

EXAMPLE 4: Using the polar form find the value of $\frac{2 + 2i}{\sqrt{3} - 3i}$.

For the numerator, $r^2 = x^2 + y^2 = 4 + 4 = 8$. Therefore $r = 2\sqrt{2}$.
$r \cdot \cos\theta = x$
$\cos\theta = \frac{2}{2\sqrt{2}} = \frac{\sqrt{2}}{2}$. Therefore $\theta = 45°$.
Thus, $2 + 2i = 2\sqrt{2}\,\text{cis}\,45°$.
For the denominator, $r^2 = x^2 + y^2 = 3 + 9 = 12$. Therefore $r = 2\sqrt{3}$.
$r \cdot \cos\theta = x$
$\cos\theta = \frac{\sqrt{3}}{2\sqrt{3}} = \frac{1}{2}$ and $\sin\theta = \frac{-3}{2\sqrt{3}} = \frac{-\sqrt{3}}{2}$
Therefore θ is in the fourth quadrant and $\theta = -60°$ or $300°$.
Thus, $\sqrt{3} + 3i = 2\sqrt{3}\,\text{cis}(-60°)$.

$\frac{2 + 2i}{\sqrt{3} - 3i} = \frac{2\sqrt{2}\,\text{cis}\,45°}{2\sqrt{3}\,\text{cis}(-60°)}$

$= \frac{\sqrt{2}}{\sqrt{3}}\text{cis}(45 - (-60))° = \frac{\sqrt{6}}{3}\text{cis}(105°)$

EXAMPLE 5: If $z = 2\text{cis}\,\frac{\pi}{9}$, what does z^5 equal?
$z^5 = 2^5 \text{cis}(5 \cdot \frac{\pi}{9}) = 32\text{cis}\,\frac{5\pi}{9}$, using DeMoivre's Theorem.

EXAMPLE 6: Find the three cube roots of i.
The complex number i is represented on the rectangular coordinate system by the point with coordinates $(0, 1)$. Therefore, the polar form of i is $1\left(\cos\frac{\pi}{2} + i \cdot \sin\frac{\pi}{2}\right)$ or $\text{cis}\,\frac{\pi}{2}$. Since the cube roots are desired, i must be represented by $\text{cis}\left(\frac{\pi}{2} + 2k\pi\right)$, where $k = 0, 1, 2$.

$\sqrt[3]{i} = 1^{1/3}\text{cis}\,\frac{1}{3}\left(\frac{\pi}{2} + 2k\pi\right)$, where $k = 0, 1, 2$

$\sqrt[3]{i} = \text{cis}\left(\frac{\pi}{6} + \frac{2k\pi}{3}\right)$, where $k = 0, 1, 2$

$\sqrt[3]{i} = \text{cis}\,\frac{\pi}{6}$ or

$\sqrt[3]{i} = \text{cis}\left(\frac{\pi}{6} + \frac{2\pi}{3}\right) = \text{cis}\,\frac{5\pi}{6}$ or

$\sqrt[3]{i} = \text{cis}\left(\frac{\pi}{6} + \frac{4\pi}{3}\right) = \text{cis}\,\frac{9\pi}{6} = \text{cis}\,\frac{3\pi}{2}$.

EXERCISES

1. A point has polar coordinates $(2, 60°)$. The same point may be represented by

   [A] $(-2, 240°)$ [B] $(2, 240°)$ [C] $(-2, 60°)$
   [D] $(2, -60°)$ [E] $(2, -240°)$

2. The graph of $r = \cos\theta$ intersects the graph of $r = \sin 2\theta$ at the points

   [A] $\left(\frac{\sqrt{3}}{2}, \frac{\pi}{6}\right), \left(\frac{\sqrt{3}}{2}, \frac{5\pi}{6}\right)$, and $\left(0, \frac{\pi}{2}\right)$
   [B] $\left(\frac{\sqrt{3}}{2}, \frac{\pi}{6}\right)$ and $\left(\frac{\sqrt{3}}{2}, \frac{11\pi}{6}\right)$
   [C] $\left(\frac{\sqrt{3}}{2}, \frac{\pi}{3}\right)$ and $\left(-\frac{\sqrt{3}}{2}, \frac{\pi}{3}\right)$
   [D] $\left(\frac{\sqrt{3}}{2}, \frac{\pi}{6}\right), \left(0, \frac{\pi}{2}\right)$, and $\left(-\frac{\sqrt{3}}{2}, \frac{5\pi}{6}\right)$
   [E] $(0, 0)$ only

3. An equation in rectangular form equivalent to $r^2 = 36\sec 2\theta$ is

   [A] $x^2 - y^2 = 36$ [B] $(x-y)^2 = 36$
   [C] $x^4 - y^4 = 36$ [D] $xy = 36$
   [E] $y = 6$

4. If $-2 - 2i\sqrt{3}$ is divided by $-1 + i\sqrt{3}$, the quotient, in trigonometric form, is

   [A] $1(\cos 240° + i \cdot \sin 240°)$
   [B] $2(\cos 120° + i \cdot \sin 120°)$
   [C] $2(\cos 135° + i \cdot \sin 135°)$
   [D] $2(\cos 150° + i \cdot \sin 150°)$
   [E] $2(\cos 210° + i \cdot \sin 210°)$

5. If $A = 2(\cos 20° + i \cdot \sin 20°)$ and $B = 3(\cos 40° + i \cdot \sin 40°)$, the product AB equals

   [A] $6 \text{cis} 20°$ [B] $6 \text{cis} 60°$ [C] $5 \text{cis} 20°$
   [D] $5 \text{cis} 60°$ [E] $5(\cos 80° + i \cdot \sin 80°)$

6. $\sqrt[6]{4\sqrt{3} + 4i}$ equals

   [A] $\sqrt{2} \left(\cos \frac{\pi}{36} - i \cdot \sin \frac{\pi}{36}\right)$
   [B] $\sqrt{2} \text{ cis } \frac{25\pi}{36}$
   [C] $\sqrt{8} \text{ cis } \frac{25\pi}{36}$
   [D] $\sqrt{2} \left(\cos \frac{13\pi}{36} - i \cdot \sin \frac{13\pi}{36}\right)$
   [E] $\sqrt{8} \text{ cis } \frac{13\pi}{36}$

MISCELLANEOUS TOPICS

CHAPTER 5

1. PERMUTATIONS AND COMBINATIONS

Any arrangement of the elements of a set in a definite order is called a *permutation*. If all n elements of a set are to be arranged, there are $n!$ (read "n factorial") ways to arrange them. $n! = n(n-1)(n-2)\cdots 3\cdot 2\cdot 1$.

EXAMPLE 1: [A] $5! = 5\cdot 4\cdot 3\cdot 2\cdot 1 = 120$
[B] $3! = 3\cdot 2\cdot 1 = 6$

If only r elements of a set that contains n elements are to be arranged, there are $\dfrac{n!}{(n-r)!}$ arrangements. $_nP_r = P(n,r) =$ the number of permutations of n elements taken r at a time. Therefore, $_nP_r = \dfrac{n!}{(n-r)!}$.

EXAMPLE 2: Evaluate each of the following:
[A] $_4P_4$ [B] $_7P_2$ [C] $_7P_5$.

[A] $_4P_4$ means "the number of permutations of 4 elements taken 4 at a time." From the original definition this is equal to $4! = 4\cdot 3\cdot 2\cdot 1 = 24$. From the definition of $_nP_r$, $P = \dfrac{4!}{(4-4)!} = \dfrac{4!}{0!}$. This will equal 24 if and only if 0! is defined to equal 1. In fact, 0! is always defined to equal 1. Therefore, using either definition, $_4P_4 = 24$.

[B] $_7P_2 = \dfrac{7!}{(7-2)!} = \dfrac{7\cdot 6\cdot 5\cdot 4\cdot 3\cdot 2\cdot 1}{5\cdot 4\cdot 3\cdot 2\cdot 1} = 7\cdot 6 = 42$

[C] $_7P_5 = \dfrac{7!}{(7-5)!} = \dfrac{7\cdot 6\cdot 5\cdot 4\cdot 3\cdot 2\cdot 1}{2\cdot 1}$
$= 7\cdot 6\cdot 5\cdot 4\cdot 3 = 2520$

EXAMPLE 3: How many four-digit lottery numbers can be drawn if there is no repetition of digits?

Method 1: This is a permutation problem:
$_{10}P_4 = \dfrac{10!}{(10-4)!}$
$= \dfrac{10!}{6!} = \dfrac{10\cdot 9\cdot 8\cdot 7\cdot 6\cdot 5\cdot 4\cdot 3\cdot 2\cdot 1}{6\cdot 5\cdot 4\cdot 3\cdot 2\cdot 1}$
$= 10\cdot 9\cdot 8\cdot 7 = 5040$

Method 2: Consider putting the digits in the grid.

Any one of the 10 digits can be put in the first space. That leaves only 9 digits for the second space, only 8 for the third, and only 7 for the fourth. Thus, there are $10\cdot 9\cdot 8\cdot 7 = 5040$ different lottery numbers.

EXAMPLE 4: In a club of 15 members, one member is to be chosen president, one secretary, and one treasurer. How many different slates of candidates can be chosen?

Method 1: The president can be any one of 15 people. After he is chosen, there are 14 remaining members who could be the secretary and only 13 members remaining who could be treasurer. Therefore, there could be $15\cdot 14\cdot 13 = 2730$ different slates of candidates.

Method 2: This is a permutation problem: $_{15}P_3 = \dfrac{15!}{(15-3)!} = \dfrac{15!}{12!} = 15\cdot 14\cdot 13 = 2730$ different slates of candidates.

In all these problems, the answer was obtained by taking the largest r factors of $n!$

EXAMPLE 5: Evaluate: [A] $_8P_3$ [B] $_5P_5$ [C] $_{12}P_2$

[A] $_8P_3 = 8\cdot 7\cdot 6 = 336$
[B] $_5P_5 = 5\cdot 4\cdot 3\cdot 2\cdot 1 = 120$
[C] $_{12}P_2 = 12\cdot 11 = 132$

If the n elements of a set are to be arranged in a circle, they may be so arranged in $\dfrac{n!}{n}$ or $(n-1)!$ ways. There are fewer such permutations because there is no beginning to a circle as there is with a line.

EXAMPLE 6: How many ways can 6 people be seated at a round table?

This is a circular permutation so there are $(6-1)! = 5! = 120$ ways.

If the circular arrangement can be viewed from either side or turned over like a bracelet, there are one-half as many permutations. Therefore, the number of permutations is $\dfrac{(n-1)!}{2}$.

If the n elements of a set are not all different, several permutations would appear to be the same. To take care of this repetition the formula for a permutation of n things where there are a of one kind and b of another kind is $\dfrac{n!}{a!b!}$.

EXAMPLE 7: How many permutations of the letters of the word TATTLETALE are there?

Of the 10 letters, there are 4 Ts, 2 Es, 2 Ls, and 2 As. Therefore, the number of permutations is $\dfrac{10!}{4!2!2!2!}$

$= \dfrac{10\cdot 9\cdot 8\cdot 7\cdot 6\cdot 5\cdot 4\cdot 3\cdot 2\cdot 1}{4\cdot 3\cdot 2\cdot 1\cdot 2\cdot 1\cdot 2\cdot 1\cdot 2\cdot 1} = 6300$ different permutations.

If a definite order in the grouping of the element of a set is *not* necessary, the grouping is called a *combination*. The number of combinations of n things taken r at a time is denoted by $_nC_r$ or $C(n,r)$ or $\binom{n}{r}$.

$\binom{n}{r} = \dfrac{_nP_r}{r!} = \dfrac{\text{the product of the largest } r \text{ factors of } n!}{r!}$

EXAMPLE 8: Set $A = \{a,b,c,d\}$. [A] List all the permutations of the elements of A taken two at a time. [B] List all the combinations of the elements of A taken two at a time.

[A] $_4P_2 = 4\cdot 3 = 12$. There should be 12 items in the permutation.

$\left.\begin{array}{llll} ab & ba & ca & da \\ ac & bc & cb & db \\ ad & bd & cd & dc \end{array}\right\}$ The permutations of A taken 2 at a time.

[B] $\binom{4}{2} = \dfrac{4\cdot 3}{2\cdot 1} = 6$. There should be 6 items in the combination.

$\left.\begin{array}{lll} ab & & \\ ac & bc & \\ ad & bd & cd \end{array}\right\}$ The combinations of A taken 2 at a time.

EXAMPLE 9: Evaluate [A] $\binom{9}{3}$ [B] $\binom{9}{6}$ [C] $\binom{8}{2}$ [D] $\binom{8}{6}$

[A] $\binom{9}{3} = \dfrac{9\cdot 8\cdot 7}{3\cdot 2\cdot 1} = 84$ [B] $\binom{9}{6} = \dfrac{9\cdot 8\cdot 7\cdot 6\cdot 5\cdot 4}{6\cdot 5\cdot 4\cdot 3\cdot 2\cdot 1} = 84$

[C] $\binom{8}{2} = \dfrac{8\cdot 7}{2\cdot 1} = 28$ [D] $\binom{8}{6} = \dfrac{8\cdot 7\cdot 6\cdot 5\cdot 4\cdot 3}{6\cdot 5\cdot 4\cdot 3\cdot 2\cdot 1} = 28$

In these examples, the answers to [A] and [B] are the same, and the answers to [C] and [D] are the same. The bottom numbers in [A] and [B] add up to the top number, 9, and the bottom numbers in [C] and [D] add up to the top number, 8. There was much less work to do when the bottom number was the smaller of the two numbers. In general $\binom{n}{r} = \binom{n}{n-r}$.

EXAMPLE 10: From a deck of 52 cards, how many different piles of 50 cards can be selected?

This is a combination problem. Evaluate $\binom{52}{50}$. Since $\binom{52}{50} = \binom{52}{2}$, it is easier to evaluate $\binom{52}{2}$, which is equal to $\dfrac{52\cdot 51}{2\cdot 1} = 1326$ different piles.

EXAMPLE 11: From a deck of 52 cards, in how many ways can a hand of 13 cards be dealt so as to contain 4 hearts and 9 spades?
Since there are 13 cards of each suit, 4 of 13 hearts and 9 of 13 spades must be selected.

$$\binom{13}{4} = \frac{13 \cdot 12 \cdot 11 \cdot 10}{4 \cdot 3 \cdot 2 \cdot 1} = 715 \text{ and } \binom{13}{9} = \binom{13}{4} = 715$$

Therefore, the number of ways this hand can be chosen is the product: $(715)(715) = 511,225$.

EXERCISES

1. $\dfrac{(5+3)!}{5! + 3!} =$

   [A] 1   [B] 56   [C] 320
   [D] 8   [E] 5040

2. Frisbees come in 5 models, 8 colors, and 3 sizes. How many Frisbees must the local dealer have on hand in order to be able to have one of each kind available?

   [A] 24   [B] 120   [C] 16   [D] 39   [E] 55

3. A craftsman has six different kinds of seashells. How many different bracelets could be constructed, if only four shells are to be used in any one bracelet?

   [A] 90   [B] 45   [C] 60
   [D] 360   [E] 180

4. How many different arrangements of the letters in the word RADAR are possible?

   [A] 120   [B] 6   [C] 30
   [D] 60   [E] 20

5. If a person is dealt two cards from a 52-card deck, how many different hands are possible if order is not important?

   [A] 104   [B] 13   [C] 1326
   [D] 103   [E] 2652

6. If $\binom{6}{x} = \binom{4}{x}$, then x equals

   [A] 5   [B] 4   [C] 11   [D] 0   [E] 1

7. How many odd numbers of three digits each can be formed from the digits 2, 4, 6, and 7, if repetition of digits is permitted?

   [A] 6   [B] 27   [C] 24   [D] 16   [E] 256

8. Given eight points in a plane, no three of which are colinear, how many lines do they determine?

   [A] 16   [B] 64   [C] 28   [D] 7   [E] 36

2. BINOMIAL THEOREM

Expanding a binomial, $(a+b)^n$, where n is a natural number, is a tedious operation for large values of n. The binomial theorem simplifies the work. For reference, consider the expansion for the first few values of n:

$$(a+b)^1 = a + b$$
$$(a+b)^2 = a^2 + 2ab + b^2$$
$$(a+b)^3 = a^3 + 3a^2b + 3ab^2 + b^3$$
$$(a+b)^4 = a^4 + 4a^3b + 6a^2b^2 + 4ab^3 + b^4$$
$$\vdots \qquad \vdots$$

Observations: There are $n + 1$ terms in each expansion.
The sum of the exponents in each term equals n.
The exponent of b is one less than the number of the term.
The coefficient of each term equals $\binom{n}{\text{either exponent}}$

EXAMPLE 1: The coefficient of the second term of $(a+b)^4$ is equal to $\binom{4}{1}$ or $\binom{4}{3}$ where the 1 is the exponent of the b and the 3 is the exponent of the a. The first exponent in each case is one. Using the fourth observation above, the first coefficient could be expressed as $\binom{n}{n}$ or $\binom{n}{0}$. Therefore, $\binom{n}{n} = \binom{n}{0} = 1$.

EXAMPLE 2: What is the third term of $(a+b)^{10}$?
The exponent of the b is $3-1$ or 2.
The exponent of the a must be 8 because the sum of the exponents, 8 and 2, must equal 10.
The coefficient is $\binom{10}{8} = \binom{10}{2} = \dfrac{10 \cdot 9}{2 \cdot 1} = 45$.
Therefore, the third term is $45a^8b^2$.

EXAMPLE 3: What is the middle term of $(a-b)^8$?
$(a-b)^8 = [a+(-b)]^8$. There are nine terms. The middle term is the fifth term. The exponent of $(-b)$ is $5-1$ or 4. The exponent of a is 4. The coefficient of the middle term is $\binom{8}{4} = \dfrac{8 \cdot 7 \cdot 6 \cdot 5}{4 \cdot 3 \cdot 2 \cdot 1} = 70$. Therefore, the middle term is $70a^4b^4$.

EXAMPLE 4: What is the fourth term of $(a-2b)^{12}$? $(a-2b)^{12} = [a+(-2b)]^{12}$. The exponent of the $(-2b)$ is 3. The exponent of a is 9. The coefficient is $\binom{12}{3} = \frac{12 \cdot 11 \cdot 10}{3 \cdot 2 \cdot 1} = 220$.
The fourth term is $220a^9(-2b)^3 = 220a^9(-8b^3) = -1760a^9b^3$.

The Binomial Theorem method of expanding $(a+b)^n$ can be enlarged to include any real number exponent, n, since the exponent of b is always a non-negative integer. Although the terminology of combinations is no longer meaningful, the symbolism is still useful. When n is a non-negative integer (greater than or equal to 3),

$$\binom{n}{3} = \frac{n(n-1)(n-2)}{3 \cdot 2 \cdot 1}.$$

Extending this process to any real number, n, gives the correct result.

EXAMPLE 5: Give the first three terms of $(a+b)^{-3}$. The exponent of b must be one less than the number of the term, and the sum of the exponents must equal -3. Thus, the expansion is

$$\binom{-3}{0}a^{-3} + \binom{-3}{1}a^{-4}b + \binom{-3}{2}a^{-5}b^2 + \ldots =$$

$$1a^{-3} + \frac{-3}{1}a^{-4}b + \frac{(-3)(-4)}{2 \cdot 1}a^{-5}b^2 + \ldots =$$

$$a^{-3} - 3a^{-4}b + 6a^{-5}b^2 - \ldots$$

EXAMPLE 6: Give the first four terms of $(a+b)^{1/2}$. The exponent of b must be one less than the number of the term, and the sum of the exponents must equal $\frac{1}{2}$. Thus, the expansion is

$$\binom{1/2}{0}a^{1/2} + \binom{1/2}{1}a^{-1/2}b + \binom{1/2}{2}a^{-3/2}b^2 +$$

$$\binom{1/2}{3}a^{-5/2}b^3 + \ldots =$$

$$1a^{1/2} + \frac{\frac{1}{2}}{1}a^{-1/2}b + \frac{\frac{1}{2}(-\frac{1}{2})}{2 \cdot 1}a^{-3/2}b^2 +$$

$$\frac{\frac{1}{2}(-\frac{1}{2})(-\frac{3}{2})}{3 \cdot 2 \cdot 1}a^{-5/2}b^3 + \ldots =$$

$$a^{1/2} + \frac{1}{2}a^{-1/2}b - \frac{1}{8}a^{-3/2}b^2 + \frac{1}{8}a^{-5/2}b^3 + \ldots$$

EXAMPLE 7: Use the Binomial Theorem to find $\sqrt[3]{9}$ to three decimal places.
Write $\sqrt[3]{9}$ as $(8+1)^{1/3}$ because 8 is a perfect cube.

$$\sqrt[3]{9} \approx \binom{1/3}{0}8^{1/3} + \binom{1/3}{1}8^{-2/3} \cdot 1 + \binom{1/3}{2}8^{-5/3} \cdot 1^2$$

$$+ \binom{1/3}{3}8^{-8/3} \cdot 1^3 + \ldots$$

$$= 8^{1/3} = 2; \; 8^{-2/3} = \frac{1}{4}; \; 8^{-5/3} = \frac{1}{32}; \; 8^{-8/3} = \frac{1}{256}$$

$$\sqrt[3]{9} \approx 1 \cdot 2 + \frac{1}{3} \cdot \frac{1}{4} + \frac{\frac{1}{3}(-\frac{2}{3})}{2 \cdot 1} \cdot \frac{1}{32}$$

$$+ \frac{\frac{1}{3}(-\frac{2}{3})(-\frac{5}{3})}{3 \cdot 2 \cdot 1} \cdot \frac{1}{256} + \ldots$$

$$\approx 2 + \frac{1}{12} - \frac{1}{9} \cdot \frac{1}{32} + \frac{5}{81} \cdot \frac{1}{256} + \ldots$$

$$\approx 2 + .08333 - .00347 + .00024$$

$$\approx 2.080$$

Since the fourth term does not affect the third decimal place only three terms were necessary to find $\sqrt[3]{9}$ to 3 decimal places.

EXERCISES

1. The middle term of $\left(\frac{1}{x} - x\right)^{10}$ is

    [A] -126 [B] 126 [C] -252
    [D] $252x^5$ [E] none of these

2. The seventh term of $\left(a^3 - \frac{1}{a}\right)^8$ is

    [A] $28a^6$ [B] $28a^{-6}$ [C] 28
    [D] $8a^5$ [E] $8a^{-5}$

3. The value of $(1.002)^7$, to the nearest ten-thousandth, is

    [A] 1.0140 [B] 1.0141 [C] 1.0143
    [D] 1.0146 [E] 1.0147

4. The coefficient of the third term of $(8x - y)^{1/3}$, after simplification, is

    [A] $-\frac{1}{9}$ [B] $-\frac{1}{288}$ [C] $\frac{5}{81}$
    [D] $\frac{5}{2592}$ [E] $-\frac{1}{576}$

3. PROBABILITY

The probability of an event happening is a number defined to be the ratio of the number of ways the event can happen successfully divided by the total number of ways the event can happen.

EXAMPLE 1: What is the probability of getting a head when a coin is flipped?
A coin can fall in one of two ways, heads or tails, and each is equally likely.
$$P(\text{head}) = \frac{\text{number of ways a head can come up}}{\text{total number of ways the coin can fall}} = \frac{1}{2}.$$

EXAMPLE 2: What is the probability of getting a 3 when one die is thrown?
A die can fall with 6 different numbers showing, and there is only one way a 3 can show.
$$P(3) = \frac{\text{number of ways a 3 can come up}}{\text{total number of ways the die can fall}} = \frac{1}{6}.$$

EXAMPLE 3: What is the probability of getting a sum of seven when two dice are thrown?
Since it is not obvious how many different throws will produce a sum of seven, or how many different ways the two dice will land, it will be useful to consider all the possible outcomes. The set of all outcomes of an experiment is called the *sample space* of the experiment. In order to keep track of the elements of the sample space in this experiment, let the first die be green and the second die be red. Since the green die can fall in one of six ways, and the red die can fall in one of six ways, there should be 6·6 or 36 elements in the sample space. The elements of the sample space are:

green	red	green	red	green	red	green	red	green	red	green	red
1	1	2	1	3	1	4	1	5	1	6	1
1	2	2	2	3	2	4	2	5	2	6	2
1	3	2	3	3	3	4	3	5	3	6	3
1	4	2	4	3	4	4	4	5	4	6	4
1	5	2	5	3	5	4	5	5	5	6	5
1	6	2	6	3	6	4	6	5	6	6	6

The circled elements of the sample space are those whose sum is seven.
$$P(7) = \frac{\text{number of successes}}{\text{total number}} = \frac{6}{36} = \frac{1}{6}$$

The probability, p, of any event is a number such that $0 \leq p \leq 1$. If $p=0$, the event cannot happen. If $p=1$, the event is sure to happen.

EXAMPLE 4: [A] What is the probability of getting a seven when one die is thrown? [B] What is the probability of getting a number less than twelve when one die is thrown?
[A] $P(7) = 0$, since a single die has only numbers 1 through 6 on its faces.
[B] $P(\# < 12) = 1$ since any face number is less than 12.

The **odds** in favor of an event happening are defined to be the ratio of the probability of an event happening successfully divided by the probability of an event not happening successfully.

EXAMPLE 5: What are the odds in favor of getting a number greater than 2 when one die is thrown?
$P(\# > 2) = \frac{4}{6} = \frac{2}{3}$ and $P(\# \not> 2) = \frac{2}{6} = \frac{1}{3}$. Therefore, the odds in favor of a number greater than
$$2 = \frac{P(\# > 2)}{P(\# \not> 2)} = \frac{\frac{2}{3}}{\frac{1}{3}} = \frac{2}{1} \text{ or } 2:1.$$

Independent events are events which have no effect on one another. Two events are defined to be independent if and only if $P(A \cap B) = P(A) \cdot P(B)$, where $A \cap B$ means the intersection of the two sets A and B. If two events are not independent, they are said to be *dependent*.

EXAMPLE 6: If two fair coins are flipped, what is the probability of getting 2 heads?
Since the flip of each coin has no effect on the outcome of any other coin, these are independent events. $P(2H) = P(H) \cdot P(H) = \frac{1}{2} \cdot \frac{1}{2} = \frac{1}{4}$.

EXAMPLE 7: When two dice are thrown, what is the probability of getting two fives?
These are independent events because the result of one die does not affect the result of the other.
$$P(2 \text{ fives}) = P(5) \cdot P(5) = \frac{1}{6} \cdot \frac{1}{6} = \frac{1}{36}$$

EXAMPLE 8: Two dice are thrown. Event A is "the sum is seven." Event B is "at least one die is a six." Are A and B independent?
From the chart in example 3, $A =$

{(1, 6), (6, 1), (2, 5), (5, 2), (3, 4), (4, 3)} and $B =$ {(1,6),(2,6),(3,6),(4,6),(5,6),(6,6),(6,1),(6,2),(6,3), (6,4),(6,5)}. Therefore, $P(A) = \frac{1}{6}$ and $P(B) = \frac{11}{36}$.

$A \cap B = \{(1,6),(6,1)\}$. Therefore, $P(A \cap B) = \frac{2}{36} = \frac{1}{18}$.

$P(A) \cdot P(B) = \frac{11}{216} \neq \frac{1}{18}$. Therefore, $P(A \cap B) \neq P(A) \cdot P(B)$ and so events A and B are dependent.

EXAMPLE 9: If the probability that John will buy a certain product is $\frac{3}{5}$, that Bill will buy that product is $\frac{2}{3}$, and that Sue will buy that product is $\frac{1}{4}$, what is the probability that at least one of them will buy the product?

Since the purchase by any one of the people does not affect the purchase by anyone else, these events are independent. The best way to approach this problem is to consider the probability that none of them buys the product.

Let $A =$ the event "John does not buy the product."
Let $B =$ the event "Bill does not buy the product."
Let $C =$ the event "Sue does not buy the product."
$P(A) = 1 - \frac{3}{5} = \frac{2}{5}$; $P(B) = 1 - \frac{2}{3} = \frac{1}{3}$; $P(C) = 1 - \frac{1}{4} = \frac{3}{4}$.

The probability that none of them buys the product
$= P(A \cap B \cap C) = P(A) \cdot P(B) \cdot P(C) = \frac{2}{5} \cdot \frac{1}{3} \cdot \frac{3}{4} = \frac{1}{10}$

Therefore, the probability that at least one of them buys the product is $1 - \frac{1}{10} = \frac{9}{10}$.

In general, the probability of event A happening or event B happening or both happening is equal to the sum of $P(A)$ and $P(B)$ less the probability of both happening. In symbols, $P(A \cup B) = P(A) + P(B) - P(A \cap B)$, where $A \cup B$ means the union of sets A and B. If $P(A \cap B) = 0$, the events are said to be *mutually exclusive*.

EXAMPLE 10: What is the probability of drawing a spade or a king from a deck of 52 cards?
Let $A =$ the event "drawing a spade."
Let $B =$ the event "drawing a king."
Since there are 13 spades and 4 kings in a deck of cards,
$P(A) = \frac{13}{52} = \frac{1}{4}$; $P(B) = \frac{4}{52} = \frac{1}{13}$
$P(A \cap B) = P(\text{drawing the king of spades}) = \frac{1}{52}$
$P(A \cup B) = P(A) + P(B) - P(A \cap B)$
$= \frac{13}{52} + \frac{4}{52} - \frac{1}{52} = \frac{16}{52} = \frac{4}{13}$.

These events are *not* mutually exclusive.

EXAMPLE 11: In a throw of two dice, what is the probability of getting a sum of seven or eleven?
Let $A =$ the event "throwing a sum of 7."
Let $B =$ the event "throwing a sum of 11."
$P(A \cap B) = 0$, so these events *are* mutually exclusive.
$P(A \cup B) = P(A) + P(B)$. From the chart in example 3,

$P(A) = \frac{6}{36}$ and $P(B) = \frac{2}{36}$.

$P(A \cup B) = \frac{6}{36} + \frac{2}{36} = \frac{8}{36} = \frac{2}{9}$.

EXERCISES

1. With the throw of two dice, what is the probability that the sum is a prime number?

   [A] $\frac{7}{18}$ [B] $\frac{5}{11}$ [C] $\frac{5}{12}$ [D] $\frac{4}{11}$ [E] $\frac{1}{2}$

2. If a coin is flipped and one die is thrown, what is the probability of getting either a head or a four?

   [A] $\frac{2}{3}$ [B] $\frac{5}{12}$ [C] $\frac{1}{12}$ [D] $\frac{7}{12}$ [E] $\frac{1}{3}$

3. Three cards are drawn from an ordinary deck of 52 cards. Each card is replaced in the deck before the next card is drawn. What is the probability that at least one of the cards was a spade?

   [A] $\frac{3}{8}$ [B] $\frac{3}{4}$ [C] $\frac{3}{52}$ [D] $\frac{9}{64}$ [E] $\frac{37}{64}$

4. A coin is tossed three times. Let $A = \{$all heads occur$\}$ and $B = \{$at least one head occurs$\}$. What is $P(A \cup B)$?

   [A] $\frac{1}{8}$ [B] $\frac{1}{4}$ [C] $\frac{1}{2}$ [D] $\frac{3}{4}$ [E] $\frac{7}{8}$

5. A class has 12 boys and 4 girls. If three students are selected at random from the class, what is the probability that they are all boys?

   [A] $\frac{1}{55}$ [B] $\frac{1}{3}$ [C] $\frac{1}{4}$ [D] $\frac{11}{15}$ [E] $\frac{11}{28}$

6. A red box contains 8 items of which 3 are defective, and a blue box contains 5 items of which 2 are defective. An item is drawn at random from each box. What is the probability that both items are non-defective?

   [A] $\frac{17}{20}$ [B] $\frac{3}{8}$ [C] $\frac{8}{13}$ [D] $\frac{3}{20}$ [E] $\frac{5}{13}$

[7] A hotel has five single rooms available. Six men and three women apply for the rooms. What is the probability that the rooms will be rented to three men and two women?

[A] $\dfrac{5}{8}$ [B] $\dfrac{10}{21}$ [C] $\dfrac{23}{112}$

[D] $\dfrac{5}{9}$ [E] $\dfrac{97}{251}$

4. SEQUENCES AND SERIES

A *sequence* is a function with a domain consisting of the natural numbers in order. A *series* is the sum of the terms of a sequence.

EXAMPLE 1: [A] $\dfrac{1}{2}, \dfrac{1}{3}, \dfrac{1}{4}, \dfrac{1}{5}, \ldots, \dfrac{1}{n+1} \ldots$ is an infinite sequence of numbers with $t_1 = \dfrac{1}{2}$, $t_2 = \dfrac{1}{3}$, $t_3 = \dfrac{1}{4}$, $t_4 = \dfrac{1}{5}$, and $t_n = \dfrac{1}{n+1}$.

[B] $2, 4, 6, \ldots, 20$ is a finite sequence of numbers with $t_1 = 2$, $t_2 = 4$, $t_3 = 6$, and $t_{10} = 20$.

[C] $\dfrac{1}{2} + \dfrac{1}{4} + \dfrac{1}{8} + \dfrac{1}{16} + \ldots + \dfrac{1}{2^n} + \ldots$ is an infinite series of numbers.

EXAMPLE 2: If $t_n = \dfrac{2n}{n+1}$, find the first five terms of the sequence.

Substituting 1, 2, 3, 4, and 5 for n, $t_1 = \dfrac{2}{2} = 1$, $t_2 = \dfrac{4}{3}$, $t_3 = \dfrac{6}{4} = \dfrac{3}{2}$, $t_4 = \dfrac{8}{5}$, and $t_5 = \dfrac{10}{6} = \dfrac{5}{3}$.

The first five terms are $1, \dfrac{4}{3}, \dfrac{3}{2}, \dfrac{8}{5}, \dfrac{5}{3}$.

EXAMPLE 3: If $a_1 = 2$ and $a_n = \dfrac{a_{n-1}}{2}$, find the first five terms of this sequence.

Since every term is expressed with respect to the immediately preceding term, this is called a *recursion formula*.

$a_1 = 2$, $a_2 = \dfrac{a_1}{2} = \dfrac{2}{2} = 1$,

$a_3 = \dfrac{a_2}{2} = \dfrac{1}{2}$, $a_4 = \dfrac{a_3}{2} = \dfrac{\frac{1}{2}}{2} = \dfrac{1}{4}$,

$a_5 = \dfrac{a_4}{2} = \dfrac{\frac{1}{4}}{2} = \dfrac{1}{8}$.

Therefore, the first five terms are $2, 1, \dfrac{1}{2}, \dfrac{1}{4}, \dfrac{1}{8}$.

A series can be abbreviated using the Greek letter sigma, Σ, which is used to represent the summation of several terms.

EXAMPLE 4: [A] The series $2 + 4 + 6 \ldots + 20 = \sum_{i=1}^{10} 2i = 110$.

[B] $\sum_{k=0}^{5} i^2 = 0^2 + 1^2 + 2^2 + 3^2 + 4^2 + 5^2 = 0 + 1 + 4 + 9 + 16 + 25 = 55$.

One of the most common sequences studied at this level is an *arithmetic sequence* (or arithmetic progression). Each term differs from the preceding term by a common difference. In general, an arithmetic sequence is denoted by:

$$t_1, t_1 + d, t_1 + 2d, t_1 + 3d, \ldots, t_1 + (n-1)d$$

where d is the common difference and $t_n = t_1 + (n-1)d$. The sum of n terms of the series constructed from an arithmetic sequence is given by the formula:

$$S_n = \dfrac{n}{2}[t_1 + t_n] \quad \text{or} \quad S_n = \dfrac{n}{2}[2t_1 + (n-1)d]$$

EXAMPLE 5: [A] Find the 28th term of the arithmetic sequence $2, 5, 8, \ldots$ [B] Express the 28 terms of the series of this sequence using sigma notation. [C] Find the sum of the first 28 terms of the series.

[A] $t_n = t_1 + (n-1)d$

$t_{28} = 2 + 27 \cdot 3 = 83$

[B] $\sum_{k=0}^{27} (3k+2)$ or $\sum_{j=1}^{28} (3j-1)$

[C] $S_n = \dfrac{n}{2}[t_1 + t_n]$

$S_{28} = \dfrac{28}{2}[2+83] = 14 \cdot 85 = 1190$

EXAMPLE 6: If $t_8 = 4$ and $t_{12} = -2$, find the first three terms.

$t_n = t_1 + (n-1)d$

$t_8 = 4 = t_1 + 7d$

$t_{12} = -2 = t_1 + 11d$

To solve these two equations for d, subtract the two equations.

$-6 = 4d$

$d = -\dfrac{3}{2}$

Substituting in the first equation: $4 = t_1 + 7\left(-\frac{3}{2}\right)$, thus

$$t_1 = 4 + \frac{21}{2} = \frac{29}{2}$$

$$t_2 = \frac{29}{2} + \left(-\frac{3}{2}\right) = \frac{26}{2} = 13$$

$$t_3 = \frac{29}{2} + 2\left(-\frac{3}{2}\right) = \frac{23}{2}$$

The first three terms are $\frac{29}{2}, 13, \frac{23}{2}$.

EXAMPLE 7: In an arithmetic series, if $S_n = 3n^2 + 2n$, find the first three terms.
When $n = 1$, $S_1 = t_1$. Therefore, $t_1 = 3(1)^2 + 2 \cdot 1 = 5$.

$$S_2 = t_1 + t_2 = 3(2)^2 + 2 \cdot 2 = 16$$
$$5 + t_2 = 16$$
$$t_2 = 11$$

Therefore, $d = 6$, which leads to a third term of 17. Thus, the first three terms are 5, 11, 17.

The terms falling between two given terms of an arithmetic sequence are called *arithmetic means*. If there is only one arithmetic mean between two given terms, it is called the *average* of the two terms.

EXAMPLE 8: Insert three arithmetic means between 1 and 9.

1, m_1, m_2, m_3, 9
t_1, t_2, t_3, t_4, t_5

Since $t_1 = 1$ and $t_5 = t_1 + (5-1) = 9$,
$$1 + 4d = 9$$
$$4d = 8$$
$$d = 2$$

Therefore, the three arithmetic means are 3, 5, and 7.

Another very common sequence studied at this level is a *geometric sequence* (or geometric progression). The ratio of any two successive terms is equal to a constant, r, called the common ratio. In general, a geometric sequence is denoted by:

$$t_1, t_1 r, t_1 r^2, \ldots, t_1 r^{n-1}, \text{ where } t_n = t_1 r^{n-1}.$$

The sum of n terms of a series constructed from a geometric sequence is given by

$$S_n = \frac{t_1(1 - r^n)}{1 - r}.$$

EXAMPLE 9: [A] Find the seventh term of the geometric sequence 1, 2, 4, ... and [B] the sum of the first seven terms.

[A] $r = \frac{t_2}{t_1} = \frac{2}{1} = 2$; $t_7 = t_1 r^{7-1}$; $t_7 = 1 \cdot 2^6 = 64$

[B] $S_7 = \frac{1(1 - 2^7)}{1 - 2} = \frac{1 - 128}{-1} = 127$

EXAMPLE 10: The first term of a geometric sequence is 64, and the common ratio is $\frac{1}{4}$. For what value of n is $t_n = \frac{1}{4}$?

$$t_n = t_1 r^{n-1}; \quad \frac{1}{4} = 64\left(\frac{1}{4}\right)^{n-1};$$

$$\frac{1}{4} = 4^3\left(\frac{1}{4}\right)^{n-1}; \quad \frac{1}{4} = \left(\frac{1}{4}\right)^{-3} \cdot \left(\frac{1}{4}\right)^{n-1};$$

$$\frac{1}{4} = \left(\frac{1}{4}\right)^{-3 + n - 1}; \quad \left(\frac{1}{4}\right)^1 = \left(\frac{1}{4}\right)^{n-4}$$

Therefore, $1 = n - 4$ and $n = 5$.

In a geometric sequence, if $|r| < 1$, the sum of the series approaches a limit as n approaches infinity. Considering the formula $S_n = \frac{a(1 - r^n)}{1 - r}$ if $|r| < 1$, the term $r^n \to 0$ as $n \to \infty$. Therefore, as long as $|r| < 1$, $\lim_{n \to \infty} S_n = \frac{a}{1 - r}$.

EXAMPLE 11: Evaluate [A] $\lim_{n \to \infty} \sum_{k=1}^{n} \frac{1}{2^k}$

[B] $\sum_{j=0}^{\infty} (-3)^{-j}$.

Both problems ask the same question: Find the limit of the sum of the geometric series as $n \to \infty$.

[A] Listing the first few terms, $\frac{1}{2} + \frac{1}{4} + \frac{1}{8} + \ldots$, it can be seen that the common ratio is $r = \frac{1}{2}$. Therefore,

$$\lim_{n \to \infty} S_n = \frac{\frac{1}{2}}{1 - \frac{1}{2}} = 1.$$

[B] Listing the first few terms, $\frac{1}{1} - \frac{1}{3} + \frac{1}{9} - \ldots$, it can be seen that the common ratio is $r = -\frac{1}{3}$. Therefore,

$$\lim_{n \to \infty} S_n = \frac{1}{1 - \left(-\frac{1}{3}\right)} = \frac{1}{\frac{4}{3}} = \frac{3}{4}.$$

EXAMPLE 12: Find the exact value of the repeating decimal $.4545\ldots$.

This can be represented by a geometric series, $.45 + .0045 + .000045 + \ldots$, with $t_1 = .45$ and $r = .01$. Since $|r| < 1$,

$$\lim_{n \to \infty} S_n = \frac{.45}{1-.01} = \frac{.45}{.99} = \frac{45}{99} = \frac{5}{11}.$$

The terms falling between two given terms of a geometric sequence are called *geometric means*. If there is only one geometric mean between the two given terms, it is called the *mean proportional* of the two terms.

EXAMPLE 13: Insert five geometric means between $\frac{1}{8}$ and 8.

$$\frac{1}{8}, \quad m_1, \quad m_2, \quad m_3, \quad m_4, \quad m_5, \quad 8$$
$$t_1, \quad t_2, \quad t_3, \quad t_4, \quad t_5, \quad t_6, \quad t_7$$

Since $t_1 = \frac{1}{8}$ and $t_7 = t_1 r^{n-1} = 8$

$$\frac{1}{8} r^6 = 8$$
$$r^6 = 64$$
$$r = \pm (64)^{1/6} = \pm (2^6)^{1/6} = \pm 2.$$

Thus, there are two sets of geometric means.

	m_1	m_2	m_3	m_4	m_5
If $r=+2$,	$\frac{1}{4}$	$\frac{1}{2}$	1	2	4
If $r=-2$,	$-\frac{1}{4}$	$\frac{1}{2}$	-1	2	-4

EXAMPLE 14: Given the sequence $2, x, y, 9$, if the first three terms form an arithmetic sequence and the last three terms form a geometric sequence, find x and y.

From the arithmetic sequence $\begin{cases} x = 2 + d \\ y = 2 + 2d \end{cases}$, substitute to eliminate the d.

$$y = 2 + 2(x - 2)$$
$$y = 2 + 2x - 4$$
$$* \quad y = 2x - 2$$

From the geometric sequence $\begin{cases} 9 = yr \\ y = xr \end{cases}$, substitute to eliminate the r.

$$9 = y \cdot \frac{y}{x}$$
$$* \quad 9x = y^2$$

Using the two equations with the * to eliminate the y gives

$$9x = (2x - 2)^2$$
$$9x = 4x^2 - 8x + 4$$
$$4x^2 - 17x + 4 = 0$$
$$(4x - 1)(x - 4) = 0$$
$$4x - 1 = 0 \text{ or } x - 4 = 0$$

Thus, $x = \frac{1}{4}$ or 4.

Substituting in $y = 2x - 2$,

if $x = \frac{1}{4}, y = -\frac{3}{2}$

if $x = 4, y = 6$.

EXERCISES

1 If $a_1 = 3$ and $a_n = n + a_{n-1}$, the sum of the first five terms is

[A] 30   [B] 17   [C] 42   [D] 45   [E] 68

2 If x, y, and z are three consecutive terms in an arithmetic sequence, which one of the following is true?

[A] $y = \frac{1}{2}(x + z)$ [B] $y = \sqrt{xz}$

[C] $y = \sqrt{x + z}$ [D] $y = z + x$

[E] $z = \frac{1}{2}(x + y)$

3 If the repeating decimal $.23\overline{737}\ldots$ is written as a fraction in lowest terms, the sum of the numerator and denominator is

[A] 245   [B] 1237   [C] 16
[D] 47   [E] 334

4 The first three terms of a geometric sequence are $\sqrt[4]{3}$, $\sqrt[8]{3}$, 1. The fourth term is

[A] $\sqrt[16]{3}$ [B] $\frac{1}{\sqrt[16]{3}}$ [C] $\frac{1}{\sqrt[8]{3}}$

[D] $\frac{1}{\sqrt[4]{3}}$ [E] $\sqrt[32]{3}$

5 By how much does the arithmetic mean between 1 and 25 exceed the positive geometric mean between 1 and 25?

[A] 5   [B] about 7.1   [C] 8
[D] 13   [E] 18

6 Let S_n equal the sum of all the prime numbers less than or equal to n. What is the largest value of n for which $S_n \leq 100$?

[A] 23   [B] 14   [C] 29   [D] 10   [E] 28

7. The units digit of $\sum_{i=1}^{10} i!$ is

[A] 0 [B] 3 [C] 5 [D] 7 [E] 9

8. $\sum_{k=1}^{100} (-1)^k \cdot k$ equals

[A] 4950 [B] 100 [C] 5050
[D] 50 [E] 0

9. In a geometric series $S_\infty = \frac{2}{7}$ and $t_1 = \frac{2}{3}$, what is r?

[A] $\frac{2}{3}$ [B] $\frac{4}{3}$ [C] $\frac{2}{7}$
[D] $-\frac{4}{3}$ [E] $-\frac{2}{7}$

5. GEOMETRY

This section contains miscellaneous topics from geometry not already covered that are apt to appear on the Level II examination.

Transformations: In a plane, a transformation slides, rotates, or stretches a geometric figure from one position to another. If a point (x,y) lies on a graph and is transformed into the point (x',y') the formula for sliding (called a *translation*) is of the form

$$\begin{cases} x' = x + r \\ y' = y + s \end{cases}.$$

The formula for stretching is of the form

$$\begin{cases} x' = hx \\ y' = kx \end{cases}.$$

The formula for rotating is beyond the scope of this book.

EXAMPLE 1: Given the equation $y = x^2 + 4x + 6$, find the equation obtained by performing the translation
$$\begin{cases} x' = x + 2 \\ y' = y - 2 \end{cases}.$$
Solving the translation equations for x and y and substituting

$$y' + 2 = (s' - 2)^2 + 4(x' - 2) + 6$$
$$y' + 2 = (x')^2 - 4x' + 4 + 4x' - 8 + 6$$
$$y' = (x')^2.$$

EXAMPLE 2: If the graph of $f(x)$ is given in the figure, what does the graph of $f(2x - 3)$ look like?

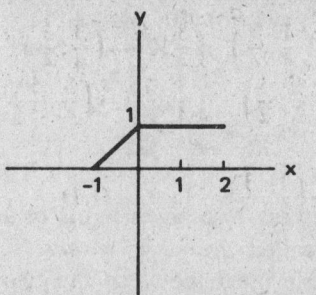

In order to avoid confusion, let the x in $f(2x - 3)$ be named x'. If $2x' - 3$ replaces the original x, the transformation relating the original graph with the desired graph is $x = 2x' - 3$. Therefore, $x' = \frac{x + 3}{2}$.
Using the graph to find y and this equation to find x' it is possible to obtain points on the desired graph.
When $x = -1$, $y = 0$ and $x' = 1$.
When $x = 0$, $y = 1$ and $x' = \frac{3}{2}$.
When $x = 1$, $y = 1$ and $x' = 2$.
When $x = 2$, $y = 1$ and $x' = \frac{5}{2}$.
Graphing the points (x', y) the following figure is obtained.

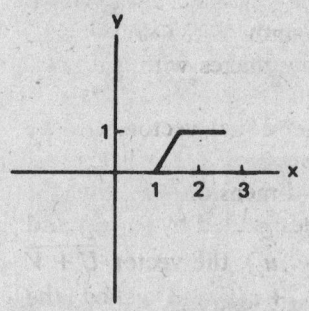

EXAMPLE 3: The shaded region in the figure is acted on by the transformation T, which transforms any point into the point $(x, x + y)$. What does this figure transform into?

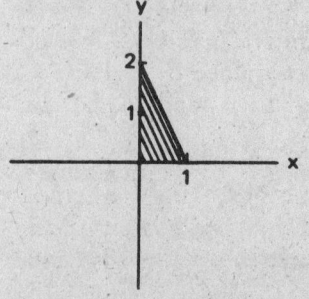

Taking points on the border of the shaded region

and transforming them by T will give an idea of the shape of the new region.

$(0,0) \xrightarrow{T} (0,0)$ $\left(\frac{1}{4},\frac{3}{2}\right) \xrightarrow{T} \left(\frac{1}{4},\frac{7}{4}\right)$ $\left(\frac{3}{4},0\right) \xrightarrow{T} \left(\frac{3}{4},\frac{3}{4}\right)$

$(0,1) \xrightarrow{T} (0,1)$ $\left(\frac{1}{2},0\right) \xrightarrow{T} \left(\frac{1}{2},\frac{1}{2}\right)$ $\left(\frac{3}{4},\frac{1}{2}\right) \xrightarrow{T} \left(\frac{3}{4},\frac{5}{4}\right)$

$(0,2) \xrightarrow{T} (0,2)$

$\left(\frac{1}{4},0\right) \xrightarrow{T} \left(\frac{1}{4},\frac{1}{4}\right)$ $\left(\frac{1}{2},1\right) \xrightarrow{T} \left(\frac{1}{2},\frac{3}{2}\right)$ $(1,0) \xrightarrow{T} (1,1)$

Vectors: A vector in a plane is defined to be an ordered pair of real numbers. A vector in space is defined to be an ordered triple of real numbers. On a coordinate system, a vector is usually represented by an arrow whose initial point is the original, and whose terminal point is at the ordered pair (or triple) that named the vector. Vector quantities always have a magnitude or *norm* (the length of the arrow) and direction (the angle the arrow makes with the positive x-axis).

All properties of 2-dimensional vectors can be extended to 3-dimensional vectors. We will express the properties in terms of 2-dimensional vectors for convenience. If vector $\vec{V}$ is designated by (v_1, v_2) and vector $\vec{U}$ is designated by (u_1, u_2), the vector $\overrightarrow{U+V}$ is designated by (u_1+v_1, u_2+v_2) and called the *resultant* of $\vec{U}$ and $\vec{V}$. The vector $-\vec{V}$ has the same magnitude as $\vec{V}$ but has a direction opposite that of $\vec{V}$. Every vector, $\vec{V}$ can be expressed in terms of any other two (three) non-parallel vectors. In many instances unit vectors parallel to the x and y-axes are used. If the vector $\vec{i}=(1,0)$ and the vector $\vec{j}=(0,1)$ (and in 3 dimensions $\vec{k}=(0,0,1)$), any vector $\vec{V}=a\vec{i}+b\vec{j}$ where a and b are real numbers. A unit vector parallel to $\vec{V}$ can be determined by dividing $\vec{V}$ by its norm, denoted by $\|\vec{V}\|$ and equal to $\sqrt{(v_1)^2+(v_2)^2}$

It is possible to determine algebraically whether two vectors are perpendicular by defining the *dot product* or *inner product* of two vectors, $\vec{V}(v_1, v_2)$ and $\vec{U}(u_1, u_2)$.

$$\vec{V} \cdot \vec{U} = v_1 u_1 + v_2 u_2$$

Notice that the dot product of two vectors is a *real number*, not a vector. Two vectors, $\vec{V}$ and $\vec{U}$, are perpendicular if and only if $\vec{V} \cdot \vec{U} = 0$.

EXAMPLE 4: If vector $\vec{V}=(2,3)$ and vector $\vec{U}=(6,-4)$: [A] What is the resultant of $\vec{U}$ and $\vec{V}$? [B] What is the norm of $\vec{U}$? [C] Express $\vec{V}$ in terms of $\vec{i}$ and $\vec{j}$. [D] Are $\vec{U}$ and $\vec{V}$ perpendicular?

[A] The resultant, $\overrightarrow{U+V}$, equals $(6+2, -4+3) = (8, -1)$.

[B] The norm of $\vec{U}$, $\|\vec{U}\| = \sqrt{36+16} = \sqrt{52} = 2\sqrt{13}$.

[C] $\vec{V} = 2\vec{i} + 3\vec{j}$. To verify this, use the definition of $\vec{i}$ and $\vec{j}$. $\vec{V} = 2(1,0) + 3(0,1) = (2,0) + (0,3) = (2,3) = \vec{V}$

[D] $\vec{U} \cdot \vec{V} = 6 \cdot 2 + (-4) \cdot 3 = 12 - 12 = 0$. Therefore, $\vec{U}$ and $\vec{V}$ are perpendicular because the dot product is equal to zero.

EXAMPLE 5: If $\vec{U}=(-1,4)$ and the resultant of $\vec{U}$ and $\vec{V}$ is $(4,5)$, find $\vec{V}$.
Let $\vec{V}=(v_1, v_2)$. The resultant $\overrightarrow{U+V}=(-1,4)+(v_1, v_2)=(4,5)$. Therefore, $(-1+v_1, 4+v_2)=(4,5)$, which implies that $-1+v_1=4$ and $4+v_2=5$. Thus, $v_1=5$ and $v_2=1$. $\vec{V}=(5,1)$.

EXAMPLE 6: Express the vector $\vec{V}=(3,-7)$ as a linear combination of (i.e., in terms of) vectors $\vec{U}=(-6,8)$ and $\vec{W}=(9,-13)$.
This question requires that two real numbers, a and b, be found such that $\vec{V} = a\vec{U} + b\vec{W}$.

$$\vec{V} = (3,-7) = a(-6,8) + b(9,-13)$$
$$(3,-7) = (-6a, 8a) + (9b, -13b)$$
$$(3,-7) = (-6a+9b, 8a-13b)$$

Therefore, $3=-6a+9b$ and $-7=8a-13b$.
Solving these two equations simultaneously gives $a=4$ and $b=3$.
Thus, $\vec{V} = 4\vec{U} + 3\vec{W}$.

Three-Dimensional Coordinate Geometry: In three dimensions, the equation of a plane is of the form $Ax+By+Cz+D=0$. The intercepts of the plane can be found by setting two of the variables equal to zero. Let $y=z=0$, and $x=-\frac{D}{A}$. The x-intercept occurs at the point $\left(-\frac{D}{A}, 0, 0\right)$. Let $x=z=0$, and $y=-\frac{D}{B}$. The y-intercept occurs at the point $\left(0, -\frac{D}{B}, 0\right)$. Let $x=y=0$, and $z=-\frac{D}{C}$.

The z-intercept occurs at the point $\left(0, 0, -\dfrac{D}{C}\right)$. The line which is the intersection of the plane and one of the coordinate planes is called a *trace*. The equation of the xy-trace is obtained by setting $z=0$ in the equation of the plane. The xy-trace is the line whose equation is $Ax + By + D = 0$.

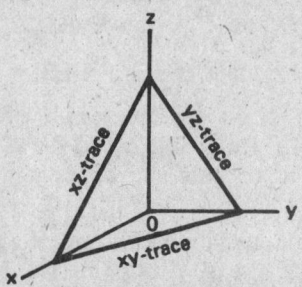

In three dimensions the equation of a line must be expressed as a set of three parametric equations:

$$x = x_0 + c_1 d$$
$$y = y_0 + c_2 d$$
$$z = z_0 + c_3 d$$

where d is the parameter, (x_0, y_0, z_0) are the coordinates of any point on the line, and c_1, c_2, and c_3 are called *direction numbers*. If (x_1, y_1, z_1) are the coordinates of any other point on the line, one set of values for the direction numbers would be $c_1 = x_1 - x_0$, $c_2 = y_1 - y_0$, and $c_3 = z_1 - z_0$. If $(c_1)^2 + (c_2)^2 + (c_3)^2 = 1$, then c_1, c_2, c_3 are called *direction cosines* and are the cosines of the three angles which the line makes with the positive x, y, and z-axes.

EXAMPLE 7: What are the traces of the plane with the equation $3x + 2y - 4z = 12$?
Set $z = 0$. The xy-trace is $3x + 2y = 12$.
Set $y = 0$. The xz-trace is $3x - 4z = 12$.
Set $x = 0$. The yz-trace is $2y - 4z = 12$.

EXAMPLE 8: [A] Find an equation of the line passing through the points whose coordinates are $(1, 2, 3)$ and $(4, -2, 6)$. [B] Find the direction cosines of the line.

[A] Let $(x_0, y_0, z_0) = (1, 2, 3)$. $c_1 = 4 - 1 = 3$, $c_2 = -2 - 2 = -4$, and $c_3 = 6 - 3 = 3$. One set of equations for the line is

$$x = 1 + 3d$$
$$y = 2 - 4d$$
$$z = 3 + 3d$$

[B] In order to find direction cosines, it is necessary to find the distance between the two points used to determine the direction numbers.

Distance $= \sqrt{(4-1)^2 + (-2-2)^2 + (6-3)^2}$
$= \sqrt{3^2 + (-4)^2 + 3^2} = \sqrt{9 + 16 + 9} = \sqrt{34}$.

With direction numbers 3, -4, and 3, the direction cosines become $\dfrac{3}{\sqrt{34}}$, $\dfrac{-4}{\sqrt{34}}$, and $\dfrac{3}{\sqrt{34}}$.

EXAMPLE 9: How far is the plane whose equation is $3x + 4y - 5z + 12 = 0$ from the origin?
Extending the formula for the distance between a point and a line to a formula for the distance between a point and a plane, gives:

$$\text{Distance} = \dfrac{|Ax_1 + By_1 + Cz_1 + D|}{\sqrt{A^2 + B^2 + C^2}}$$

In this problem $(x_1, y_1, z_1) = (0, 0, 0)$ and $A = 3$, $B = 4$, $C = -5$.

$$\text{Distance} = \dfrac{|3 \cdot 0 + 4 \cdot 0 - 5 \cdot 0 + 12|}{\sqrt{3^2 + 4^2 + (-5)^2}} = \dfrac{|12|}{\sqrt{9 + 16 + 25}}$$
$$= \dfrac{12}{\sqrt{50}} = \dfrac{12}{5\sqrt{2}} = \dfrac{6\sqrt{2}}{5}.$$

Solid Figures: A solid of revolution is obtained by taking a plane figure and rotating it about some line in the plane that does not intersect the figure to form a solid figure.

EXAMPLE 10: If the segment of the line $y = -2x + 2$ which lies in the first quadrant (Figure a) is rotated about the y-axis, a cone is formed (Figure b). What is the volume of the cone?

$$V = \dfrac{1}{3} \pi r^2 h$$
$$V = \dfrac{1}{3} \pi \cdot 1^2 \cdot 2 = \dfrac{2\pi}{3}$$

Many three-dimensional problems concern situations where it is necessary to picture the solid figures and their relationship to one another.

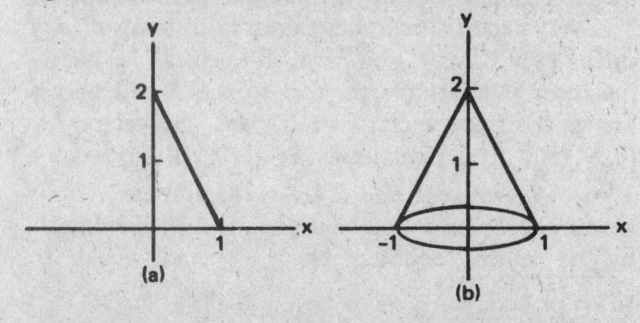

EXAMPLE 11: What figure is formed by the set of points at a fixed distance, d, from a line and at the same time $2d$ units from a point, P, on the line?
The set of points d units from the line is a cylinder of radius d, and with the line as the central axis of

the cylinder. The set of points $2d$ units from P is a sphere with center P and radius $2d$. Since the radius of the sphere is greater than that of the cylinder, the set of points satisfying the conditions is the intersection of the cylinder piercing the sphere. This intersection consists of two parallel circles, with centers on the line, perpendicular to the line, and equidistant from the fixed point P.

EXAMPLE 12: In the preceding example, how far apart are the two parallel circles?
A partial picture of the situation is shown below.

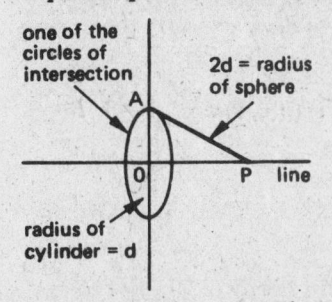

The distance OP is $\frac{1}{2}$ the distance between the circles, and it is a side of a right triangle whose other two sides are known. Using the Pythagorean Theorem:

$$\overline{OP}^2 + d^2 = (2d)^2$$
$$\overline{OP}^2 = 4d^2 - d^2$$

Therefore, $OP = \sqrt{3}\,d$.
The distance between the circles is $2\sqrt{3}\,d$.

EXAMPLE 13: A cone is inscribed in a cylinder. What is the ratio of the volume of the cone to the volume of the cylinder?
The radius of the base and the height of the cone and the cylinder are equal. The formula for the volume of the cone is $V_1 = \frac{1}{3}\pi r^2 h$. The formula for the volume of the cylinder is $V_2 = \pi r^2 h$. Therefore,

$$\frac{V_1}{V_2} = \frac{\frac{1}{3}\pi r^2 h}{\pi r^2 h} = \frac{\frac{1}{3}}{1} = \frac{1}{3}.$$

EXERCISES

1 In the figure, the graph of $f(x)$ has two transformations performed on it. First it is rotated 180°, and then it is reflected about the x-axis. Which of the following is the equation of the resulting curve?

[A] $f(-x)$ [B] $f(x)$ [C] $-f(x)$
[D] $f(x-1)$ [E] none of these

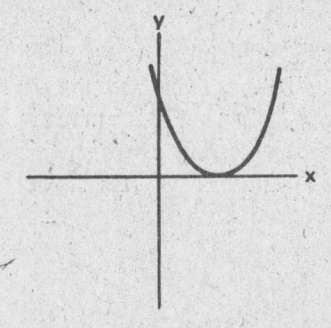

2 If (x,y) represents a point on the graph of $y = x + 2$, which of the following could be a portion of the graph of the set of points $(x, \sqrt{y}\,)$?

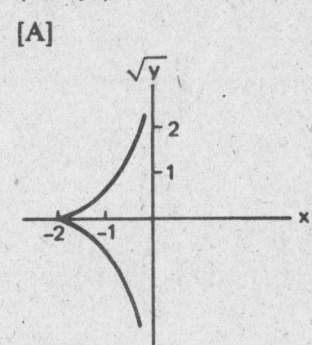

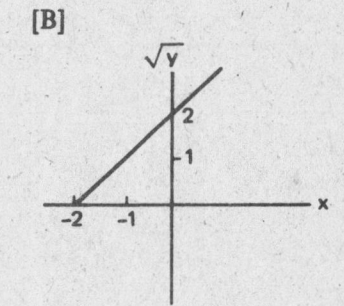

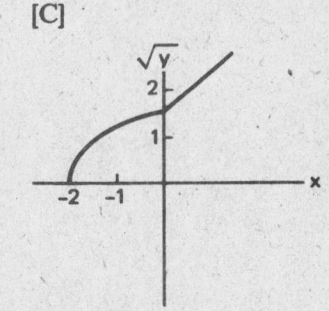

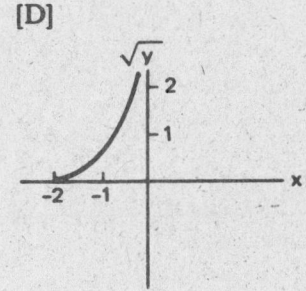

[A]

[B]

[C]

[D]

[E]

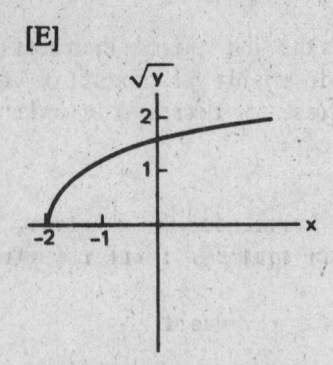

[3] In the figure, S is the set of points in the shaded region. Which of the following represents the set of all points $(x+y, y)$ where (x,y) is a point in S?

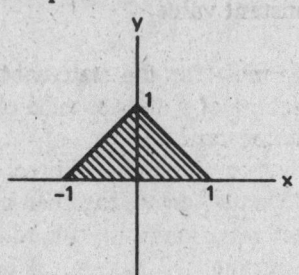

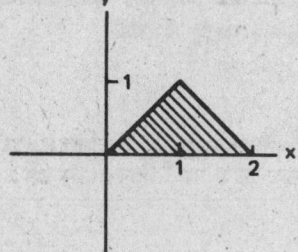

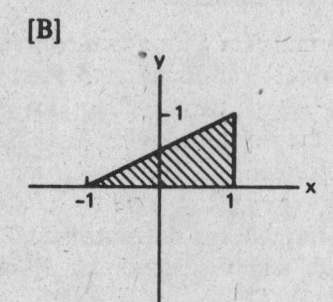

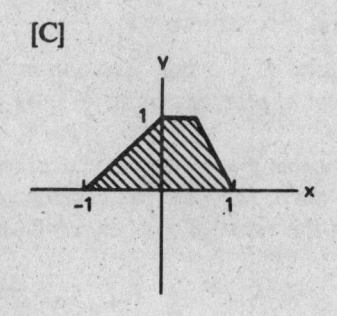

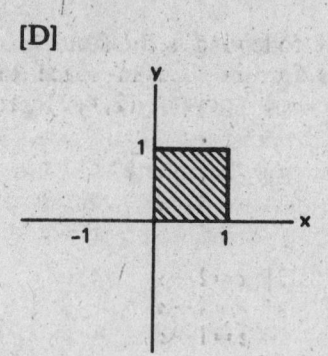

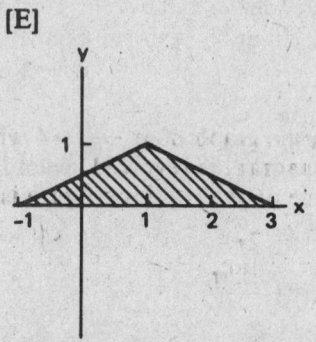

[A]

[B]

[C]

[D]

[E]

[4] If $\vec{V} = 2\vec{i} + 3\vec{j}$ and $\vec{U} = \vec{i} - 5\vec{j}$, the resultant vector of $2\vec{U} + 3\vec{V}$ equals

[A] $3\vec{i} - 2\vec{j}$ [B] $5\vec{i} + \vec{j}$ [C] $7\vec{i} - 9\vec{j}$
[D] $8\vec{i} - \vec{j}$ [E] $2\vec{i} + 3\vec{j}$

[5] A unit vector perpendicular to the vector $\vec{V} = (3, -4)$ is

[A] $(4, 3)$ [B] $\left(\frac{3}{5}, \frac{4}{5}\right)$
[C] $\left(-\frac{3}{5}, -\frac{4}{5}\right)$ [D] $\left(-\frac{4}{5}, -\frac{3}{5}\right)$
[E] $\left(-\frac{4}{5}, \frac{3}{5}\right)$

[6] The plane $2x + 3y - 4z = 5$ intersects the x-axis at the point where the x-coordinate is

[A] 2   [B] 2.5   [C] $1\frac{2}{3}$
[D] -1.25 [E] 5

[7] The distance between two points in space, $P_1(x, -1, -1)$ and $P_2(3, -3, 1)$, is 3. Find the possible values of x.

[A] 1 & 2 only   [B] 2 & 3 only
[C] -2 & -3 only   [D] 2 & 4 only
[E] -2 & -4 only

[8] Which of the following is a trace of the plane $3x + 5y - 7z + 8 = 0$?

[A] $3x + 5y = 8$ [B] $3x + 7z + 8 = 0$
[C] $5y - 7z = 8$ [D] $7z + 3x = 8$
[E] $7z - 5y = 8$

[9] Which one of the following is the set of parametric equations for the line in space that passes through the points $(2, 4, 1)$ and $(-1, 5, 2)$?

[A] $x = -1 + 2d$
$y = 5 + 4d$
$z = 2 + d$

[B] $x = -1 + d$
$y = 5 - d$
$z = 2 - d$

[C] $x = 2 - d$
$y = 4 + 5d$
$z = 1 + 2d$

[D] $x = 2 - 3d$
$y = 4 + d$
$z = 1 + d$

[E] $x = 2 + d$
$y = 4 + 9d$
$z = 1 + 3d$

[10] If the portion of the graph of $x^2 + y^2 = 4$ which lies in the first quadrant is revolved around the x-axis, the volume of the resulting figure is

[A] π [B] 2π [C] 4π
[D] $\dfrac{16\pi}{3}$ [E] $\dfrac{32\pi}{3}$

[11] A right circular cone whose base radius is 12 is inscribed in a sphere of radius 13. What is the volume of the cone?

[A] 720π [B] $\dfrac{2197\pi}{3}$ [C] 864π
[D] 1440π [E] 2592π

[12] Given a right circular cylinder such that the volume has the same numerical value as the total surface area, the smallest integral value for the radius of the cylinder is

[A] 1   [B] 2   [C] 3
[D] 4   [E] cannot be determined

[13] A square pyramid is inscribed in a cylinder whose base radius is 4 and height is 9. What is the volume of the pyramid?

[A] 48 [B] 144 [C] 288 [D] 96 [E] 108

[14] The set of points in space equidistant from two fixed points is

[A] one point   [B] one line   [C] a plane
[D] a hyperbola   [E] none of these

6. VARIATION

In a *direct variation* the ratio of the variables is a constant. The first variable mentioned is in the numerator, and all others form a product in the denominator. Thus, as the first variable increases (or decreases), the second variable (or product of variables) must also increase (or decrease) in order to maintain a constant value.

EXAMPLE 1: Give a formula for the statement, "x varies directly as y, the square of z and the square root of a."
$\dfrac{x}{yz^2\sqrt{a}} = K$, where K is a constant.

In an *inverse variation* the product of all variables is equal to a constant. Thus, as the first variable increases (or decreases) the second variable (or product of variables) must decrease (or increase) in order to maintain a constant value.

EXAMPLE 2: Give a formula for the statement, "x varies inversely as the square of y and the cube of z."
$xy^2z^3 = K$, where K is a constant.

In any variation that combines both direct and inverse variation the formula is put together using the procedure of a direct variation (a ratio) and then an inverse variation (a product). The result is equal to a constant.

EXAMPLE 3: Give a formula for the statement, "z varies directly as x and inversely as y."
$\dfrac{z \cdot y}{x} = K$, where K is a constant.

EXAMPLE 4: Give a formula for the statement, "a varies as b and the square of c and inversely as x and the square root of y."
$\dfrac{ax\sqrt{y}}{bc^2} = K$, where K is a constant.

In general, the formula for a *combined variation* is set up as a ratio (direct variation) until there is some indication that what follows is an inverse variation (usually just the word *inversely*). From that point on a product is formed.

EXAMPLE 5: Give a formula for the statement, "T varies jointly as L, P, and the square of D and inversely as R and Q."
$\dfrac{T \cdot RQ}{LPD^2} = K$, where K is a constant.

In all these examples K is called the *constant of variation* or the *constant of proportionality*.

EXAMPLE 6: If x varies as y and inversely as z, and $x = 2$ when $y = 4$ and $z = 3$, [A] find the constant of variation, and [B] find the value of z when $x = 8$ and $y = 2$.

The formula is $\dfrac{xz}{y} = K$.

[A] $\dfrac{2 \cdot 3}{4} = K = \dfrac{3}{2}$ = the constant of variation.

[B] $\dfrac{8z}{2} = K = \dfrac{3}{2}$

$8z = 3$

Therefore, $z = \dfrac{3}{8}$.

EXAMPLE 7: If x varies as y and z^2, and inversely as the square root of w, what is the effect on y when x is doubled, z is halved, and w is multiplied by 4?

The formula is $\dfrac{x\sqrt{w}}{yz^2} = K$.

For all practical purposes, the quickest way to solve a problem of this type is to arbitrarily choose values for each variable and compute K. For example, let $x = 1$, $y = 3$, $z = 2$, and $w = 4$ (a perfect square since $\sqrt{w}$ is needed).

$$\dfrac{1\sqrt{4}}{3 \cdot 2^2} = K = \dfrac{1 \cdot 2}{3 \cdot 4} = \dfrac{1}{6}$$

The new values of the variables are $x = 2$, $z = 1$, and $w = 16$.

Let $y = y_2$ just to emphasize a new value of y.

$$\dfrac{2\sqrt{16}}{y_2 \cdot 1} = \dfrac{1}{6}$$
$$\dfrac{2 \cdot 4}{y_2} = \dfrac{1}{6}$$
$$y_2 = 48$$

The original value of y was 3. The new value of y is 48. Therefore, for any values of $x, y, z,$ and w, when x is doubled, z is halved, and w is multiplied by 4, y is multiplied by 16.

EXERCISES

1. If x varies directly as y, which of the following statements must be true?
   I. Their product is a constant.
   II. Their sum is a constant.
   III. Their quotient is a constant.

   [A] I only [B] II only [C] III only
   [D] I & II only [E] II & III only

2. If $\{(x,y): (4,3), (-2,a)\}$ consists of pairs of numbers that belong to the relationship "y varies inversely as x^2," what does a equal?

   [A] $-\dfrac{3}{2}$ [B] -6 [C] $-\dfrac{2}{3}$
   [D] 12 [E] $\dfrac{3}{4}$

3. D varies as R and inversely as the square of M. If R is divided by 3 and M is doubled, what is the effect on D?

   [A] unchanged
   [B] divided by 12
   [C] multiplied by 12
   [D] divided by 6
   [E] multiplied by 6

7. LOGIC

There are three major statements in logic:

I. A *conjunction* is a statement of the form "A and B" ($A \wedge B$). It is considered to be true if A and B are both true. The *negation* of a conjunction, denoted by $(A \wedge B)'$, or $\sim(A \wedge B)$, is equivalent to the statement "the negation of A *or* the negation of B," which is denoted by $A' \vee B'$ or $(\sim A) \vee (\sim B)$.

II. A *disjunction* is a statement of the form "A or B" ($A \vee B$). It is considered to be true if either A or B or both are true. The *negation* of a disjunction, denoted by $(A \vee B)'$, or $\sim(A \vee B)$, is equivalent to the statement "the negation of A *and* the negation of B," which is denoted by $A' \wedge B'$ or $(\sim A) \wedge (\sim B)$.

III. An *implication* is a statement of the form "if A, then B" ($A \rightarrow B$). It is considered to be true except when A is true and B is false. The negation of an implication, denoted by $(A \rightarrow B)'$, or $\sim(A \rightarrow B)$, is equivalent to the statement "A and the negation of B," which is denoted by $A \wedge B'$ or $A \wedge (\sim B)$. There are many forms of an implication that are all equivalent. They are:

> if A, then B
> A implies B
> A only if B
> B, if A
> A is sufficient for B
> B is necessary for A

The double implication "A if and only if B," denoted by "A iff B" or $A \leftrightarrow B$, is equivalent to the statement "if A, then B and if B, then A."

Associated with every implication, $A \rightarrow B$, are three other statements:
1. The converse is $B \rightarrow A$.
2. The inverse is $A' \rightarrow B'$.

3. The contrapositive is $B' \to A'$, or $(\sim B) \to (\sim A)$, which is equivalent to the original implication $A \to B$. (If "$A \to B$" is true, then "$B' \to A'$" is also true. If "$A \to B$" is false, then "$B' \to A'$" is also false.)

The negation of the statement "for all A" is "for some not-A." The negation of the statement "for some A" is "for all not-A."

If a statement is always true it is called a *tautology*.

EXAMPLE 1: State the negation of each of the following:
[A] All men are mortal. [B] For some x, $x = 2$. [C] Some boys play tennis. [D] All cats are not dogs.

[A] Some men are not mortal. [B] For all x, $x \ne 2$. [C] All boys do not play tennis. [D] Some cats are dogs.

EXAMPLE 2: State the negation of $(p \wedge q) \to p$.
The negation is $(p \wedge q) \wedge p'$, which is always false because there is no way that both p and p' can both be true.

EXAMPLE 3: Is the statement, "I will drive only if I am wrong," true or false if it is known that I am always wrong?
Let $p = I$ *will drive* and $q = I$ *am wrong*. The statement is equivalent to $p \to q$, which is true for all cases except when p is true and q is false. Since it is known that q is true, the statement itself must always be true.

EXAMPLE 4: What conclusion can be drawn from the following statements?

The boy is handsome, or he is short and fat.
If he is short, then he is blond.
He is not blond.

Let $H = $ *he is handsome*, let $S = $ *he is short*, let $F = $ *he is fat*, and let $B = $ *he is blond*. The three statements become:

$$H \vee (S \wedge F)$$
$$S \to B$$
$$B'$$

Since B' is true, B is false. The only way that $S \to B$ can be true when B is false is when S is also false. Since S is false, $S \wedge F$ is false regardless of whether F is true or false. Since $S \wedge F$ is false, the only way $H \vee (S \wedge F)$ can be true is if H is true. Therefore, the conclusion is, "The boy is handsome."

EXERCISES

1. The statement $(p \vee q) \to p$ is *false* if
   [A] p is true and q is true
   [B] p is true and q is false
   [C] p is false and q is true
   [D] p is false and q is false
   [E] the statement is a tautology

2. Which of the following is equivalent to the statement "Having equal radii is necessary for two circles to have equal areas"?
   I. Having equal areas is sufficient for two circles to have equal radii.
   II. Two circles have equal areas only if they have equal radii.
   III. Having equal radii implies that two circles have equal areas.

   [A] I only   [B] III only
   [C] I and II only   [D] II and III only
   [E] I, II, and III

3. Given the statement, "If $x = 2$, then $x^2 = 4$," the negation of this statement is
   [A] $x \ne 2$ and $x^2 \ne 4$
   [B] $x = 2$ and $x^2 \ne 4$
   [C] $x \ne 2$ or $x^2 = 4$
   [D] $x \ne 2$ and $x^2 = 4$
   [E] $x \ne 2$ or $x^2 \ne 4$

4. Given the statements:
   I. Some numbers are not prime.
   II. No primes are squares.
   If *some* means "at least one," it can be concluded from I and II that

   [A] some numbers are squares
   [B] some squares are not numbers
   [C] some numbers are not squares
   [D] no numbers are squares
   [E] none of these is a conclusion of I and II

5. In a particular town the following facts are true:
   I. Some smarties do not smoke cigars.
   II. All men smoke cigars.
   A necessary conclusion is

   [A] some smarties are men
   [B] some smarties are not men
   [C] no smartie is a man
   [D] some men are not smarties
   [E] no man is a smartie

6. Given the statement, "The student will not pass the course only if he does not come to class," which one of the following can be concluded?

[A] If the student does not pass the course, then he probably missed too many classes.
[B] If a student comes to class, then he may pass the course.
[C] If a student comes to class, then he will pass the course.
[D] If a student passes the course, then he came to class.
[E] If a student does not come to class, then he will not pass the course.

7 The contrapositive of $p \rightarrow q'$ is

[A] $p' \rightarrow q$ [B] $p' \rightarrow q'$ [C] $q' \rightarrow p$
[D] $q \rightarrow p'$ [E] $q' \rightarrow p'$

8. STATISTICS

One final subject that may appear is statistics, and only the most elementary topics would be considered. Among those would be measures of central tendency (averages), frequency distributions, and a simple measure of dispersion (range).

Descriptive statistics consists of methods for describing numerical information in an organized fashion. Probably the most common measure of central tendency, the arithmetic mean can be determined regardless of whether the data is organized or not. The arithmetic mean is obtained by adding up all the data and dividing by the number of pieces of data. The mean of the 18 pieces of data in Table 1 is

$$\frac{\Sigma \text{ scores}}{18} = \frac{1123}{18} \approx 62.39$$

In order to draw other conclusions from the test scores in Table 1, they would have to be ordered into a frequency distribution (Table 2). A frequency distribution consists of the data organized in a table (usually from smallest value to largest value) with the number of times each value occurs.

Many things about the data can be easily determined from a frequency distribution. Some of them are:

1. The *range* of the data is the difference between the largest and smallest values. The range of this data is 12. $(67 - 55 = 12)$.
2. The *mode* of the data is the value which occurs most often. The mode of this data is 62 (There are four 62's).
3. The *median* of the data is the middle score after the data have been ordered. (If there is an even number of scores, the mean of the two middle scores is the median.) The median of this data is 62.5 $\frac{(62 + 63}{2} = 62.5)$.

Although any one of the mean, median, and mode could correctly be called the average of the data, when average is mentioned it usually refers to the mean.

EXAMPLE 1: Consider this frequency distribution:

Data	Frequency
0	2
1	3
2	5
3	8
4	2

Find the value of the mode, median, and mean.

Mode = 3

Median = $\frac{2 + 3}{2}$ = 2.5

Mean = $\frac{0 + 3 + 10 + 24 + 8}{20} = \frac{45}{20} = 2.25$

EXAMPLE 2: If the set of data 1, 2, 3, 1, 5, 7, x is to have a mode and x is equal to one of the other elements of the set, what value must x have?

Since there are two 1's and one of every other value, x must equal 1. If it were to equal any of the other numbers there would be two values with two elements. Thus, the set of data would not have a mode.

EXAMPLE 3: Given the set of data: 15, 24, 28, 32, 35, x. x is the largest element. If the range is 35, what is the value of x?

TABLE 1

67	62	65
62	57	59
67	60	65
66	62	65
63	55	64
63	59	62

TABLE 2

Data (N)	Frequency (f)
55	1
57	1
59	2
60	1
62	4
63	2
64	1
65	3
66	1
67	2

Since the range is the difference between the largest element and the smallest element, $x - 15 = 35$. Therefore $x = 50$.

EXAMPLE 4: Given a set of integers: 1, 2, 3, 3, 4, 1, x. If the median is 2, what is the value of x?

Since this set of data has seven elements, the median is the middle number after the data have been ordered. Since the median is 2, the value of x must be any integer ≤ 2.

EXAMPLE 5: Given the set of integers: 1, 2, 3, 3, 4, 1, x. If the mean = median = 2, what is the value of x?

The mean $= \dfrac{1+2+3+3+4+1+x}{7} = \dfrac{14+x}{7} = 2$.

$14 + x = 14$. Therefore, $x = 0$.

EXERCISES

[1] In statistics, when the average is mentioned, it refers to
[A] mean [B] median [C] mode
[D] mean or median [E] mean, median, or mode

[2] If the range of a set of integers is 2 and the mean is 50, which of the following statements must be true?
I. The mode is 50.
II. The median is 50.
III. There are exactly 3 pieces of data.
[A] Only I [B] Only II [C] Only III
[D] I and II [E] I, II & III

[3] If the range of the set of data 1, 1, 2, 2, 3, 3, 3, x is equal to the mean and x is an integer, then x must be
[A] -1 [B] -2 [C] 0 [D] 1
[E] There are no values of x that satisfy the stated conditions.

[4] In the following frequency distribution, if the mean is to be equal to one of the elements of data, how many fours must there by?

Data	Frequency
0	1
2	3
3	2
4	?
5	1

[A] 0 [B] 2 [C] 3 [D] 4 [E] 5

[5] If the mean, median, and mode are calculated from the data in this frequency distribution, which of the following statements is true?

Data	Frequency
2	4
4	3
6	2
8	2

[A] mode < median < mean
[B] mean = median
[C] mode < mean < median
[D] mean < median < mode
[E] median < mode < mean

9. ODDS AND ENDS

Each year one or two problems appear on the Level II examination using topics that do not fall into one of the categories above. Also, occasionally, definitions are made within problems, and then a question is asked using these definitions. Usually these questions are quite easy if the student takes the time to read them carefully.

EXAMPLE 1: If a 2 by 2 determinant $\begin{vmatrix} a & b \\ c & d \end{vmatrix}$ is defined to equal $ad - bc$, what does x equal if the determinant $\begin{vmatrix} 2 & 3 \\ x & 5 \end{vmatrix} = 0$?

$$2 \cdot 5 - 3x = 0$$
$$3x = 10$$
$$x = \frac{10}{3}$$

EXAMPLE 2: If the operation $*$ is defined as follows: $a*b = ab - b$, what does $3*5$ equal?
$3*5 = 3 \cdot 5 - 5 = 15 - 5 = 10$

EXAMPLE 3: If the operations $*$ and $\#$ are defined such that $a*b = 2a - b$ and $a\#b = \dfrac{a}{b} + b$, does $(3*2)\#4 = 3*(2\#4)$? Explain.
$(3*2)\#4 = (2 \cdot 3 - 2)\#4 = 4\#4 = \dfrac{4}{4} + 4 = 5$.
$3*(2\#4) = 3*(\dfrac{2}{4} + 4) = 3*4\dfrac{1}{2} = 2 \cdot 3 - 4\dfrac{1}{2} = 6 - 4\dfrac{1}{2} = 1\dfrac{1}{2}$.
Therefore, they are not equal.

EXAMPLE 4: Follow the sequence of instructions and tell what number(s) will be printed.
1. Let $x = 2$.

2. Let $y = x+2$.
3. If $x \cdot y < 10$, then print the value of x, replace x by $x+3$ and return to step 2.
   If $x \cdot y > 10$, then stop.
Starting off with $x=2$ and $y=4$, $x \cdot y = 8 < 10$. Print 2, replace x by 5 and $y=7$. Then $x \cdot y = 35 \not< 10$. Stop. Thus, the only number that is printed is 2.

EXAMPLE 5: $P(A|B)$ is the probability of event A happening after event B has already happened. Event A is "a red card is drawn from a deck of 52 cards." Event B is "a black card drawn from a deck of 52 cards." What does $P(A|B)$ equal?

Since event B is given as having been accomplished, a black card has been drawn from the deck, and there are only 51 cards left in the deck: 26 red cards and 25 black cards. Therefore, $P(A|B) = \dfrac{26}{51}$.

EXAMPLE 6: If $f^*(x)$ is defined to be $\dfrac{[f(x)]^2 - f(x)}{x}$ and $f(x) = x^x$, what does $f^*(2)$ equal?

$f^*(2) = \dfrac{(2^2)^2 - 2^2}{2} = \dfrac{4^2 - 4}{2} = \dfrac{16-4}{2} = 6$.

EXAMPLE 7: If $\begin{Bmatrix} ax+by=c \\ dx+ey=f \end{Bmatrix}$ are solved simultaneously, and $x=5$, $y=2$, and $\begin{vmatrix} a & b \\ d & e \end{vmatrix} = 2$, what does $\begin{vmatrix} c & b \\ f & e \end{vmatrix}$ equal?

Using determinants to solve a pair of equations,

$$x = \dfrac{\begin{vmatrix} c & b \\ f & e \end{vmatrix}}{\begin{vmatrix} a & b \\ d & e \end{vmatrix}} \text{ and } y = \dfrac{\begin{vmatrix} a & c \\ d & f \end{vmatrix}}{\begin{vmatrix} a & b \\ d & e \end{vmatrix}}.$$

In this case, $x = 5 = \dfrac{\begin{vmatrix} c & b \\ f & e \end{vmatrix}}{2}$.

Therefore, $\begin{vmatrix} c & b \\ f & e \end{vmatrix} = 10$.

EXAMPLE 8: At the dog pound there are 47 dogs. 16 are large dogs, 18 are brown dogs, and 20 are neither. How many large, brown dogs are there?

Let L = the set of large dogs, let B = the set of brown dogs, and let N = the set of dogs that are neither large nor brown. A Venn diagram is helpful.

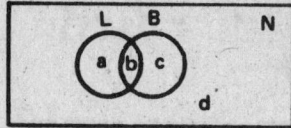

All dogs at the dog pound

Label the number of dogs in each set with a variable, a, b, c, or d. Thus,

$$\begin{aligned} a+b+c+d &= 47 \\ a+b &= 16 \\ b+c &= 18 \\ d &= 20 \end{aligned}$$

Multiplying the top equation by -1 and adding all four equations results in $b=7$.
Therefore, there are 7 large, brown dogs.

EXERCISES

1 A *harmonic sequence* is a set of numbers such that their reciprocals, taken in order, form an arithmetic sequence. If S_n = the sum of the first n terms of the harmonic sequence, and the first three terms are 2, 3, and 6, then S_4 equals

[A] 11   [B] 12   [C] 1
[D] 22   [E] cannot be determined

2 If the pair of equations $\begin{Bmatrix} 2x+3y=4 \\ 5x+2y=8 \end{Bmatrix}$ is solved using determinants, the determinant appearing in the denominator of the solution has a value of

[A] -11 [B] -4 [C] -32
[D] 4   [E] 16

3 In the statement "$a \equiv b \pmod{n}$" b is the remainder when a is divided by n. What is the smallest value of n such that $125 \equiv 7 \pmod{n}$?

[A] 47   [B] 118   [C] 19
[D] 8   [E] 59

4 If $*$ is a binary operation defined by $a*b = \dfrac{a+b}{2}$ and $\#$ is a binary operation defined by $a \# b = \sqrt{ab}$, for what values of a and b does $a*b = a\#b$?

[A] all real numbers   [B] no real numbers
[C] 0 only   [D] 1 only
[E] only when $a=b$

5 A binary operation $*$ is defined on the set of ordered pairs of real numbers as follows: $(a,b)*(x,y) = (a+x, b-y)$. If $(1,2)*(1,1) = (p,q)*(1,2)$, then (p,q) equals

[A] $(1,1)$ [B] $(1,2)$ [C] $(1,3)$
[D] $(0,2)$ [E] $(2, \dfrac{1}{2})$

6 Carry out the following instructions in order and find what the first number printed is.
1. Let $a = x = 2$.
2. Go to step 4.
3. Replace x by the value of $a+2$.
4. If $a < 6$, replace a by $x+2$ and go to step 3, otherwise print the value of x.

[A] 10 [B] 8 [C] 6 [D] 12 [E] 2

7 If darts are thrown at a target like the one in the figure (and every dart must hit the figure), what is the probability that the first dart hits outside the circle?

[A] $\dfrac{4-\pi}{4}$ [B] $\dfrac{\pi}{4}$ [C] $\dfrac{1}{\pi}$

[D] $\dfrac{\pi-1}{4}$ [E] $\dfrac{\pi-2}{2}$

8 In a group of 100 birds, 85 had a long beak, 45 had gray feathers, and 38 had both a long beak and gray feathers. How many did not have a long beak and did not have gray feathers?

[A] 15 [B] 55 [C] 62 [D] 8 [E] 30

SUMMARY OF FORMULAS
CHAPTER 6

CHAPTER 2:
Polynomial Functions

LINEAR FUNCTIONS:

General form of the equation: $Ax + By + C = 0$

Slope-intercept form: $y = mx + b$, where m represents the slope and b the y-intercept

Point-slope form: $y - y_1 = m(x - x_1)$, where m represents the slope and (x_1, y_1) are the coordinates of some point on the line

Slope: $m = \dfrac{y_1 - y_2}{x_1 - x_2}$, where (x_1, y_1) and (x_2, y_2) are the coordinates of two points

Parallel lines have equal slopes.

Perpendicular lines have slopes that are negative reciprocals.

If m_1 and m_2 are the slopes of two perpendicular lines, $m_1 \cdot m_2 = -1$

Distance between two points with coordinates (x_1, y_1) and $(x_2, y_2) =$
$$\sqrt{(x_1 - x_2)^2 + (y_1 - y_2)^2}$$

Coordinates of the midpoint between two points =
$$\left(\dfrac{x_1 + x_2}{2}, \dfrac{y_1 + y_2}{2}\right)$$

Distance between a point with coordinates (x_1, y_1) and a line, $Ax + By + C = 0$, is
$$\dfrac{|Ax_1 + By_1 + C|}{\sqrt{A^2 + B^2}}$$

If θ is the angle between two lines: $\operatorname{Tan}\theta = \dfrac{m_1 - m_2}{1 + m_1 m_2}$, where m_1 and m_2 are the slopes of the two lines.

QUADRATIC FUNCTIONS:

General quadratic equation: $ax^2 + bx + c = 0$

General Quadratic Formula:
$$x = \dfrac{-b \pm \sqrt{b^2 - 4ac}}{2a}$$

General quadratic function: $y = ax^2 + bx + c$

Coordinates of vertex: $\left(-\dfrac{b}{2a}, c - \dfrac{b^2}{4a}\right)$

Axis of symmetry equation: $x = -\dfrac{b}{2a}$

Sum of zeros (roots) $= -\dfrac{b}{a}$

Product of zeros (roots) $= \dfrac{c}{a}$

Nature of zeros (roots):
 If $b^2 - 4ac < 0$, 2 complex numbers
 If $b^2 - 4ac = 0$, 2 equal real numbers
 If $b^2 - 4ac > 0$, 2 unequal real numbers

CHAPTER 3:
Trigonometric Functions

$\sin\theta = \dfrac{\text{opposite}}{\text{hypotenuse}}$ $\cos\theta = \dfrac{\text{adjacent}}{\text{hypotenuse}}$

$\tan\theta = \dfrac{\text{opposite}}{\text{adjacent}}$ $\cot\theta = \dfrac{\text{adjacent}}{\text{opposite}}$

$\sec\theta = \dfrac{\text{hypotenuse}}{\text{adjacent}}$ $\csc\theta = \dfrac{\text{hypotenuse}}{\text{opposite}}$

$\pi^R = 180°$

Length of arc in circle of radius r and central angle θ is given by $r\theta^R$

Area of sector of circle of radius r and central angle θ is given by $\dfrac{1}{2}r^2\theta^R$

TRIGONOMETRIC REDUCTION FORMULAS:

1. $\sin^2 x + \cos^2 x = 1$
2. $\tan^2 x + 1 = \sec^2 x$ } Pythagorean identities
3. $\cot^2 x + 1 = \csc^2 x$
4. $\sin(A+B) = \sin A \cdot \cos B + \cos A \cdot \sin B$
5. $\sin(A-B) = \sin A \cdot \cos B - \cos A \cdot \sin B$
6. $\cos(A+B) = \cos A \cdot \cos B - \sin A \cdot \sin B$
7. $\cos(A-B) = \cos A \cdot \cos B + \sin A \cdot \sin B$ } sum and difference formulas
8. $\tan(A+B) = \dfrac{\tan A + \tan B}{1 - \tan A \cdot \tan B}$
9. $\tan(A-B) = \dfrac{\tan A - \tan B}{1 + \tan A \cdot \tan B}$
10. $\sin 2A = 2\sin A \cdot \cos A$
11. $\cos 2A = \cos^2 A - \sin^2 A$
12. $ = 2\cos^2 A - 1$ } double angle formulas
13. $ = 1 - 2\sin^2 A$
14. $\tan 2A = \dfrac{2\tan A}{1 - \tan^2 A}$
15. $\sin \dfrac{1}{2}A = \pm\sqrt{\dfrac{1-\cos A}{2}}$
16. $\cos \dfrac{1}{2}A = \pm\sqrt{\dfrac{1+\cos A}{2}}$
17. $\tan \dfrac{1}{2}A = \pm\sqrt{\dfrac{1-\cos A}{1+\cos A}}$ } half angle formulas
18. $\phantom{\tan \dfrac{1}{2}A} = \dfrac{1-\cos A}{\sin A}$
19. $\phantom{\tan \dfrac{1}{2}A} = \dfrac{\sin A}{1+\cos A}$

In any $\triangle ABC$:

Law of sines: $\dfrac{\sin A}{a} = \dfrac{\sin B}{b} = \dfrac{\sin C}{c}$

Law of cosines: $a^2 = b^2 + c^2 - 2bc\cdot\cos A$

Area $= \dfrac{1}{2}bc\cdot\sin A$

CHAPTER 4:
Miscellaneous Relations and Functions

GENERAL QUADRATIC EQUATION IN TWO VARIABLES:

$$Ax^2 + Bxy + Cy^2 + Dx + Ey + F = 0$$

If $B^2 - 4AC < 0$ and $A = C$, graph is a circle
If $B^2 - 4AC < 0$ and $A \neq C$, graph is an ellipse
If $B^2 - 4AC = 0$, graph is a parabola
If $B^2 - 4AC > 0$, graph is a hyperbola

CIRCLE:

$(x-h)^2 + (y-k)^2 = r^2$

Center (h,k) Radius $= r$

ELLIPSE:

$\dfrac{(x-h)^2}{a^2} + \dfrac{(y-k)^2}{b^2} = 1$, major axis horizontal

$\dfrac{(x-h)^2}{b^2} + \dfrac{(y-k)^2}{a^2} = 1$, major axis vertical

where $a^2 = b^2 + c^2$. Coordinates of center (h,k).
Vertices: $\pm a$ units along major axis from center.
Foci: $\pm c$ units along major axis from center.
Minor axis: perpendicular to major axis at center.
Length $= 2b$.
Eccentricity $= \dfrac{c}{a}$

Length of latus rectum $= \dfrac{2b^2}{a}$

HYPERBOLA:

$\dfrac{(x-h)^2}{a^2} - \dfrac{(y-k)^2}{b^2} = 1$, transverse axis horizontal

$\dfrac{(y-k)^2}{a^2} - \dfrac{(x-h)^2}{b^2} = 1$, transverse axis vertical

where $c^2 = a^2 + b^2$. Coordinates of center (h,k).
Vertices: $\pm a$ units along the transverse axis from center.
Foci: $\pm c$ units along the transverse axis from center.

Conjugate axis: perpendicular to transverse axis at center.
Eccentricity $= \dfrac{c}{a}$

Length of latus rectum $= \dfrac{2b^2}{a}$

Asymptotes: Slopes =

$\pm \dfrac{b}{a}$, if transverse axis horizontal

$\pm \dfrac{a}{b}$, if transverse axis vertical

PARABOLA:

$(x-h)^2 = 4p(y-k)$, opens up or down—axis of symmetry vertical

$(y-k)^2 = 4p(x-h)$, opens to the side—axis of symmetry horizontal

Coordinates of the vertex: (h,k)

Equation of axis of symmetry:
  $x = h$, if vertical
  $y = k$, if horizontal

Focus: p units along the axis of symmetry from vertex

Equation of directrix:
  $y = -p$, if axis of symmetry vertical
  $x = -p$, if axis of symmetry horizontal

Eccentricity $= 1$

Length of latus rectum $= 4p$

EXPONENTS:

$$x^a \cdot x^b = x^{a+b} \qquad \dfrac{x^a}{x^b} = x^{a-b}$$

$$(x^a)^b = x^{ab}$$

$$x^0 = 1 \qquad x^{-a} = \dfrac{1}{x^a}$$

LOGARITHMS:

$\log_b(p \cdot q) = \log_b p + \log_b q \qquad \log_b\left(\dfrac{p}{q}\right) = \log_b p - \log_b q$

$\log_b p^x = x \cdot \log_b p \qquad \log_b 1 = 0$

$\log_b p = \dfrac{\log_a p}{\log_a b} \qquad \log_b b = 1$

$b^{\log_b p} = p$

$\text{Log}_b N = x$ if and only if $b^x = N$

ABSOLUTE VALUE:

if $x \geq 0$, then $|x| = x$
if $x < 0$, then $|x| = -x$

GREATEST INTEGER FUNCTION:

$[x] = i$, where i is an integer and $i \leq x < i+1$

POLAR COORDINATES:

$x = r \cdot \cos\theta$
$y = r \cdot \sin\theta$
$x^2 + y^2 = r^2$

DE MOIVRE'S THEOREM:

If
$$z_1 = x_1 + y_1 i = r_1(\cos\theta_1 + i \cdot \sin\theta_1) = r_1 \text{cis}\,\theta_1$$
and
$$z_2 = x_2 + y_2 i = r_2(\cos\theta_2 + i \cdot \sin\theta_2) = r_2 \text{cis}\,\theta_2:$$

1. $z_1 \cdot z_2 = r_1 \cdot r_2 (\cos(\theta_1 + \theta_2) + i \cdot \sin(\theta_1 + \theta_2)) =$
   $r_1 \cdot r_2 \cdot \text{cis}(\theta_1 + \theta_2)$

2. $\dfrac{z_1}{z_2} = \dfrac{r_1}{r_2}(\cos(\theta_1 - \theta_2) + i \cdot \sin(\theta_1 - \theta_2)) =$
   $\dfrac{r_1}{r_2} \text{cis}(\theta_1 - \theta_2)$

3. $z^n = r^n(\cos n\theta + i \cdot \sin n\theta) = r^n \text{cis}\, n\theta$

4. $z^{1/n} = r^{1/n}\left(\cos\dfrac{\theta + 2\pi k}{n} + i \cdot \sin\dfrac{\theta + 2\pi k}{n}\right)$
   $= r^{1/n} \text{cis}\,\dfrac{\theta + 2\pi k}{n}$, where k is an integer taking on values from 0 up to $n-1$.

CHAPTER 5:
Miscellaneous Topics

PERMUTATIONS:

$${}_n P_r = \dfrac{n!}{(n-r)!}$$, where $n! = n(n-1)(n-2)\ldots 3 \cdot 2 \cdot 1$

Circular permutation (around a table, etc.) of n elements $= (n-1)!$

Circular permutation (beads on a bracelet, etc.) of n elements $= \dfrac{(n-1)!}{2}$

Permutations of n elements with a repetitions and with b repetitions $= \dfrac{n!}{a!\,b!}$

COMBINATIONS:

$$\binom{n}{r} = \dfrac{n!}{(n-r)!\,r!} = \dfrac{{}_nP_r}{n!}$$

BINOMIAL THEOREM:

There are $n+1$ terms in $(a+b)^n$.
The sum of the exponents in each term is n.
The exponent on the b is one less than the number of the term.
The coefficient of each term $= \binom{n}{\text{either exponent}}$

PROBABILITY:

$P(\text{event}) = \dfrac{\text{number of ways to get a successful result}}{\text{total number of ways of getting any result}}$

Independent events: $P(A \cap B) = P(A) \cdot P(B)$
Mutually exclusive events: $P(A \cap B) = 0$
and $P(A \cup B) = P(A) + P(B)$

SEQUENCES AND SERIES:

Arithmetic Sequence (or Progression)
nth term $= t_n = t_1 + (n-1)d$
Sum of n terms $=$
$S_n = \dfrac{n}{2}(t_1 + t_n)$
$= \dfrac{n}{2}[2t_1 + (n-1)d]$
Geometric Sequence (or Progression)
nth term $= t_n = t_1 r^{n-1}$

Sum of n terms $= S_n = \dfrac{t_1(1-r^n)}{1-r}$

If $|r| < 1$, $S_\infty = \lim\limits_{n \to \infty} S_n = \dfrac{t_1}{1-r}$

VECTORS:

If $\vec{V} = (v_1, v_2)$ and $\vec{U} = (u_1, u_2)$,
$\vec{V} + \vec{U} = (v_1 + u_1, v_2 + u_2)$
$\vec{V} \cdot \vec{U} = v_1 u_1 + v_2 u_2$
Two vectors are perpendicular if and only if $\vec{V} \cdot \vec{U} = 0$.

GEOMETRY:

Distance between two points with coordinates (x_1, y_1, z_1) and (x_2, y_2, z_2) equals

$$\sqrt{(x_1 - x_2)^2 + (y_1 - y_2)^2 + (z_1 - z_2)^2}.$$

Distance between a point with coordinates (x_1, y_1, z_1) and a plane with equation $Ax + By + Cz + D = 0$ equals

$$\dfrac{|Ax_1 + By_1 + Cz_1 + D|}{\sqrt{A^2 + B^2 + C^2}}$$

DETERMINANTS:

$\begin{vmatrix} a & b \\ c & d \end{vmatrix} = ad - bc$

ANSWERS TO PRACTICE EXERCISES

CHAPTER 7

ANSWER KEY

CHAPTER 1 INTRODUCTION TO FUNCTIONS
PART 1.1 DEFINITION
1. A   2. E   3. A

PART 1.2 FUNCTION NOTATION
1. B   3. A   5. C
2. D   4. E   6. C

PART 1.3 INVERSE FUNCTIONS
1. E   3. B
2. A   4. E

PART 1.4 ODD AND EVEN FUNCTIONS
1. D   3. B
2. D   4. C

PART 1.5 MULTI-VARIABLE FUNCTIONS
1. B   2. C   3. B

CHAPTER 2 POLYNOMIAL FUNCTIONS
PART 2.2 LINEAR FUNCTIONS
1. C   4. D   7. B
2. E   5. C   8. E
3. D   6. B   9. D

PART 2.3 QUADRATIC FUNCTIONS
1. B   3. B   5. B
2. C   4. E   6. E

PART 2.4 HIGHER DEGREE POLYNOMIAL FUNCTIONS
1. D   5. C   9. B
2. C   6. C   10. E
3. A   7. A   11. C
4. C   8. C   12. D

PART 2.5 INEQUALITIES
1. C   2. B   3. C

CHAPTER 3 TRIGONOMETRIC FUNCTIONS
PART 3.1 DEFINITIONS
1. A   3. D   5. A
2. B   4. B

PART 3.2 ARCS AND ANGLES
1. C   3. D
2. D   4. C

PART 3.3 SPECIAL ANGLES
1. C   3. B   5. D
2. E   4. B

PART 3.4 GRAPHS
1. C 4. C 6. D
2. E 5. C 7. D
3. C

PART 3.5 IDENTITIES
1. D 5. A 8. B
2. D 6. A 9. A
3. E 7. B 10. B
4. E

PART 3.6 INVERSE FUNCTIONS
1. B 4. B 7. B
2. D 5. B 8. E
3. A 6. B

PART 3.7 TRIANGLES
1. D 5. C 8. D
2. E 6. D 9. C
3. A 7. E 10. A
4. D

CHAPTER 4 MISCELLANEOUS RELATIONS AND FUNCTIONS
PART 4.1 CONIC SECTIONS
1. D 3. C 5. E
2. C 4. D 6. D

PART 4.2 EXPONENTIAL AND LOGARITHMIC FUNCTIONS
1. C 3. B 5. B
2. D 4. B 6. E

PART 4.3 INEQUALITIES
1. C 2. B 3. C

PART 4.4 ABSOLUTE VALUE
1. B 3. A 5. B
2. E 4. A

PART 4.5 GREATEST INTEGER FUNCTION
1. C 2. C 3. C

PART 4.6 RATIONAL FUNCTIONS
1. A 3. A 5. B
2. B 4. C

PART 4.7 PARAMETRIC EQUATIONS
1. B 2. D 3. A

PART 4.8 POLAR COORDINATES
1. A 3. A 5. B
2. D 4. B 6. B

CHAPTER 5 MISCELLANEOUS TOPICS
PART 5.1 PERMUTATIONS AND COMBINATIONS
1. C 4. C 7. D
2. B 5. C 8. C
3. B 6. D

PART 5.2 BINOMIAL THEOREM
1. C 3. B
2. C 4. B

PART 5.3 PROBABILITY
1. C 4. E 6. B
2. D 5. E 7. B
3. E

PART 5.4 SEQUENCES AND SERIES
1. D 4. C 7. B
2. A 5. C 8. D
3. A 6. A 9. D

PART 5.5 GEOMETRY
1. A 6. B 11. C
2. E 7. D 12. C
3. B 8. E 13. D
4. D 9. D 14. C
5. D 10. D

PART 5.6 VARIATION
1. C 2. D 3. B

PART 5.7 LOGIC
1. C 4. E 6. C
2. C 5. B 7. D
3. B

PART 5.8 STATISTICS
1. E 3. D 5. A
2. B 4. D

PART 5.9 ODDS AND ENDS
1. E 4. E 7. A
2. A 5. C 8. D
3. E 6. A

ANSWER EXPLANATIONS FOR PRACTICE EXERCISES

CHAPTER 1 INTRODUCTION TO FUNCTIONS

Part 1.1 Definition

1. **A** Either (3,2) or (3,1) must be removed so 3 will be paired with only one number.

2. **E** For each value of x there is only one value for y in each case. Therefore, all three are functions.

3. **A** Since division by zero is forbidden, x cannot equal 2.

Part 1.2 Function Notation

1. **B** $f(-2)=3(-2)^2-2(-2)+4=20$.

2. **D** $g(2)=3^2=9$. $f(g(2))=f(9)=31$.

3. **A** In order to get from $f(x)$ to $f(g(x))$, the x^2 must become a $4x^2$. Therefore the answer must contain a $2x$ since $(2x)^2=4x^2$.

4. **E** $g(x)$ cannot equal 0. Therefore $x \neq \pm 1$.

5. **C** Since $f(2)$ implies that $x=2$, $g(f(2))=2$. Therefore, $g(f(2))=3(f(2))+2=2$. Therefore, $f(2)=0$.

6. **C** $p(a)=0$ implies $4a-6=0$.

Part 1.3 Inverse Functions

1. **E** If $y=2x-3$, the inverse is $x=2y-3$, which equals $y=\dfrac{x+3}{2}$.

2. **A** By definition.

3. **B** The inverse is $\{(2,1),(3,2),(4,3),(1,4),(2,5)\}$, which is not a function because of (2,1) and (2,5). Therefore, the domain of the original function must lose either 1 or 5.

4. **E** If this line were reflected in the line $y=x$ to get its inverse, the slope would be less than 1 and the y-intercept would be less than zero. Only possibilities are D or E.

Part 1.4 Odd and Even Functions

1. **D** Only I and III remain unchanged when reflected about the y-axis.

2. **D** Only II and III remain unchanged when rotated 180° about the origin.

3. **B** Only I and II remain unchanged when reflected about the y-axis or rotated 180° about the origin.

4. **C** This is the only one that does not satisfy one of the two tests.

Part 1.5 Multi-variable Functions

1. **B** $g(4)=16$. $f(3,16)=9+32-8=33$.

2. **C** $f(-1,1,-1)=(-1)^2+(1)^2+2(1)^2=4$.

3. **B** $g(1,2)=1^2-2^2=-3$. $f(-3,3)=3(-3)+2(3)=-3$.

CHAPTER 2 POLYNOMIAL FUNCTIONS

Part 2.2 Linear Functions

1. **C** Slope $=\dfrac{-3-(-2)}{-2-3}=\dfrac{1}{5}$.

2. **E** $y=-\dfrac{2}{3}x-\dfrac{5}{12}$. The slope is $-\dfrac{2}{3}$.

3. **D** $y=\dfrac{3}{5}x+\dfrac{8}{5}$. Slope of this line is $\dfrac{3}{5}$. Slope of perpendicular is $-\dfrac{5}{3}$.

4. **D** If (x,y) represents any point on the line, the slope =
$$\dfrac{y-3}{x-1}=\dfrac{-2-3}{5-1}.$$
Therefore $y=-\dfrac{5}{4}x+\dfrac{17}{4}$. y-intercept $=\dfrac{17}{4}$.

5. **C** Slope of the line $=\dfrac{3-4}{-2-1}=\dfrac{1}{3}$. Therefore the slope of a perpendicular line is -3. Midpoint of the segment is $\left(\dfrac{1-2}{2},\dfrac{4+3}{2}\right)=\left(\dfrac{1}{2},\dfrac{7}{2}\right)$. Equation of the line is $\dfrac{\left(y-\dfrac{7}{2}\right)}{\left(x-\dfrac{1}{2}\right)}=-3$. So $3x+y-2=0$.

6. **B** Length $=\sqrt{(3+2)^2+(-5-4)^2}=\sqrt{25+81}=\sqrt{106}$.

7. **B** $y=-\dfrac{2}{3}x+\dfrac{8}{3}$. Therefore slope of a parallel line $=-\dfrac{2}{3}$.

8. **E** Distance $= \frac{|12(m) + 5(2m) - 1|}{\sqrt{144+25}} = 5$.
   $|22m-1| = 65$. $22m-1=65$ or $-22m+1=65$.
   Therefore, $m=3$ or $-\frac{32}{11}$.

9. **D** Slope of first line is $\frac{2}{3}$. Slope of second line is 2.
   $$\tan\theta = \frac{2-\frac{2}{3}}{1+\frac{4}{3}} = \frac{4}{7}$$

Part 2.3 Quadratic Functions

1. **B** Complete the square. $y = 2(x^2+2x+1) - 5 - 2 = 2(x+1)^2 - 7$. Vertex $(-1,-7)$.

2. **C** Complete the square. $y = -1(x^2+4x+4)+5+4 = -1(x+2)^2+9$. This is a parabola with vertex at $(-2,9)$ which opens down. Therefore, the range is $\{y : y \leq 9\}$.

3. **B** Complete the square. $y = 2\left(x^2+\frac{3}{2}x+\frac{9}{16}\right) - 6 - \frac{9}{8} = 2\left(x+\frac{3}{4}\right)^2 - \frac{57}{8}$. Axis of symmetry is $x+\frac{3}{4} = 0$.

4. **E** $2x^2 + x - 6 = (2x-3)(x+2) = 0$. Zeros $\frac{3}{2}$ and -2.

5. **B** Sum of zeros $= -\frac{b}{a} = -\frac{-6}{3} = 2$.

6. **E** $b^2 - 4ac = 4 - 4\cdot 1 \cdot 3 = -8 < 0$.

Part 2.4 Higher Degree Polynomial Functions

1. **D** By Descartes' Rule of Signs, there can be 2 or 0 positive real roots and 1 negative real root.

2. **C** $P(0) = -2$. $P(1) = 30$. Therefore, Answer C.

3. **A** Since the degree of the polynomial is an even number, both ends of the graph go off in the same direction. Since $P(x)$ increases without bound as x increases without bound, $P(x)$ increases without bound as x decreases without bound.

4. **C** Since $h(x) = -h(-x)$, it is the only odd function.

5. **C** Rational roots of the form $\frac{p}{q}$, where p is a factor of 12 and q is a factor of 2. $\frac{p}{q} \in \left\{\pm 12, \pm 6, \pm 4, \pm 3, \pm 2, \pm 1, \pm\frac{3}{2}, \pm\frac{1}{2}\right\}$. Total 16.

6. **C** By Descartes' Rule of Signs, only 1.

7. **A** Sum of the roots $= 3$. 1 and 2 are roots. Therefore third root is zero. Product of the roots $= 4b$. Therefore, $b = 0$.

8. **C** Using synthetic division with i and $-i$ results in a depressed equation of $x^2+2x-4=0$, whose roots are $-1 \pm \sqrt{5}$.

9. **B** By Descartes' Rule of Signs, only 1.

10. **E** Sum of roots $= -\frac{b}{a} = \frac{2}{8}$.

11. **C** Using synthetic division with 3 gives $K=4$ and a depressed equation of $3x^2+4$.

12. **D** $1-i$ is also a root. Multiply $(x+1)(x-(1+i))(x-(1-i))$, which are the factors that produced the three roots.

Part 2.5 Inequalities

1. **C** $3x^2 - x - 2 = (3x+2)(x-1) = 0$ when $x = -\frac{2}{3}$ or 1. Numbers between these satisfy the original inequality.

2. **B** The smallest number possible is -84 and the largest is -18.

3. **C** $x^2 - 16x + 48 = (x-4)(x-12) = 0$ when $x = 4$ or 12. Numbers between these satisfy the original inequality.

CHAPTER 3 TRIGONOMETRIC FUNCTIONS

Part 3.1 Definitions

1. **A** Reference angle is 40°. Cos in quadrant IV is positive.

2. **B** See corresponding figure.

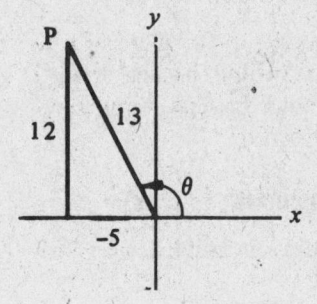

3. **D** θ in quadrant II since sec < 0 and sin > 0.

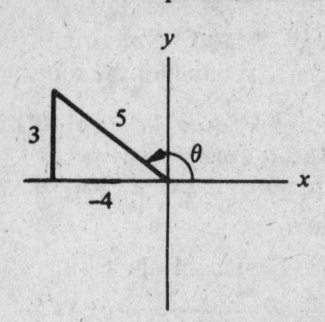

4. **B** Cofunctions of complementary angles are equal. $x - 30 + x = 90$ finds a reference angle of $60°$ for x.

5. **A** α is in quadrant II and $\sin \alpha$ is positive. β is in quadrant IV and $\sin \beta$ is negative.

Part 3.2 Arcs and Angles

1. **C** $\dfrac{30}{\pi} = \dfrac{x}{180°}$. Therefore $x = \dfrac{5400°}{\pi}$.

2. **D** $s = r\theta$. $2\pi = r\theta$. $A = \dfrac{1}{2}r^2\theta$. $6\pi = \dfrac{1}{2}r^2\theta = \dfrac{1}{2}r(r\theta)$ $= \dfrac{1}{2}r(2\pi)$. $r = 6$.

3. **D** $C = 2\pi r = 16$. $r = \dfrac{8}{\pi}$. $A = \dfrac{1}{2}r^2\theta = \dfrac{1}{2}\left(\dfrac{8}{\pi}\right)^2 \cdot \left(\dfrac{3\pi}{2}\right) = \dfrac{48}{\pi}$.

4. **C** $40° = \dfrac{2\pi^R}{9}$. $s = r\theta = 1 \cdot \dfrac{2\pi}{9}$.

Part 3.3 Special Angles

1. **C** $-60°$ is in quadrant IV where $\tan < 0$.

2. **E** $-135°$ is in quadrant III with reference angle $45°$. Tan > 0 in quadrant III. Therefore $\tan(-135°) = 1$. Reference angle for $315°$ is $45°$. Cot < 0 in quadrant IV. Therefore, cot $315° = -1$.

3. **B** The terms equal, in order, $-1, -\dfrac{1}{2}, -1, 1$.
$-1 - \left(-\dfrac{1}{2}\right) - (-1) + 1 = \dfrac{3}{2}$.

4. **B** The terms equal, in order, $\dfrac{2\sqrt{3}}{3}, \dfrac{-\sqrt{3}}{1}, \dfrac{-\sqrt{2}}{2}$.

5. **D** Sin $300° = \dfrac{-\sqrt{3}}{2}$. Choices A and B are ruled out because they are positive. Choices C and E are ruled out because they are cosines with the same reference angle ($60°$) as sin $300°$.

Part 3.4 Graphs

1. **C** Period $= \dfrac{2\pi}{2} = \pi$. Point P is $\dfrac{1}{4}$ of the way through period. Amplitude is 1.

2. **E** Amplitude $= \dfrac{1}{2}$. Period $= \pi$. Graph shifted $\dfrac{1}{2}$ unit up. Graph looks like a cosine graph reflected about x-axis.

3. **C** Graph has amplitude of 4 and period of 4π.

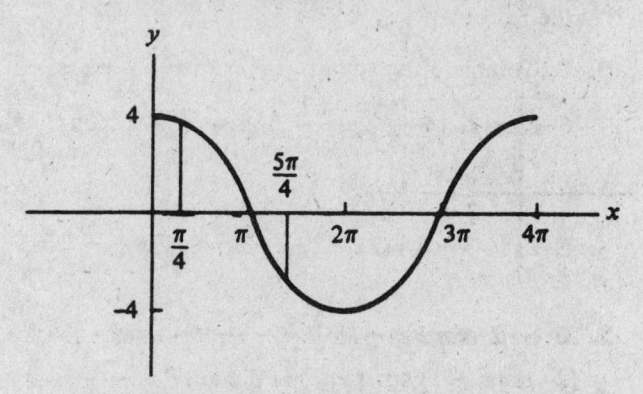

4. **C** Multiply and divide by 2.
$$f(x) = 2\left(\dfrac{\sqrt{3}}{2}\cos x + \dfrac{1}{2}\sin x\right)$$
$$= 2(\sin 60° \cos x + \cos 60° \sin x)$$
$$= 2\sin(60° + x).$$

5. **C** Period $= \dfrac{2\pi}{P} = \dfrac{2\pi}{3}$.

6. **D** Period $= 6\pi$. Curve reaches its maximum when $x = \dfrac{6\pi}{4}$, which is not in the allowable set of x values. Therefore, maximum is reached when $x = \dfrac{\pi}{2}$. Maximum $= \sin \dfrac{1}{3} \cdot \dfrac{\pi}{2} = \dfrac{1}{2}$.

7. **D** Period $= \dfrac{2\pi}{M} = 4\pi$ (from the figure). $M = \dfrac{1}{2}$. Phase shift for a sine curve in the figure is $-\pi$. Therefore, $\dfrac{1}{2}x + N = 0$ when $x = -\pi$.

Part 3.5 Identities, Equations, and Inequalities

1. **D** $\cos 2x = 2\cos^2 x - 1$. Equation becomes
$$2\cos^2 x + \cos x = 0.$$
$$(2\cos x + 1)\cos x = 0.$$

2. **D** $2\sin\theta\cdot\cos\theta = \frac{1}{2}$ becomes $\sin 2\theta = \frac{1}{2}$. $2\theta = 30°, 150°, 390°, 510°$. $\theta = 15°, 75°, 195°, 255°$.

3. **E** $4x = 60°, 240°, 420°, 600°, 780°, 960°, 1140°, 1320°$. $x = 15°, 60°, 105°, 150°, 195°, 240°, 285°, 330°$.

4. **E** $\tan 2\theta = \frac{2\tan\theta}{1-\tan^2\theta} = \frac{2\left(\frac{4}{3}\right)}{1-\left(\frac{16}{9}\right)}$. Since θ can be in quadrants I or IV $\tan 2\theta = \pm\frac{24}{7}$.

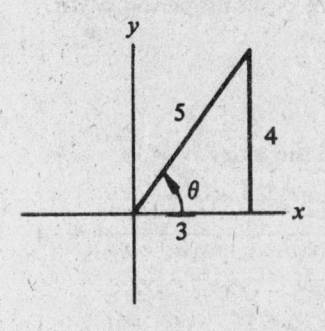

5. **A** $1 - 2\sin^2\theta = -\sin\theta$. $2\sin^2\theta - \sin\theta - 1 = 0$. $(2\sin\theta + 1)(\sin\theta - 1) = 0$. $\sin\theta = -\frac{1}{2}$ or $\sin\theta = 1$.

6. **A** $74 = 2(37)$. $\sin 74° = \sin 2(37°)$. $2\sin 37° \cdot \cos 37° = 2\left(\frac{3}{5}\right)\left(\frac{4}{5}\right) = \frac{24}{25}$.

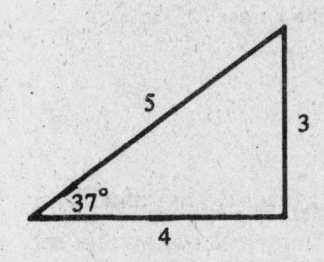

7. **B** $\cot(A+B) = \frac{1}{\tan(A+B)}$
$$= \frac{1-\tan A\cdot\tan B}{\tan A + \tan B}$$
$$= \frac{1-\frac{1}{\cot A\cdot\cot B}}{\frac{1}{\cot A}+\frac{1}{\cot B}}$$

8. **B** This is part of the formula for $\tan(A+B) = \tan(140° + 70°) = \tan 210° = \frac{\sqrt{3}}{3}$.

9. **A** $\cos^4 40° - \sin^4 40°$
$= (\cos^2 40° + \sin^2 40°)(\cos^2 40° - \sin^2 40°)$
$= 1(\cos 2(40°)) = \cos 80°$.

10. **B** A is in quadrant II.
$$\tan 2A = \frac{2\tan A}{1-\tan^2 A} = \frac{2\left(-\frac{2}{\sqrt{5}}\right)}{1-\frac{4}{5}} = -4\sqrt{5}.$$

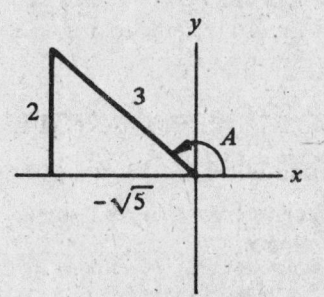

11. **D** To find answer to $\sin 2x < \sin x$, solve the associated equation, $\sin 2x = \sin x$. $\sin 2x - \sin x = 2\sin x\cdot\cos x - \sin x = \sin x(2\cos x - 1) = 0$.
$\sin x = 0$ or $\cos x = \frac{1}{2}$. So $x = 0, \frac{\pi}{3}, \pi, \frac{5\pi}{3}, 2\pi$.
Regions where inequality is satisfied are only $\frac{\pi}{3} < x < \pi$ or $\frac{5\pi}{3} < x < 2\pi$.

Part 3.6 Inverse Functions

1. **B** $2x^2 - 2x = \cos\frac{2\pi}{3} = -\frac{1}{2}$. $4x^2 - 4x + 1 = 0$. $(2x-1)^2 = 0$.

2. **D** $\sin\frac{7\pi}{6} = -\frac{1}{2}$. $\arcsin\left(-\frac{1}{2}\right)$ is an angle in quadrant IV.

3. **A** $\arcsin 1 = \frac{\pi}{2}$ and $\arcsin(-1) = -\frac{\pi}{2}$. I is true. $\arccos 1 = 0$ and $\arccos(-1) = \pi$. II is false. Since Arccos is defined in quadrants I & II part III is false.

4. **B** If the side of a cube is 1, its diagonal $= \sqrt{3}$ and the diagonal of the base is $\sqrt{2}$. Therefore, $\theta = \arcsin\frac{\sqrt{3}}{3}$.

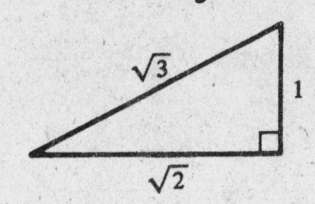

5. **B** Let $\theta = \text{Arcsin}\left(-\frac{3}{5}\right)$. $\sin\theta = -\frac{3}{5}$. Therefore, $\tan\theta = -\frac{3}{4}$.

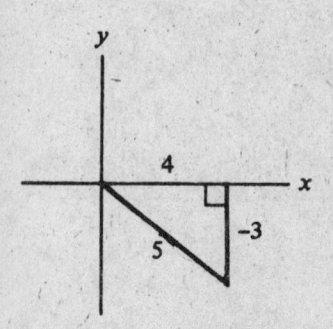

6. **B** Let $A = \text{Arcsin}\left(-\frac{5}{13}\right)$ and $B = \text{Arccos}\left(-\frac{3}{5}\right)$. $-90° < A < 0°$ and $90° < B < 180°$, so $0° < A+B < 180°$. Since cosine has different signs in quadrants I & II, evaluate

$$\cos(A+B) = \frac{12}{13} \cdot \frac{-3}{5} - \frac{-5}{13} \cdot \frac{4}{5} = \frac{-16}{65} < 0.$$

Therefore $A+B$ is in quadrant II.

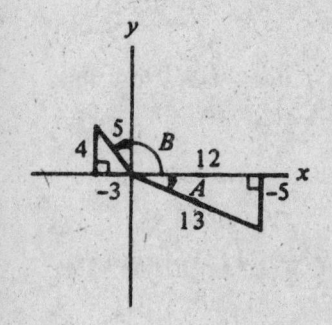

7. **B** $-1 \leq \cos\theta \leq 1$. Thus choice B is not defined.

8. **E** $3x = \text{Arccos}\left(\frac{1}{2}\right)$, so $x = \frac{1}{3}\text{Arccos}\left(\frac{1}{2}\right)$.

Part 3.7 Triangles

1. **D** Law of sines: $\frac{\sin B}{8} = \frac{\frac{1}{2}}{4\sqrt{2}}$. $\sin B = \frac{\sqrt{2}}{2}$. $B = 45°$ or $135°$. Therefore, $C = 105°$ or $15°$.

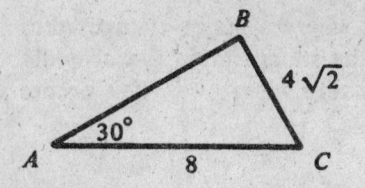

2. **E** Law of sines: $\frac{\sin C}{8} = \frac{\frac{1}{2}}{6}$. $\sin C = \frac{2}{3} \approx .67$. $\sin 45° = \frac{\sqrt{2}}{2} \approx .7$. Since sine is an increasing function in the first quadrant, $0° < C < 45°$. Angles in the second quadrant greater than $135°$ use these values of C as reference angles.

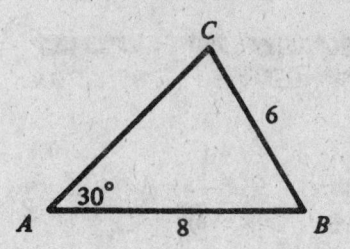

3. **A** The angles are $15°$, $45°$, and $120°$. Let c be longest side, b next. $\frac{\sin 120°}{c} = \frac{\sin 45°}{b}$. $\frac{c}{b} = \frac{\sin 120°}{\sin 45°}$.

4. **D** Law of cosines. Let the sides be 4, 5, and 6. $16 = 25 + 36 - 60 \cos A$. $\cos A = \frac{45}{60} = \frac{3}{4}$.

5. **C** Law of cosines: $d^2 = 36 + 64 - 96\cos 120°$. $d^2 = 148$.

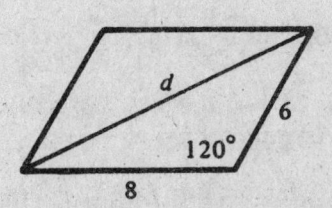

6. **D** Altitude to base $= 8 \sin 60° = 4\sqrt{3}$. Therefore, $4\sqrt{3} < a < 8$.

7. **E** $A = 45°$. Law of sines: $\frac{\sin 45°}{a} = \frac{\sin 30°}{10}$. Therefore $a = 10\sqrt{2}$.

8. **D** Area $= \frac{1}{2}ab\sin C$. $24\sqrt{3} = \frac{1}{2} \cdot 6 \cdot 16 \sin C$. $\sin C = \frac{\sqrt{3}}{2}$.

9. **C** Area $= \frac{1}{2}ab\sin C$. $12\sqrt{3} = \frac{1}{2} \cdot 6 \cdot 8 \sin C$. $\sin C = \frac{\sqrt{3}}{2}$. $C = 60°$ or $120°$. Use law of cosines with $60°$, then with $120°$.

10. **A** In I the altitude $= 12 \cdot \frac{1}{2} = 6$, $6 < c < 12$, so 2 triangles. In II $b > 12\sqrt{2}$, so only 1 triangle. In III the altitude $= 12 \cdot \frac{\sqrt{3}}{2} > 5\sqrt{3}$, so no triangles.

CHAPTER 4 MISCELLANEOUS RELATIONS AND FUNCTIONS

Part 4.1 Conic Sections

1. **D** Complete the squares: $5(x^2 - 4x + 4) + 4(y^2 + 2y + 1) = -4 + 20 + 4$. $5(x-2)^2 + 4(y+1)^2 = 20$. $\frac{(x-2)^2}{4} + \frac{(y+1)^2}{5} = 1$. $a^2 = 5$, $b^2 = 4$, therefore $c^2 = 1$. Therefore, the foci are 1 unit above and below the center, which is at $(2, -1)$.

2. **C** See figure given.

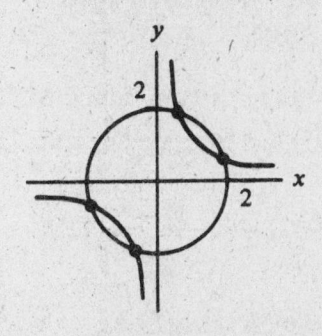

3. **C** Vertex at $(2,0)$, focus at $(2,1)$, directrix $y = -1$.

4. **D** $\frac{x^2}{4} - \frac{y^2}{3} = 1$. Asymptotes $y = \pm \frac{\sqrt{3}}{2} x$.

5. **E** $x^2 - (2y+3)^2 = 0$, factors into $(x - 2y - 3)(x + 2y + 3) = 0$, which breaks into $x - 2x - 3 = 0$ or $x + 2y + 3 = 0$, which are the equations of two intersecting lines.

6. **D** Equation of a semi-circle with radius 2. $A = \frac{1}{2} \pi r^2 = 2\pi$.

Part 4.2 Exponential and Logarithmic Functions

1. **C** $x^a \cdot x^{a^2+a} \cdot x^{a-a^2} = x^{a+a^2+a-a^2} = x^{3a}$.

2. **D** Let $\log_8 3 = y$ and $\log_2 3^x = y$. $8^y = 3$. $2^{3y} = 3$. $2^y = 3^{1/3}$. From the second equation $2^y = 3^x$. Therefore, $3^x = 3^{1/3}$, so $x = \frac{1}{3}$.

3. **B** $\text{Log}_{10}(10m^2) = \log_{10} 10 + 2\log_{10} m = 1 + 2 \cdot \frac{1}{2} = 2$.

4. **B** $b^a = 5$, $b^c = 2.5 = 5^x$ using the relationship between logs and exponents: $b^{ax} = 5^x = b^c$. Therefore $ax = c$ and $x = \frac{c}{a}$.

5. **B** $f\left(\frac{2}{x}\right) + f(x) = \log_2\left(\frac{2}{x}\right) + \log_2 x$
$= \log_2 2 - \log_2 x + \log_2 x = 1$.

6. **E** If $b > 1$, Choice **B** is the answer. If $b < 1$, Choice **C** is the answer. Since no restriction was put on b, the correct answer is **E**.

Part 4.3 Absolute Value

1. **B** If $2x - 1 \geqslant 0$, the equation becomes $2x - 1 = 4x + 5$ and $x = -3$. However, $2x - 1 \geqslant 0$ implies $x \geqslant \frac{1}{2}$ so the -3 will not work. If $2x - 1 < 0$ the equation becomes $-2x + 1 = 4x + 5$ and $x = -\frac{2}{3}$. Since $2x - 1 < 0$ implies that $x < \frac{1}{2}$, $x = -\frac{2}{3}$ is the only root.

2. **E** x must be more than 1 unit from 2 but less than 4 units from 2 (including 1 and 4)

3. **A** The figure is a square $2\sqrt{2}$ on a side. The area is 8.

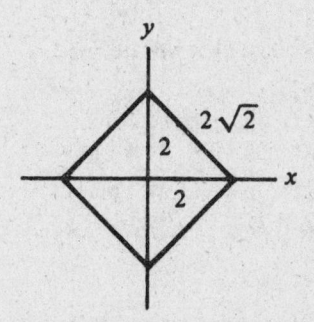

4. **A** Since $f(x)$ must $= f(|x|)$, the graph must be symmetric about the y-axis. The only graph to do that is Choice **A**.

5. **B** Since the point where a major change takes place is at $(1,1)$, the statement in the absolute value should equal zero when $x = 1$. This occurs only in Choice **B**.

Part 4.4 Greatest Integer Function

1. **C** If $N < 1$, the value should be 15 and if $1 < N < 2$ the value should be 28. The only answer that satisfies this is **C**.

2. **C** Since $f(x) =$ an integer by definition, the answer is **C**.

3. **C** Plotting a few numbers between 0 and 2 leads to a rather steep line going down to the right.

x	0	.5	1	1.5	2
$f(x)$	0	-1	-2	-3	-4

Part 4.5 Rational Functions

1. **A** Factor: $\dfrac{(x-1)(x+1)(x^2+1)}{(x-1)(x^2+x+1)}$. Substitute 1 for x and the fraction equals $\dfrac{4}{3}$.

2. **B** Factor and reduce the fraction which becomes $3x+2$. As x approaches zero this approaches 2.

3. **A** Factor and reduce: $\dfrac{(x-2)(x^2+2x+4)}{(x-2)(x+2)(x^2+4)}$. Substitute 2 for x and the fraction equals $\dfrac{3}{8}$.

4. **C** Divide numerator and denominator through by x^2. As $x \to \infty$, the fraction approaches $\dfrac{5}{3}$.

5. **B** Factor and reduce: $\dfrac{(3x+1)(x-1)}{(3x+1)(3x-1)}$. Therefore one asymptote is $3x-1=0$ or $x=\dfrac{1}{3}$. As $x \to \infty$, $y \to \dfrac{1}{3}$. Therefore $y=\dfrac{1}{3}$.

Part 4.6 Parametric Equations

1. **B** Complete the square on the x equation: $x=\left(t^2+t+\dfrac{1}{4}\right)-\dfrac{1}{4}=\left(t+\dfrac{1}{2}\right)^2-\dfrac{1}{4}$. This represents a parabola that opens up with vertex at $\left(-\dfrac{1}{2}, -\dfrac{1}{4}\right)$. Therefore, $x \geq -\dfrac{1}{4}$.

2. **D** $\dfrac{y}{2}=\cos t$. So $\dfrac{y^2}{4}=\cos^2 t$. Adding this to $x=\sin^2 t$ gives $\dfrac{y^2}{4}+x=\cos^2 t+\sin^2 t=1$. Since $0 \leq x \leq 1$ because $0 \leq \sin^2 t \leq 1$, this can only be a portion of the parabola given by the equation $y^2+4x=4$.

3. **A** Removing the parameter in I by squaring and adding gives $x^2+y^2=1$, which is a circle of radius 1. Substituting x for t in the y equation of II and squaring gives $x^2+y^2=1$, but $y \geq 0$ so this is only a semicircle. Squaring and substituting x^2 for s in the y equation of III gives $x^2+y^2=1$, but $x \geq 0$ so this is only a semicircle.

Part 4.7 Polar Coordinates

1. **A** The angle must be either coterminal with $60°$ or $180°$ away. If coterminal, r must equal 2. No such points are given. $180°$ away gives $+240°$ or $-60°$. In either case r must equal -2.

2. **D** Set the two equations equal and use the double angle formula: $\cos\theta = 2\sin\theta \cdot \cos\theta$. $2\sin\theta \cdot \cos\theta - \cos\theta = 0$. Therefore $\cos\theta(2\sin\theta - 1)=0$, so $\cos\theta=0$ or $\sin\theta=\dfrac{1}{2}$.

3. **A** $\sec 2\theta = \dfrac{1}{\cos 2\theta} = \dfrac{1}{2\cos^2\theta - 1}$. Substituting in $r^2 = \dfrac{36}{2\cos^2\theta - 1}$ gives $r^2 = \dfrac{36}{2 \cdot \dfrac{x^2}{r^2} - 1}$.

$r^2 = \dfrac{36 r^2}{2x^2 - r^2}$. Dividing by r^2 gives $1 = \dfrac{36}{2x^2 - x^2 - y^2}$. $x^2 - y^2 = 36$.

4. **B** $-2-2\sqrt{3}\,i = 4\,\text{cis}\,240°$.

$-1+\sqrt{3}\,i = 2\,\text{cis}\,120°$.

$\dfrac{-2-2\sqrt{3}\,i}{-1+\sqrt{3}\,i} = \dfrac{4}{2}\,\text{cis}(240° - 120°)$

$= 2\,\text{cis}\,120°$.

5. **B** $AB = 2 \cdot 3\,\text{cis}(20° + 40°) = 6\,\text{cis}\,60°$.

6. **B** $4\sqrt{3}+4i = 8\,\text{cis}\,\dfrac{\pi}{6}$. Sixth root of this $= \sqrt{2}\,\text{cis}\,\dfrac{1}{6}\left(\dfrac{\pi}{6}+2\pi k\right)$, where $k=0, 1, 2, 3, 4, 5$.

CHAPTER 5 MISCELLANEOUS TOPICS

Part 5.1 Permutations and Combinations

1. **C** $\dfrac{(5+3)!}{5!+3!} = \dfrac{8!}{5!+3!}$

$= \dfrac{8 \cdot 7 \cdot 6 \cdot 5 \cdot 4 \cdot 3 \cdot 2 \cdot 1}{(5 \cdot 4 \cdot 3 \cdot 2 \cdot 1)+(3 \cdot 2 \cdot 1)}$

$= \dfrac{40320}{120+6} = 320$.

2. **B** Number available $= 5 \cdot 8 \cdot 3 = 120$.

3. **B** There are $\binom{6}{4} = 15$ different groups of 4 shells out of the 6 available. In each of these groups the circular permutation can be turned over∴ # of bracelets $= 15 \cdot \frac{3!}{2} = 45$.

4. **C** Permutation with repetitions: $\frac{5!}{2!2!} = 30$.

5. **C** A combination: $\binom{52}{2} = \frac{52 \cdot 51}{2 \cdot 1} = 1326$.

6. **D** x must be less than 4. The only value that works is 0.

7. **D** Only the 7 can go in the units place. Any one of the four numbers can go in the other two places. Number of numbers $= 4 \cdot 4 \cdot 1 = 16$.

8. **C** In order to draw lines through 2 points each, there are 8 points to be chosen 2 at a time. Number of lines $= \binom{8}{2} = \frac{8 \cdot 7}{2 \cdot 1} = 28$.

Part 5.2 Binomial Theorem

1. **C** Middle term is the 6th term, which
$$= \binom{10}{5} x^5 \left(-\frac{1}{x}\right)^2 = \frac{10 \cdot 9 \cdot 8 \cdot 7 \cdot 6}{5 \cdot 4 \cdot 3 \cdot 2 \cdot 1} = -252.$$

2. **C** 7th term $=$
$$\binom{8}{6}(a^3)^2 \left(-\frac{1}{a}\right)^6 = \frac{8 \cdot 7}{2 \cdot 1} \cdot a^6 \cdot \frac{1}{a^6} = 28.$$

3. **B** $(1.002)^7 = (1+.002)^7 \approx 1^7 + 7 \cdot 1^6 \cdot (.002) + \binom{7}{2} \cdot$
$1^5 \cdot (.002)^2 \approx 1 + .014 + \frac{7 \cdot 6}{2 \cdot 1} \cdot 1 \cdot (.000004)$
$= 1.014 + .000084 = 1.014084 \approx 1.0141$.

4. **B** Third term $=$
$$\binom{1/3}{2} \cdot (8x)^{-5/3}(-y)^2 = \frac{\frac{1}{3} \cdot \frac{-2}{3}}{2 \cdot 1}$$
$$\cdot \frac{1}{32} x^{-5/3} y^2 = -\frac{1}{9} \cdot \frac{1}{32} x^{-5/3} y^2.$$
Coefficient of third term is $-\frac{1}{288}$.

Part 5.3 Probability

1. **C** One way to get a 2, 2 ways to get a 3, 4 ways to get a 5, 6 ways to get a 7, 2 ways to get an 11. 15 successes out of 36 elements in the sample space. P (prime) $= \frac{15}{36} = \frac{5}{12}$.

2. **D** Probability of getting neither is $\frac{1}{2} \cdot \frac{5}{6} = \frac{5}{12}$. Therefore, probability of getting either is $1 - \frac{5}{12} = \frac{7}{12}$.

3. **E** Probability that none of the cards was a spade $= \frac{39}{52} \cdot \frac{39}{52} \cdot \frac{39}{52} = \frac{3}{4} \cdot \frac{3}{4} \cdot \frac{3}{4} = \frac{27}{64}$. Probability that one was a spade $= 1 - \frac{27}{64} = \frac{37}{64}$.

4. **E** The only situation when neither of these sets is satisfied is when 3 tails appear. $P(A \cup B) = \frac{7}{8}$.

5. **E** If all 3 are boys that means zero are girls.
$$P(3B) = \frac{\binom{12}{3}\binom{4}{0}}{\binom{16}{3}},$$ where $\binom{12}{3}$ is the number of ways 3 of 12 boys may be selected. $\binom{4}{0}$ is the number of ways 0 of the 4 girls can be selected. $\binom{16}{3}$ is the number of ways the 3 of 16 can be selected. Therefore, $P(3B) = \frac{11}{28}$.

6. **B** Probability of both being *non*-defective
$$= \frac{5}{8} \cdot \frac{3}{5} = \frac{3}{8}.$$

7. **B** $\binom{6}{3}$ is the number of ways the 3 men can be selected. $\binom{3}{2}$ is the number of ways the 2 women can be selected. $\binom{9}{5}$ is the total number of ways people can be selected to fill the 5 rooms. Probability of 3 men and 2 women
$$= \frac{\binom{6}{3}\binom{3}{2}}{\binom{9}{5}} = \frac{10}{21}.$$

Part 5.4 Sequences and Series

1. **D** $a_2 = 5, a_3 = 8, a_4 = 12, a_5 = 17$. Therefore $S_5 = 45$.

2. **A** $y = x + d, z = y + d$. Substituting for d gives $y = x + (z - y)$. Therefore, $y = \frac{1}{2}(x + z)$.

3. **A** $.23\overline{737} = .2 + (.037 + .00037 + .0000037 + \ldots)$, which is $.2 +$ and infinite geometric series with common ratio of $.01$.

$S_n = .2 + \frac{.037}{.99} = \frac{2}{10} + \frac{37}{990} = \frac{235}{990} = \frac{47}{198}$.

The sum of numerator and denominator is 245.

4. **C** Terms are $3^{1/4}$, $3^{1/8}$, 1. Common ratio = $3^{-1/8}$. Therefore the fourth term is $1 \cdot 3^{-1/8} = 3^{-1/8}$.

5. **C** Arithmetic mean = $\frac{1+25}{2} = 13$. Geometric mean = $\sqrt{1 \cdot 25} = 5$. Their difference is 8.

6. **A** List the primes and add them until sum reaches close to 100. 2, 3, 5, 7, 11, 13, 17, 19, 23. Sum equals 100.

7. **B** Adding just units digits of $i!$ gives $1+2+6+4+0+0+0+\cdots$. Sum of units digits for any $i!$ with $i > 4 = 13$.

8. **D** The first few terms are $-1+2-3+4-\cdots+100$. Split up the even numbers and the odd numbers, giving $(2+4+\cdots+100)-(1+3+\cdots+99)$: two arithmetic series with 50 terms and a difference of 2. The sum

$= \frac{50}{2}(2+100) - \frac{50}{2}(1+99) = 2550 - 2500 = 50$.

9. **D** $\frac{2}{7} = \frac{\frac{2}{3}}{1-r}$. $2-2r = \frac{14}{3}$. $6-6r = 14$. $r = -\frac{4}{3}$.

Part 5.5 Geometry

1. **A** The resulting graph is the same as the given one except that it is situated to the left of the y-axis instead of to the right. Therefore, it is the graph of $f(-x)$.

2. **E** Using the equation to determine y, make up a table of values for $(x, \sqrt{y})$.

x	0	1	2	3	4	-2	-1
y	2	3	4	5	6	0	1
$\sqrt{y}$	$\sqrt{2}$	$\sqrt{3}$	$\sqrt{4}$	$\sqrt{5}$	$\sqrt{6}$	0	1

This leads to graph in Choice E.

3. **B** Setting up a table of a few representative values leads to Choice B.

x	-1	$-\frac{1}{2}$	$-\frac{1}{2}$	0	0	0	$\frac{1}{2}$	$\frac{1}{2}$	1
y	0	0	$\frac{1}{2}$	0	$\frac{1}{2}$	1	0	$\frac{1}{2}$	0
x+y	-1	$-\frac{1}{2}$	0	0	$\frac{1}{2}$	1	$\frac{1}{2}$	1	1

4. **D** $2\vec{U} = 2\vec{i} - 10\vec{j}$ and $3\vec{V} = 6\vec{i} + 9\vec{j}$.

$2\vec{U} + 3\vec{V} = 8\vec{i} - \vec{j}$.

5. **D** The dot product must equal zero. Choices B and D satisfy. However, only D is a unit vector.

6. **B** When a plane intersects the x-axis, the y and z-coordinates are zero. Therefore, $2x = 5$ and $x = \frac{5}{2}$.

7. **D** $3 = \sqrt{(x-3)^2 + (-1+3)^2 + (-1-1)^2}$. Square both sides. $9 = x^2 - 6x + 9 + 4 + 4$. $x^2 - 6x + 8 = 0$. $(x-4)(x-2) = 0$. $x = 2$ or 4.

8. **E** In order to find a trace, set one variable to zero and check to see if the result is an answer.

9. **D** Direction numbers must be multiples of $2-(-1)$, $4-5$, $1-2$ or $3, -1, -1$.

10. **D** If the quarter-circle is revolved about the x-axis, a hemisphere of radius 2 is formed. $A = \frac{2}{3}\pi r^3 = \frac{16\pi}{3}$.

11. **C** The altitude is 18.

$V = \frac{1}{3}\pi r^2 h = \frac{1}{3}\pi \cdot 144 \cdot 18 = 864\pi$.

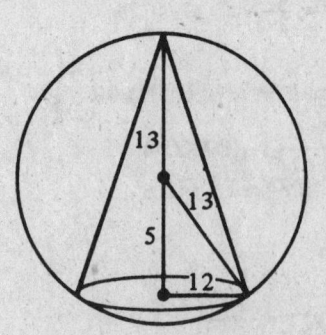

12. **C** $V = \pi r^2 h$. Total area = $2\pi r^2 + 2\pi rh$. $\pi r^2 h = \pi r(2r + 2h)$, which gives $rh = 2r + 2h$. Therefore, $h = \frac{2r}{r-2}$. Since h must be positive, the smallest integral value of r is 3.

13. **D** If the base radius is 4, the side of the square base is $4\sqrt{2}$. Volume of pyramid = $\frac{1}{3}$ (area of base) $h = \frac{1}{3} \cdot 32 \cdot 9 = 96$.

14. **C** In space, the set of points would be the perpendicular bisector plane of the segment joining the two fixed points.

Part 5.6 Variation

1. **C** $\frac{x}{y} = K$ by definition of direct variation.
2. **D** $yx^2 = K$. Substitute $(4,3)$ to find $K = 48$. $a \cdot 4 = 48$.
3. **B** $\frac{DM^2}{R} = K$. Let $D=1$, $M=2$, and $R=3$ (arbitrary choices), then $K = \frac{4}{3}$. Part 2 makes $R=1$ and $M=4$ and $K = \frac{4}{3}$. Thus, $D = \frac{1}{12}$. Therefore, the original value of $D(1)$ was divided by 12.

Part 5.7 Logic

1. **C** It is false only when $(p \vee q)$ is true and p is false. If p is false $(p \vee q)$ is true only when q is true.
2. **C** The statement "q is necessary for p" is equivalent to "p is sufficient for q" and "p only if q," but *not* "q implies p."
3. **B** The negation of "if p, then q" is "p and not q."
4. **E** Counter-examples of each answer can be found.
5. **B** Since "some smarties do not smoke cigars" and "all men smoke cigars," "some smarties are not men."
6. **C** The statement "p', only if c'" is equivalent to "if p', then c'," which is equivalent to its contrapositive "if c, then p."
7. **D** By definition the contrapositive is $q \to p'$.

Part 5.8 Statistics

1. **E** Although arithmetic mean is usually what is meant, anyone of the three is considered an average.
2. **B** Since the values are integers, the range is 3, and the mean is 50, the possible numbers in the set can only be 49, 50, and 51.
    (I) The set could consist of n 49's and n 51's, so 50 would not even be an element of the set. Therefore, the mode is not necessarily 50.
    (II) Since the mean is 50, there must be the same number of 49's as 51's. Thus, the median is 50.
    (III) Obviously false.
3. **D** Mean $= \frac{2+4+9+x}{8} = \frac{15+x}{8}$. The range could equal $x-1$ if $x<3$, or $3-x$ if $x<1$; or 2 if $1 < x < 3$. Therefore, $\frac{15+x}{8} = x-1$ or $\frac{15+x}{8} = 3-x$ or $\frac{15+x}{8} = 2$. $7x = 22$ or $x=1$ or $x=1$. Since x is an integer, the answer can only be 1.
4. **D** Mean $= \frac{6+6+4x+5}{7+x} = 2$; $\frac{17+4x}{7+x} = 2$; $17 + 4x = 14 + 2x$; $x = -\frac{3}{2}$. Mean $= \frac{17+4x}{7+x} = 3$; $17 + 4x = 21 + 3x$; $x = 4$; a positive integer; correct. The mean can be checked against 4 and 5 in similar fashion to show that only 4 fours satisfies the condition of the problem.
5. **A** mode $= 2$; median $=$ middle number $=4$; mean $= \frac{8+12+12+16}{11} = 4\frac{4}{11}$. Therefore, mode $<$ median $<$ mean.

Part 5.9 Odds and Ends

1. **E** Since 2, 3, 6 are three terms of a harmonic sequence, $\frac{1}{2}, \frac{1}{3}, \frac{1}{6}$ are three terms of an arithmetic sequence. The fourth term of the arithmetic sequence is 0 since the difference is $\frac{1}{6}$. Since the reciprocal of zero does not exist, four terms of the harmonic sequence do not exist.
2. **A** The denominator determinant is made up of the coefficients of the variables. It is $\begin{vmatrix} 2 & -3 \\ 5 & 2 \end{vmatrix} = 4 - (+15) = -11$.
3. **E** The division can be written in the form $n \cdot Q + 7 = 125$, so $n \cdot Q = 118 = 2 \cdot 59$. To get a remainder of 7, n must be greater than 7. Since 59 is a prime, the answer is E.
4. **E** $\frac{a+b}{2} = \sqrt{ab}$. Multiplying by 2 and squaring gives
$$a^2 + 2ab + b^2 = 4ab. \quad a^2 - 2ab + b^2 = 0.$$
$$(a-b)^2 = 0. \quad a = b.$$
5. **C** $(1,2)*(1,1) = (2,1)$ and $(p,q)*(1,2) = (p+1, q-2)$. Therefore $p+1=2$ and $q-2=1$.
6. **A** Keeping track of the values of a and x:

step	1	4	3	4	3
a	2	4		8	
x	2		6		10

Print 10.

7. **A** Probability of hitting the target is 1. Probability of hitting inside the circle = $\frac{\text{area of circle}}{\text{area of square}} = \frac{4\pi}{16}$. Probability of hitting outside the circle is $1 - \frac{4\pi}{16}$.

8. **D**
$$a + b + c + d = 100$$
$$b + c = 85$$
$$c + d = 45$$
$$c = 38$$

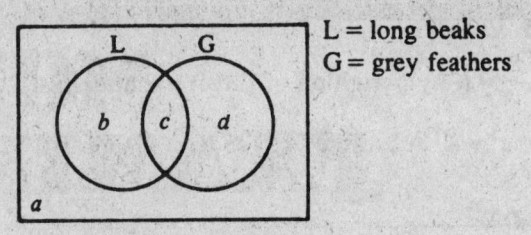

L = long beaks
G = grey feathers

Substituting the value of c, then b and d, gives $a = 8$.

MODEL EXAMINATIONS

SECTION 3

MODEL TEST 1

50 questions
1 hour

Directions: For each of the following problems, decide which is the best of the choices given. Then blacken the corresponding space on the answer sheet on page 137. Answers may be found on page 143, and solutions begin on page 144.

Notes: (1) Figures that accompany problems in this test are intended to provide information useful in solving the problems. They are drawn as accurately as possible EXCEPT when it is stated in a specific problem that its figure is not drawn to scale. All figures lie in a plane unless otherwise indicated.

(2) Unless otherwise specified, the domain of a function f is assumed to be the set of all real numbers x for which $f(x)$ is a real number.

1. The center of the circle $x^2+y^2-10y-36=0$ is

   [A] (0,6)   [B] (0,-11)   [C] (5,0)
   [D] (0,-5)   [E] (0,5)

2. If the sequence $\frac{1}{3}$, 1,... is to be a geometric sequence, the third term must be

   [A] 3   [B] $\frac{5}{3}$ [C] $-\frac{1}{3}$ [D] $\frac{2}{3}$ [E] 6

3. The value of $\frac{453!}{450!\ 3!}$ is

   [A] greater than 10^{100}
   [B] between 10^{10} and 10^{100}
   [C] between 10^5 and 10^{10}
   [D] between 10 and 10^5
   [E] less than 10

4. If $f(x)=2x+3$ and $g(x)=x^2-1$, then $f(g(2))=$

   [A] 48   [B] 9   [C] 7   [D] 5   [E] 3

5. $\left(-\frac{1}{64}\right)^{2/3}$ equals

   [A] $\frac{1}{16}$ [B] $-\frac{1}{16}$ [C] 16
   [D] -16 [E] $16i$

6. The slope of a line perpendicular to the line whose equation is $\frac{x}{3}-\frac{y}{4}=1$ would be

   [A] $\frac{1}{4}$ [B] $-\frac{4}{3}$ [C] $-\frac{3}{4}$ [D] $\frac{4}{3}$
   [E] -3

7 The volume of the region between two concentric spheres of radius 2 and 5 is

[A] 21π [B] 9π [C] 117π
[D] 156π [E] 36π

8 If $16^x = 4$ and $5^{x+y} = 625$, then y equals

[A] 2 [B] 5 [C] $\frac{25}{2}$ [D] $\frac{7}{2}$ [E] 1

9 If A is the angle formed by the line $2y = 3x + 7$ and the x-axis, then $\tan A$ equals

[A] 3 [B] $\frac{3}{2}$ [C] 9 [D] 0 [E] -1

10 If $(x-4)^2 + 4(y-3)^2 = 16$ is graphed, the sum of the distances from any fixed point on the curve to the two foci is

[A] 4 [B] 8 [C] 12 [D] 16 [E] 32

11 If the parameter is eliminated from the equations $x = t^2 + 1$ and $y = 2t$, then the relation between x and y is

[A] $y = x - 1$ [B] $y = 1 - x$
[C] $y^2 = x - 1$ [D] $y^2 = (x-1)^2$
[E] $y^2 = 4x - 4$

12 Let $f(x)$ be a polynomial function: $f(x) = x^5 + \ldots$. If $f(1) = 0$ and $f(2) = 0$, then $f(x)$ is divisible by

[A] $x - 3$ [B] $x^2 - 2$
[C] $x^2 + 2$ [D] $x^2 - 3x + 2$
[E] $x^2 + 3x + 2$

13 If $f(x) = 2$ for all real numbers x, then $f(x+2)$ equals

[A] 0 [B] 2 [C] 4
[D] x [E] cannot be determined

14 In right triangle ABC, $AB = 10$, $BC = 8$, $AC = 6$. The sine of angle A is

[A] $\frac{4}{3}$ [B] $\frac{3}{4}$ [C] $\frac{4}{5}$ [D] $\frac{5}{4}$ [E] $\frac{3}{5}$

15 If $f(x) = |x| + [x]$, the value of $f(-2.5) + f(1.5)$ is

[A] 3 [B] 1 [C] -2 [D] 1.5 [E] 2

16 The number of terms in the expansion of $(2x^2 - 3y^{1/3})^7$ is

[A] 6 [B] 8 [C] 1 [D] 7 [E] 9

17 For what values of k does the graph of $\frac{(x-2k)^2}{1} - \frac{(y-3k)^2}{3} = 1$ pass through the origin?

[A] 0 only [B] 1 only [C] ± 1
[D] $\pm\sqrt{5}$ [E] no value

18 In the figure, $\angle A = 120°$, $a = \sqrt{6}$, and $b = 2$. How large is $\angle C$?

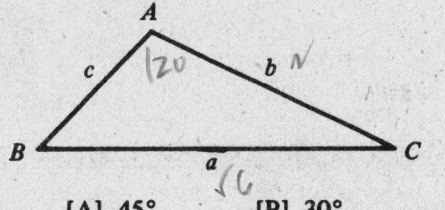

[A] 45° [B] 30° [C] 15°
[D] 10° [E] none of these

19 The remainder obtained when $3x^4 + 7x^3 + 8x^2 - 2x - 3$ is divided by $x + 1$ is

[A] 5 [B] 0 [C] -3 [D] 3 [E] 13

20 The ellipse $4x^2 + 8y^2 = 64$ and the circle $x^2 + y^2 = 7$ intersect at points where the y-coordinate is

[A] $\sqrt{10}$ [B] 3 [C] $\pm\sqrt{10}$
[D] ± 3 [E] they do not intersect

21 Each term of a sequence, after the first, is inversely proportional to the term preceding it. If the first two terms are 2 and 6, what is the twelfth term?

[A] 2 [B] 6 [C] $2 \cdot 3^{11}$
[D] 46 [E] cannot be determined

22 If the function $g(x)$ represents the slope of the line tangent to the graph of the function $f(x)$, shown below, at each point (x,y), which of the following could be the graph of $g(x)$?

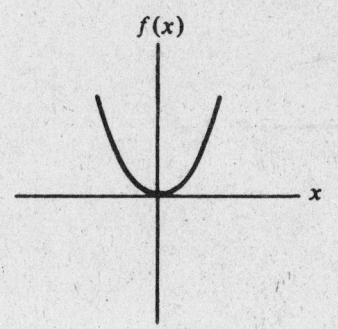

[A]

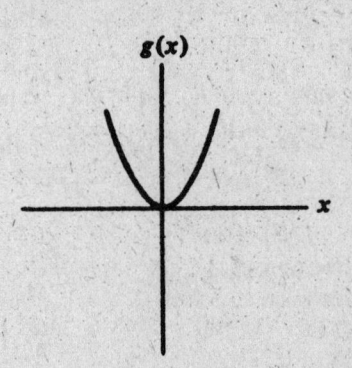

  g(x) (parabola opening upward)

[B]

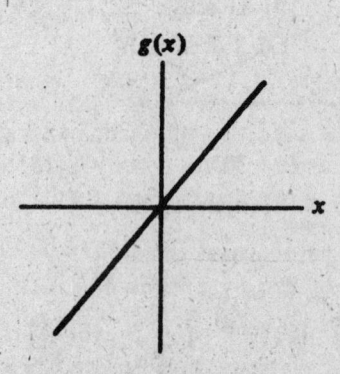

  g(x) (line through origin)

[C]

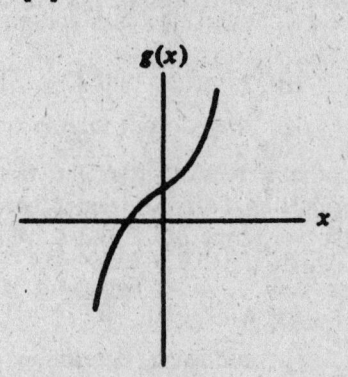

  g(x) (cubic-like curve through origin)

[D]

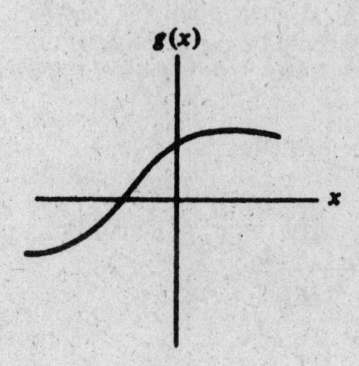

  g(x) (S-shaped curve leveling off)

[E]

  g(x) (S-shaped curve leveling off)

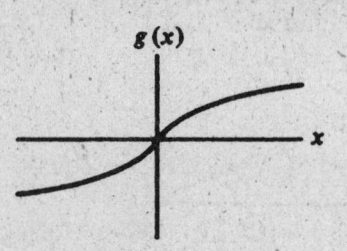

23 The graph of $r = \sec\theta$ is

[A] a point [B] a straight line
[C] an ellipse [D] a hyperbola
[E] none of these

24 If $f(x) = 2x + 3$ where $-2 \leq x \leq 4$, what is the range of $f(|x|)$?

[A] $-2 \leq y \leq 4$ [B] $2 \leq y \leq 4$
[C] $-1 \leq y \leq 11$ [D] $1 \leq y \leq 11$
[E] $3 \leq y \leq 11$

25 A sector of a circle $0 - \overset{\frown}{AB}$ with a central angle of $\dfrac{8\pi}{5}$ and radius of 5 is bent to form a cone with vertex at 0. What is the volume of the cone that is formed?

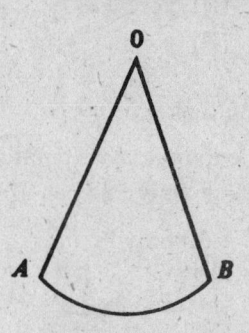

[A] 48π [B] 36π [C] 24π
[D] 16π [E] 100π

26 $\dfrac{\tan 15°}{\cot 75°}$ equals

[A] 0 [B] 1 [C] -1 [D] ∞ [E] $\sqrt{3}$

27 Which of the following is equivalent to $\cos\theta$ when $0 \leq \theta \leq \pi$?

[A] $\sqrt{\dfrac{\sec^2\theta - 1}{\sec^2\theta}}$ [B] $\dfrac{\tan\theta}{\sec\theta}$

[C] $\dfrac{\cot\theta}{\csc\theta}$ [D] $\sqrt{1 - \cos^2\theta}$

[E] $\dfrac{\sec\theta \cdot \cos\theta}{\csc\theta}$

28. In $\triangle ABC$, $a=x$, $b=3x+2$, $c=\sqrt{12}$, and $\angle C=60°$. Find x.

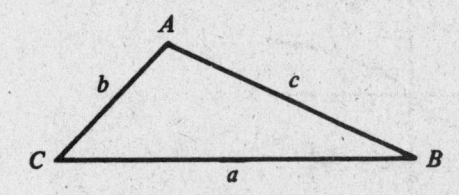

[A] $\dfrac{4}{7}$ [B] 2 [C] $\dfrac{4}{7}$ and 2

[D] $\dfrac{1}{2}$ [E] $\dfrac{2\sqrt{3}}{2}$

29. In the equation $3x^2+kx+54=0$, one root is twice the other root. The value(s) of k is (are)

[A] 3 [B] ± 3 [C] -27
[D] ± 27 [E] 6

30. If $x^2+3x+2<0$ and $f(x)=x^2-3x+2$, then

[A] $0<f(x)<6$ [B] $f(x)>\dfrac{3}{2}$
[C] $f(x)>12$ [D] $f(x)>0$
[E] $6<f(x)<12$

31. $f=\{(x,y):y=4\}$, $g=\{(x,y):x^2-y^2=1\}$, and $h=\{(x,y):x=3\}$. Which of these is a function?

[A] f only [B] g only [C] h only
[D] f and g only [E] g and h only

32. If point $P(3a,a)$ is six units from the line $5x-12y-3=0$, then a equals

[A] 27 only [B] -25 only
[C] 25 or -27 [D] ± 27
[E] -25 or 27

33. What is the set of points in space equidistant from two vertices of an equilateral triangle and 2 inches from the third vertex?

[A] a circle [B] a line segment
[C] two points [D] a parabola
[E] two parallel lines

34. If $f(x)=3x^2+4x+5$, what must the value of k equal so that the graph of $f(x-k)$ is symmetric to the y-axis?

[A] $\dfrac{2}{3}$ [B] $-\dfrac{2}{3}$ [C] 0
[D] -4 [E] $-\dfrac{4}{3}$

35. Which of the following could be a possible root of $4x^3-px^2+qx-6=0$?

[A] $\dfrac{1}{6}$ [B] 4 [C] $\dfrac{2}{3}$ [D] $\dfrac{3}{2}$ [E] $\dfrac{4}{3}$

36. $(x+2)^3-3(x+2)^2+3(x+2)-1$ equals

[A] $(x+1)^3$ [B] $(x-1)^3$ [C] $(x+2)^3$
[D] x^3 [E] $(x+3)^3$

37. If $f(x)=\cos x$ and $g(x)=2x+1$, which of the following are even functions?
  I. $f(x)\cdot g(x)$
  II. $f(g(x))$
  III. $g(f(x))$

[A] I only [B] II only
[C] III only [D] I & II only
[E] II & III only

38. At the end of a meeting all participants shook hands with each other. Twenty-eight handshakes were exchanged. How many people were at the meeting?

[A] 14 [B] 7 [C] 8
[D] 28 [E] 56

39. If the roots of the equation $3x^4-5x^3+8x^2-4x+2=0$ are multiplied in pairs and these products are then added, the sum will equal

[A] $\dfrac{5}{3}$ [B] $-\dfrac{5}{3}$ [C] $\dfrac{8}{3}$
[D] $-\dfrac{8}{3}$ [E] $\dfrac{4}{3}$

40. If $f(x)=\dfrac{2x-2}{x-1}$ and $f^2(x)=f(f(x)), f^3(x)=f(f^2(x))\ldots f^n(x)=f(f^{n-1}(x))$, where n is an integer greater than 1, what values of n are there such that $f^n(x)=f(x)$?

[A] all values of n
[B] all even values of n
[C] all odd values of n
[D] no values of n
[E] $\{n:n=3k$, where k is a positive integer$\}$

41. $\operatorname{Sin}\left(\operatorname{Arctan}\dfrac{1}{3}\right)$ equals

[A] $\dfrac{3\sqrt{10}}{10}$ [B] $\dfrac{\sqrt{10}}{10}$ [C] $\dfrac{1}{3}$
[D] $\dfrac{\sqrt{2}}{4}$ [E] $\dfrac{1}{2}$

42 A cylinder whose base radius is three is inscribed in a sphere of radius five. What is the difference between the volume of the sphere and the volume of the cylinder?

[A] $112\frac{2}{3}\pi$ [B] $94\frac{2}{3}\pi$ [C] 28π

[D] $142\frac{2}{3}\pi$ [E] 428π

43 If determinants are used to solve the equations $\begin{cases} 3x+21=1 \\ x-3y=2 \end{cases}$ simultaneously, then y equals

[A] $\dfrac{\begin{vmatrix} 3 & 1 \\ 1 & 2 \end{vmatrix}}{\begin{vmatrix} 3 & 2 \\ 1 & -3 \end{vmatrix}}$ [B] $\dfrac{\begin{vmatrix} 3 & 2 \\ 1 & -3 \end{vmatrix}}{\begin{vmatrix} 3 & 1 \\ 1 & 2 \end{vmatrix}}$

[C] $\dfrac{\begin{vmatrix} 1 & 2 \\ 3 & 1 \end{vmatrix}}{\begin{vmatrix} 3 & 2 \\ 1 & -3 \end{vmatrix}}$ [D] $\dfrac{\begin{vmatrix} 3 & 2 \\ 1 & -3 \end{vmatrix}}{\begin{vmatrix} 1 & 2 \\ 2 & -3 \end{vmatrix}}$

[E] $\dfrac{\begin{vmatrix} 1 & 2 \\ 2 & -3 \end{vmatrix}}{\begin{vmatrix} 3 & 2 \\ 1 & -3 \end{vmatrix}}$

44 If $f(x)=i$, where i is an integer such that $i \leq x < i+1$, and $g(x)=f(x)-2x$, then the period of $g(x)$ is

[A] $\dfrac{1}{2}$ [B] 1   [C] 2

[D] -2 [E] none of these

45 The figure could be the graph of

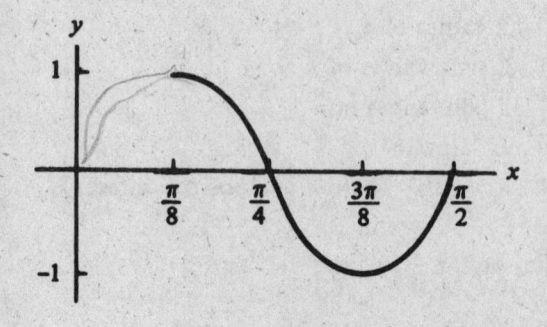

[A] $y=\sin\dfrac{1}{4}x$

[B] $y=\cos\left(4x-\dfrac{\pi}{2}\right)$

[C] $y=\sin 2x \cdot \cos 2x$

[D] $y=-\sin 4x$

[E] $y=\cos\left(\dfrac{1}{4}x-2\pi\right)$

46 What is the smallest positive angle that will make $5-\sin\left(x+\dfrac{\pi}{6}\right)$ a maximum?

[A] $\dfrac{\pi}{3}$ [B] $\dfrac{2\pi}{3}$ [C] $\dfrac{\pi}{2}$

[D] $\dfrac{4\pi}{3}$ [E] $\dfrac{5\pi}{3}$

47 If $f(x)=\begin{cases} \dfrac{5}{x-2}, & \text{when } x \neq 2 \\ k, & \text{when } x=2 \end{cases}$, what must the value of k equal in order that $f(x)$ be a continuous function?

[A] 2   [B] 5   [C] 0   [D] -2

[E] no value of k will make $f(x)$ a continuous function

48 Given the set of data 8, 12, 12, 15, 18. What is the range of this set of data?

[A] 12   [B] 18   [C] 13   [D] 15   [E] 10

49 $f(x)=ax+b$. Which of the following make $f(x)=f^{-1}(x)$?

I. $a=-1$, $b=$ any real number
II. $a=1$, $b=0$
III. $a=$ any real number, $b=0$

[A] I only   [B] II only   [C] III only

[D] I & II only   [E] I & III only

50 If $\dfrac{1-\cos\theta}{\sin\theta}=\dfrac{\sqrt{3}}{3}$, then θ equals

[A] 15°   [B] 30°   [C] 45°

[D] 60°   [E] 75°

MODEL TEST 2

50 questions
1 hour

Directions: For each of the following problems, decide which is the best of the choices given. Then blacken the corresponding space on the answer sheet on page 138. Answers may be found on page 143, and solutions begin on page 146.

Notes: (1) Figures that accompany problems in this test are intended to provide information useful in solving the problems. They are drawn as accurately as possible EXCEPT when it is stated in a specific problem that its figure is not drawn to scale. All figures lie in a plane unless otherwise indicated.

(2) Unless otherwise specified, the domain of a function f is assumed to be the set of all real numbers x for which $f(x)$ is a real number.

1. If $f(x) = \dfrac{x-2}{x^2-4}$, for what value(s) of x does the graph of $f(x)$ have a vertical asymptote?

   [A] −2, 0, and 2 [B] −2 and 2
   [C] 2   [D] 0   [E] −2

2. If a regular square pyramid has x pairs of parallel edges, then x equals

   [A] 1 [B] 2 [C] 4 [D] 8 [E] 12

3. The radius of the sphere $x^2 + y^2 + z^2 + 2x - 2y = 10$ is

   [A] $2\sqrt{2}$ [B] 2   [C] $2\sqrt{3}$
   [D] $3\sqrt[3]{2}$ [E] none of these

4. Every sixth degree polynomial with real coefficients must have k real roots; k is

   [A] 0   [B] 2   [C] 6
   [D] at least 1   [E] at most 1

5. If the roots of the equation $x^2 + bx + c = 0$ are r and s, the value of $\dfrac{(r+s)^2}{rs}$ in terms of b and c is

   [A] $\dfrac{b^2}{c^2}$ [B] $\dfrac{c^2}{b}$ [C] $\dfrac{b^2}{c}$
   [D] $-\dfrac{b}{c}$ [E] $-\dfrac{b^2}{c}$

6. The maximum value of $6 \cdot \sin x \cdot \cos x$ is

   [A] $\dfrac{1}{3}$ [B] 1 [C] 3 [D] 6 [E] $\dfrac{3\sqrt{3}}{2}$

7. If $f(r, \theta) = r \cos \theta$, $f\left(2, \dfrac{\pi}{3}\right) =$

   [A] $\dfrac{2\pi}{3}$ [B] 1 [C] 3 [D] $\dfrac{\pi}{6}$ [E] $\dfrac{1}{2}$

99

8. The sum of the zeros of $f(x) = x^3 - 3x^2 - 4x + 12$ is

[A] 3 [B] −3 [C] 7 [D] −7 [E] 12

9. $i^{14} + i^{15} + i^{16} + i^{17} =$

[A] 1 [B] $2i$ [C] $1-i$ [D] 0
[E] $2+2i$

10. If the graph of $y = \sin 2x$ is drawn for all values of x between 10° and 350°, this graph crosses the x-axis

[A] zero times [B] one time
[C] two times [D] three times
[E] six times

11. The range of the function $y = x^{-2/3}$ is

[A] $y < 0$ [B] $y > 0$ [C] $y \geq 0$
[D] $y \leq 0$ [E] all real numbers

12. A particular sphere has the property that its surface area has the same numerical value as its volume. What is the radius of this sphere?

[A] 1 [B] 2 [C] 3 [D] 4 [E] 6

13. A unit vector parallel to the vector $\vec{V} = (2, -3, 6)$ is the vector

[A] $(-2, 3, -6)$ [B] $(6, -3, 2)$
[C] $\left(-\frac{2}{7}, \frac{3}{7}, -\frac{6}{7}\right)$ [D] $\left(\frac{2}{7}, \frac{3}{7}, \frac{6}{7}\right)$
[E] $\left(-\frac{2}{\sqrt{31}}, -\frac{3}{\sqrt{31}}, \frac{6}{\sqrt{31}}\right)$

14. The pendulum on a clock swings through an angle of one radian and the tip sweeps out an arc of 12 inches. How long is the pendulum?

[A] 6 inches [B] 12 inches [C] 24 inches
[D] $\frac{12}{\pi}$ inches [E] $\frac{24}{\pi}$ inches

15. If $f(x) = \frac{1}{3}x + 2$ and $g(f(x)) = x$, then $g(x) =$

[A] $-\frac{1}{3}x - 2$ [B] $\frac{3}{x+6}$
[C] $3x - 6$ [D] $3x - 2$
[E] $\frac{1}{3}x - 2$

16. Which of the following is the solution set for $x(x-3)(x+2) > 0$?

[A] $x < -2$ [B] $-2 < x < 3$
[C] $-2 < x < 3$ or $x > 3$
[D] $x < -2$ or $0 < x < 3$
[E] $-2 < x < 0$ or $x > 3$

17. If the function $g(x)$ represents the slope of the line tangent to the graph of the function $f(x)$, shown at right, at each point (x,y), which of the following could be the graph of $g(x)$?

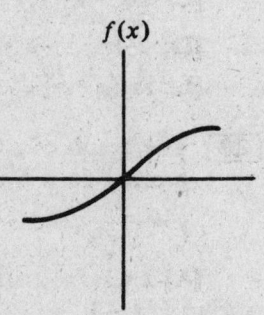

[A] 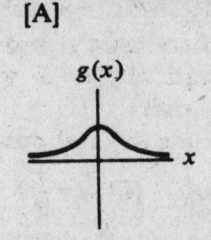 [B]

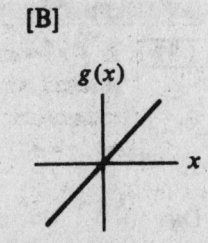

[C] 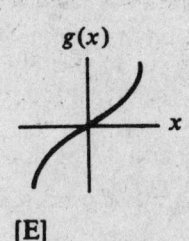 [D]

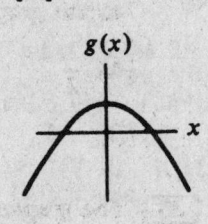

[E]

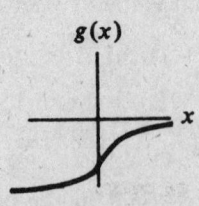

18. The polar coordinates of a point of intersection of the circle $r = \sin \theta$ and the line $r = \dfrac{1}{\sin \theta + \cos \theta}$ are

[A] $\left(\dfrac{\sqrt{2}}{2}, \dfrac{5\pi}{4}\right)$ [B] $\left(-1, \dfrac{\pi}{2}\right)$

[C] $\left(\dfrac{\sqrt{2}}{2}, \dfrac{\pi}{4}\right)$ & $\left(-1, \dfrac{\pi}{2}\right)$

[D] $\left(-\dfrac{\sqrt{2}}{2}, \dfrac{5\pi}{4}\right)$ & $\left(1, \dfrac{3\pi}{2}\right)$

[E] $\left(\dfrac{\sqrt{2}}{2}, \dfrac{\pi}{4}\right)$ & $\left(-1, \dfrac{3\pi}{2}\right)$

19 Log $(a^2 - b^2)$ equals

[A] $\log a^2 - \log b^2$
[B] $\log \dfrac{a^2}{b^2}$
[C] $\log \dfrac{a+b}{a-b}$
[D] $2 \cdot \log a - 2 \cdot \log b$
[E] $\log(a+b) + \log(a-b)$

20 The graph of $xy - 4x - 2y - 4 = 0$ can be expressed as a set of parametric equations. If $y = \dfrac{4t}{t-3}$ and $x = f(t)$, then $f(t) =$

[A] $t+1$ [B] $t-1$ [C] $3t-3$
[D] $\dfrac{t-3}{4t}$ [E] $\dfrac{t-3}{2}$

21 If $f(x) = ax^2 + bx + c$, how must a and b be related so that the graph of $f(x-3)$ will be symmetric about the y-axis?

[A] $a = b$ [B] $b = 0$, a is any real number
[C] $b = 3a$ [D] $b = 6a$ [E] $a = \dfrac{1}{9}b$

22 If the graph of $x + 2y + 3 = 0$ is perpendicular to the graph of $ax + 3y + 2 = 0$, then a equals

[A] $\dfrac{3}{2}$ [B] $-\dfrac{3}{2}$ [C] 6
[D] -6 [E] $\dfrac{2}{3}$

23 The graphs of $y = 2 \cdot \log_b x$ and $y = \log_b 2x$ intersect in

[A] only one point
[B] only 2 points
[C] no points
[D] more than 2 points
[E] an infinite number of point

24 In how many ways can a committee of four be selected from nine men so as to always include a particular man?

[A] 84 [B] 70 [C] 48 [D] 56 [E] 126

25 Which of the following is equivalent to $\sin(A + 30°) + \cos(A + 60°)$ for all values of A?

[A] $\sin A$ [B] $\cos A$
[C] $\sqrt{3} \cdot \sin A + \cos A$ [D] $\sqrt{3} \cdot \sin A$
[E] $\sqrt{3} \cdot \cos A$

26 Sin 15° equals

[A] $\dfrac{\sqrt{6} - \sqrt{2}}{2}$ [B] $\dfrac{\sqrt{6} + \sqrt{2}}{4}$
[C] $\dfrac{\sqrt{2 + \sqrt{3}}}{2}$ [D] $\dfrac{\sqrt{2 - \sqrt{3}}}{2}$
[E] $\sqrt{\dfrac{2 + \sqrt{3}}{2}}$

27 When $\left(1 - \dfrac{1}{x}\right)^{-6}$ is expanded, the sum of the last three coefficients is

[A] 10 [B] 11 [C] 16
[D] -11 [E] cannot be determined

28 If L varies inversely as the square of D, what is the effect on D when L is multiplied by four?

[A] multiplied by $\dfrac{3}{2}$
[B] multiplied by 4
[C] multiplied by 2
[D] divided by 2
[E] none of these

29 The contrapositive of $(p \lor q) \to q$ is

[A] $q' \to (p \land q)'$
[B] $(p \lor q)' \to q'$
[C] $q' \to (p' \land q')$
[D] $q \to (p' \lor q')$
[E] $q' \to (p' \lor q')$

30 If $f(x, y, z) = 2x + 3y - z$ and $f(a, b, 0) = f(0, a, b)$, then $\dfrac{b}{a}$ equals

[A] 4 [B] $\dfrac{1}{2}$ [C] 1 [D] $\dfrac{1}{4}$ [E] 2

31 If $f(x) = \dfrac{x}{x-1}$ and $f^2(x) = f(f(x))$, $f^3(x) = f(f^2(x)) \ldots f^n(x) = f(f^{n-1}(x))$, where n is a positive integer greater than 1, what is the smallest value of n such that $f^n(x) = f(x)$?

[A] 2 [B] 3 [C] 4
[D] 6 [E] no value of n works

32 The set of points satisfying the inequalities $y > |x| - 3$ and $2y > x + 6$ lies entirely in quadrants

[A] I & II [B] II & III
[C] III & IV [D] I & IV
[E] II & IV

33. Which of the following could represent the inverse of the function graphed at right?

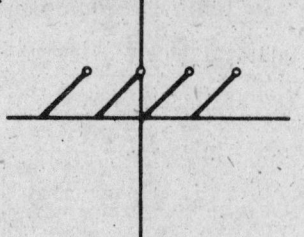

[A]

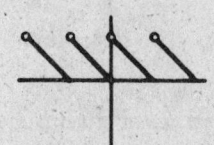

[B]

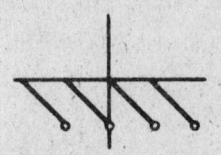

[C]

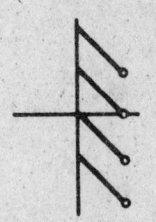

[D]

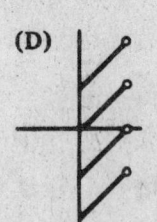

[E]

34. If f is a linear function and f(1) = −2 and f(−2) = 1 and f(x) = 1.5, what is x?

[A] −1.5 [B] −.5 [C] 0
[D] −2.5 [E] −1

35. Which of the following could be a term in the expansion of $(p-q)^{18}$?

[A] $816p^{15}q^3$ [B] $-816p^{15}q^3$
[C] $816p^{16}q^2$ [D] $-816p^{16}q^2$
[E] $-816p^4q^{14}$

36. If the domain of $f(x) = 3x^2 + 2$ is $\{x: -2 \leq x \leq 2\}$, then $f(x)$ has a minimum value when x equals

[A] −2 [B] 0 [C] 2
[D] $\frac{\sqrt{6}}{3}$ [E] $-\frac{\sqrt{6}}{3}$

37. If $\dfrac{\sin x + \cos \dfrac{\pi}{3}}{\cos \dfrac{5\pi}{6} - \sin 270°} = 0$ and $90° < x < 270°$, then x equals

[A] 120° [B] 150° [C] 180°
[D] 210° [E] 240°

38. A coin is tossed three times. Given that at least one head appears, what is the probability of exactly two heads appearing?

[A] $\frac{3}{8}$ [B] $\frac{3}{7}$ [C] $\frac{3}{4}$ [D] $\frac{5}{8}$ [E] $\frac{7}{8}$

39. Which of the following is the equation of the circle with center at the origin and tangent to the line with equation $3x - 4y = 10$?

[A] $x^2 + y^2 = 2$
[B] $x^2 + y^2 = 4$
[C] $x^2 + y^2 = 3$
[D] $x^2 + y^2 = 5$
[E] $x^2 + y^2 = 10$

40. If $f(x) = 3 - 2x + x^2$, then $\dfrac{f(x+t) - f(x)}{t}$ equals

[A] $t^2 + 2xt - 2t$
[B] $x^2t^2 - 2xt + 3$
[C] $t + 2x - 2$
[D] $2x - 2$
[E] none of these

41. If $x + y = 90°$, which of the following must be true?

[A] $\cos x = \cos y$
[B] $\sin x = -\sin y$
[C] $\tan x = \cot y$
[D] $\sin x + \cos x = 1$
[E] $\tan x + \cot y = 1$

42 If $\vec{V}=(3,-5)$ and $\vec{U}=(-7,2)$, what is the magnitude of the resultant of $\vec{V}$ and $\vec{U}$?

[A] $(-4,-3)$ [B] $(-21,-10)$
[C] 5   [D] $\sqrt{34}+\sqrt{53}$
[E] $\sqrt{87}$

43 Three consecutive terms, in order, of an arithmetic sequence are $x+2$, $2x+3$, $5x-6$. x equals

[A] 1   [B] -1 [C] 5   [D] 0   [E] $\frac{5}{4}$

44 If $f(x)=x^3$ and $g(x)=x^2+1$, which of the following are odd functions?
   I. $f(x) \cdot g(x)$
   II. $f(g(x))$
   III. $g(f(x))$

[A] I only   [B] II only   [C] III only
[D] II & III only   [E] I, II, & III

45 The graph of the equation $y=x^3+5x+1$

[A] does not intersect the x-axis
[B] intersects the x-axis in one and only one point
[C] intersects the x-axis in exactly three points
[D] intersects the x-axis in more than three points
[E] intersects the x-axis in exactly two points

46 If $f(x)=\sin x$ and $g(x)=\cos x$ over the interval $(0, 2\pi)$, and $M(f,g)$ is defined to be the maximum of f and g, then the graph of $m(f,g)$ would look like which one of the following?

[A]

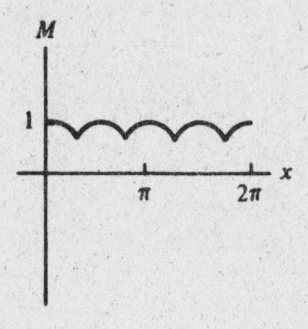

[B]

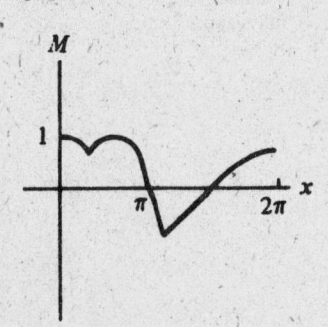

[C]

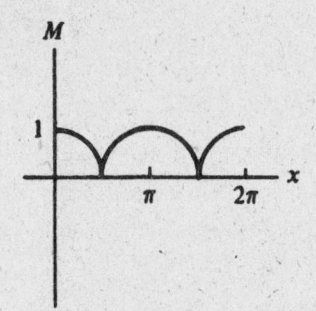

[D]

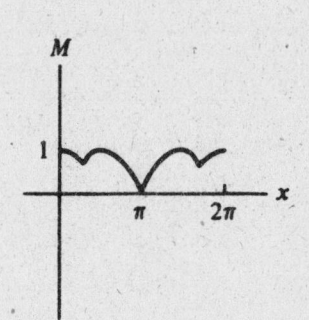

[E]

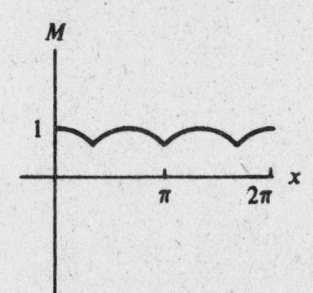

47 If the mean of the set of data 1, 2, 3, 1, 2, 5, x is 3, what is the value of x?

[A] 7   [B] $4\frac{2}{3}$ [C] 2
[D] 4   [E] $2\frac{3}{7}$

48 In $\triangle JKL$, $\sin L = \frac{1}{3}$, $\sin J = \frac{3}{5}$, and $JK = 5$ inches. The length of KL, in inches, is

[A] $\frac{25}{9}$ [B] 25   [C] 15
[D] 12   [E] 9

49 In $\triangle ABC$, $a = 1$, $b = 4$, and $\angle C = 60°$. The length of c is

[A] $\sqrt{21}$ [B] 13   [C] $\sqrt{13}$
[D] 21   [E] 3

50 If (x,y) represents a point on the graph of $y = 2x + 1$, which of the following could be a portion of the graph of the set of points (x, y^2)?

[A]

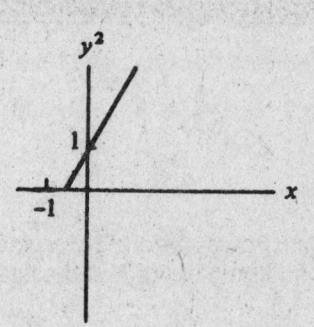

[B]

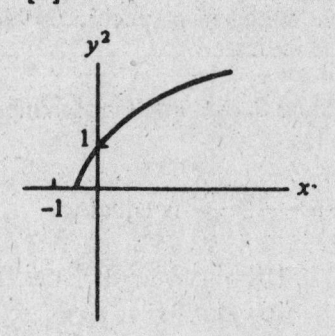

[C]

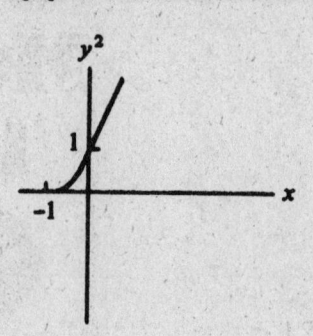

[D]

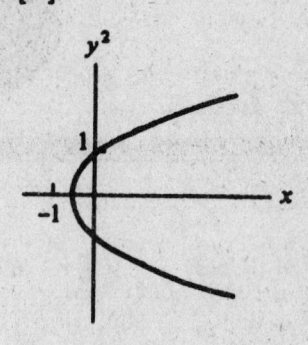

[E]

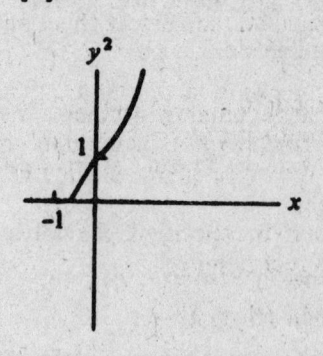

MODEL TEST 3

50 questions
1 hour

<u>Directions:</u> For each of the following problems, decide which is the best of the choices given. Then blacken the corresponding space on the answer sheet on page 138. Answers may be found on page 143, and solutions begin on page 148.

<u>Notes:</u> (1) Figures that accompany problems in this test are intended to provide information useful in solving the problems. They are drawn as accurately as possible EXCEPT when it is stated in a specific problem that its figure is not drawn to scale. All figures lie in a plane unless otherwise indicated.

(2) Unless otherwise specified, the domain of a function f is assumed to be the set of all real numbers x for which $f(x)$ is a real number.

1. If $f(x) = 2^x$, then $f(\log_2 5) =$
   [A] 2^5 [B] 2^2 [C] $2 \cdot \log_2 5$ [D] 2
   [E] 5

2. A sphere is tangent to two distinct parallel planes. The set of all points equidistant from the two planes that intersect the sphere is
   [A] a circle
   [B] 2 points
   [C] a plane
   [D] a straight line segment
   [E] the empty set

3. If $f(x) = \sqrt{x-1}$, the domain of $f^{-1}(x)$ would be
   [A] all real numbers [B] $x > 1$
   [C] $x \geq 1$ [D] $x > 0$
   [E] $x \geq 0$

4. The expression $\dfrac{1 - \cos 4x}{2}$ is equal to
   [A] $\sin^2 6x$ [B] $\sin^2 2x$ [C] $\cos^2 8x$
   [D] $\cos^2 4x$ [E] $\sin^2 8x$

5. If $g(x) = x^3 - x^2 - x - 1$, then the value of $g(-1)$ is
   [A] 0 [B] -1 [C] -2 [D] -3 [E] -4

6. The smallest positive value of x satisfying the equation $\tan 5x = -1$ is
   [A] $9°$ [B] $18°$ [C] $27°$
   [D] $45°$ [E] $135°$

7. If $\log A = .2222$ and $\log B = .3333$, then the value of $\log (\sqrt{A} \cdot B^2)$ is
   [A] .0741 [B] .1111 [C] .5555
   [D] .7777 [E] .9999

8. The number of positive values of x less than $360°$ which satisfy the equation $\sin \frac{1}{2}x = \cos x$ is

[A] 0 [B] 1 [C] 2 [D] 3 [E] 4

9. For each value of θ, $\sin(90° + \theta)$ equals

[A] $\sin \theta$ [B] $\cos \theta$ [C] $-\sin \theta$
[D] $-\cos \theta$ [E] none of these

10. If $f(x) = x^2 + bx + c$ for all x, and if $f(-3) = 0$ and $f(1) = 0$, then $b + c$ equals

[A] -5 [B] -1 [C] 1 [D] 5 [E] 0

11. If $3x^3 - x^2 + 12x - 4 = (x - 2i)(3x - 1)Q(x)$ for all real numbers x, then $Q(x)$ equals

[A] $x - 2$ [B] $x + 2$ [C] $x - 2i$
[D] $x + 2i$ [E] $x + i$

12. A point in space has coordinates $(7, -3, 4)$. How far is the point from the x-axis?

[A] 5 [B] 7 [C] $\sqrt{58}$
[D] $\sqrt{65}$ [E] $\sqrt{74}$

13. If $P(x)$ is a sixth degree polynomial and $P(1) = 3$ and $P(-1) = -2$, what is the value of the remainder when $P(x)$ is divided by $x - 1$?

[A] -2 [B] 2 [C] -3
[D] 3 [E] cannot be determined

14. If f represents an even function, which of the following is also an even function?

I. $g(x) = f(x + 1)$
II. $h(x) = f(x) + 1$
III. $k(x) = f^{-1}(x)$

[A] I only [B] II only
[C] III only [D] II and III
[E] I and III

15. If $f(x) = 4x + 2$ for all x, then $f^{-1}\left(\frac{1}{2}\right) =$

[A] 4 [B] $\frac{1}{16}$ [C] $\frac{1}{4}$ [D] $-\frac{3}{8}$ [E] -4

16. The sides of a triangle are 2 in., 3 in., and 4 in. The cosine of the angle opposite the 3-inch side is

[A] $\frac{2}{3}$ [B] $\frac{11}{16}$ [C] $-\frac{1}{4}$ [D] $\frac{1}{4}$ [E] $\frac{5}{6}$

17. The set of points (x, y) which satisfy $(x - 3)(y + 2) > 0$ lies in quadrant(s)

[A] I & IV only
[B] II only
[C] III & IV only
[D] I, III, & IV only
[E] I, II, III, & IV

18. A particle travels in a circular path at 50 cm/min. If it traverses an arc of $30°$ in 30 seconds, what is the radius of the circular path?

[A] $\frac{9000}{\pi}$ [B] $\frac{150}{\pi}$ [C] $\frac{75}{\pi}$

[D] 50 [E] $\frac{300}{\pi}$

19. If the line passing through the point (a, b) forms a right isosceles triangle with the x and y-axes, the area of the triangle is

[A] $\frac{(a + b)^2}{2}$ [B] $\frac{a^2 - b^2}{2}$

[C] $\frac{a^2 + b^2}{2}$ [D] $\frac{1}{2}ab$

[E] cannot be determined

20. The graph of $y = x + \sin x$ intersects the x-axis n times. n equals

[A] 0 [B] 1 [C] 2
[D] 4 [E] more than 4

21. A cube is inscribed in a sphere and a smaller sphere is inscribed in the cube. What is the ratio of the volume of the small sphere to the volume of the large sphere?

[A] $\frac{1}{2}$ [B] $\frac{1}{3}$ [C] $\frac{\sqrt{3}}{3}$ [D] $\frac{\sqrt{3}}{9}$ [E] $\frac{\sqrt{2}}{2}$

22. If x and y are real numbers and $y = \sqrt{4 - x^2}$, what is the minimum value of y?

[A] 2 [B] 0 [C] -2
[D] -4 [E] $-\infty$

Model Test 3

23 If $\log(\cos\theta)=p$, then $\log(\sec\theta)$ equals

[A] $-p$ [B] $1-p$ [C] $\dfrac{1}{p}$
[D] $-\dfrac{1}{p}$ [E] p

24 A purse contains five different coins (penny, nickel, dime, quarter, half dollar). How many different sums of money can be made using one or more coins?

[A] 32   [B] 10   [C] 5
[D] 120   [E] none of these

25 If the graph at right represents the function f defined on the interval $[-2,2]$ which of the following could represent the graph of $y=f(2x)$?

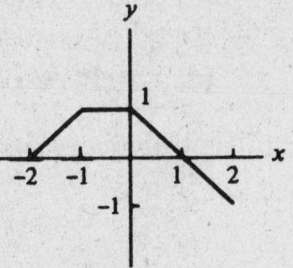

[A]

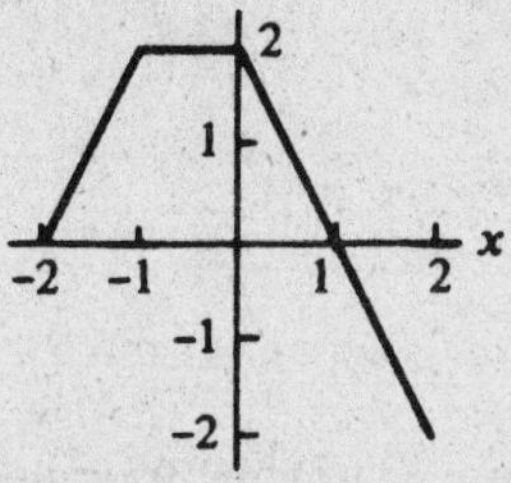

[B]

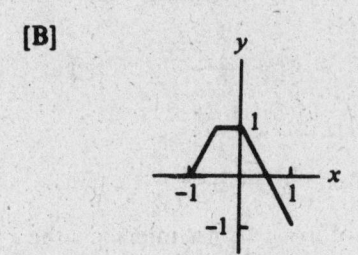

[C]

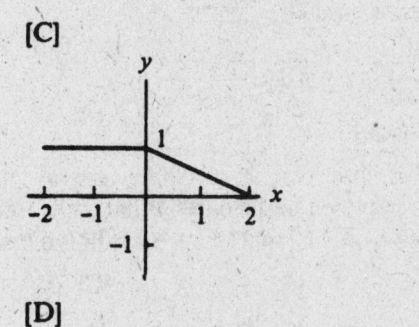

[D]

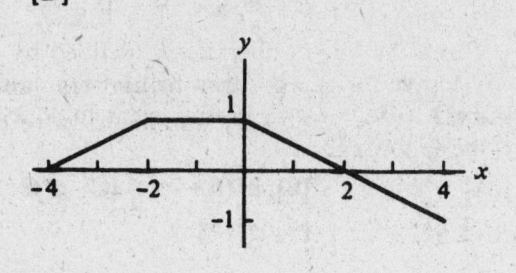

[E]

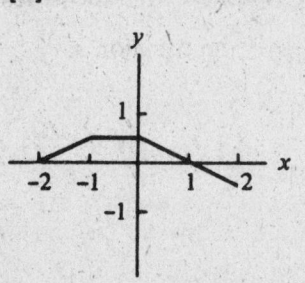

26 An equation in polar form equivalent to $x^2+y^2-4x+2=0$ is

[A] $r=4\cos\theta+2$
[B] $r^2=4\cos\theta$
[C] $4r=\cos\theta$
[D] $r^2-4r\cos\theta+2=0$
[E] $r^2=4\cos\theta+2$

27 If $\begin{cases} x=t^3+9 \\ y=\dfrac{3}{4}t^3+7 \end{cases}$ represents a line, what is the y-intercept?

[A] $\dfrac{3}{4}$ [B] 9   [C] 16
[D] $-\dfrac{27}{4}$ [E] $\dfrac{1}{4}$

28 If θ is the angle between the segment PQ and the x-axis, then $\sin\theta$ equals

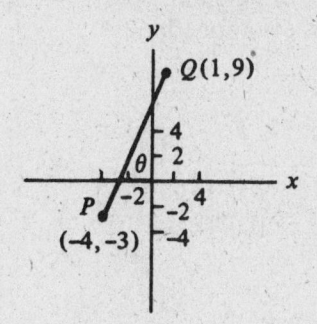

[A] $\dfrac{5}{13}$ [B] $\dfrac{1}{2}$ [C] $\dfrac{5}{12}$
[D] $\dfrac{2\sqrt{5}}{15}$ [E] $\dfrac{12}{13}$

29. If $f(x)=3x-5$ and $g(y)=y^2-1$, then $f(g(z))$ equals

[A] $(3z-5)^2-1$
[B] $3z^2-8$
[C] $3z^3-5z^2-3z+5$
[D] z^2+3z-6
[E] $9z^2-30z-26$

30. If $[x]$ is defined to represent the greatest integer less than or equal to x, and $f(x)=\left|x-[x]-\frac{1}{2}\right|$, the graph of $f(x)$ is discontinuous at

[A] no values of x
[B] all integer values of x
[C] all even integer values of x
[D] all odd integer values of x
[E] all odd multiples of $\frac{1}{2}$

31. A cone is inscribed in a hemisphere of radius r such that the base of the cone coincides with the base of the hemisphere. What is the ratio of the volume of the cone to the volume of the hemisphere?

[A] $\frac{r}{2}$ [B] $\frac{r}{3}$ [C] $\frac{2r}{3}$ [D] $\frac{1}{2}$ [E] $\frac{1}{3}$

32. If (x,y) represents a point on the graph of $y=x+2$, which of the following could be a portion of the graph of the set of points $(\sqrt{x},y)$?

[A]

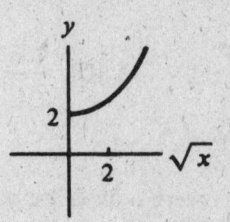

[B]

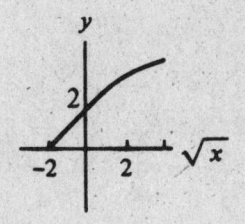

[C]

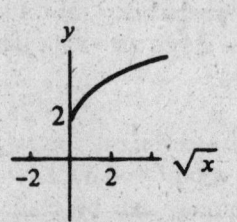

[D]

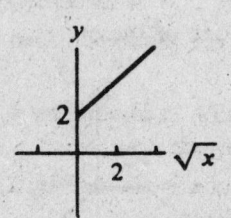

[E]

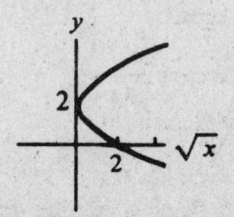

33. If $f(x)=x^7-97x^6-199x^5+99x^4-2x+190$, $f(99)$ equals

[A] 16 [B] 10 [C] 4
[D] -2 [E] -8

34. If $x_0=1$, then $\sum_{i=1}^{n}\left(x_{i-1}+\frac{1}{2}\right)$ equals

[A] $\frac{n+2}{2}$ [B] $\frac{n-1}{2}$
[C] $\frac{n^2-1}{4}$ [D] $\frac{n^2+5n}{4}$
[E] $\frac{n^2+3n}{4}$

35. If $ax^3+bx^2+cx+3=0$ when $x=-1$, what is the value of ax^3-bx^2+cx+3 when $x=1$?

[A] -6 [B] -3 [C] 0
[D] 3 [E] 6

36. Given the binary operation * defined by $a*b=\log_a b$ for all positive numbers a and b, $a\neq 1$, $(a*b)+(a*c)$ is equivalent to which of the following?

[A] $a*b^2$ [B] $a*(b+c)$ [C] $a*(bc)$
[D] $ab*ac$ [E] $a*(b*c)$

37 A basket contains 10 apples, of which 5 are rotten. What is the probability that a person who buys 4 apples will get no rotten ones?

[A] $\dfrac{1}{2}$ [B] $\dfrac{2}{5}$ [C] $\dfrac{2}{25}$
[D] $\dfrac{1}{4032}$ [E] $\dfrac{1}{42}$

38 If h varies as V and inversely as the square of r, which of the following is true?

[A] If r is increased by 2, V is increased by 4.
[B] If both r and h are doubled, then V is doubled.
[C] If r is doubled and h is divided by 4, then V remains unchanged.
[D] If r is doubled and h is divided by 2, then V remains unchanged.
[E] None of these is true.

39 Arcsin .8 + Arccos .8 equals

[A] Arcsin $\dfrac{24}{25}$ [B] Arccos $\dfrac{24}{25}$
[C] Arcsin 1 [D] Arccos 1
[E] Arctan .8

40 Two cards are drawn from a regular deck of 52 cards. What is the probability that the cards are an ace and a ten?

[A] $\dfrac{8}{51}$ [B] $\dfrac{8}{663}$ [C] $\dfrac{1}{13}$
[D] $\dfrac{1}{2500}$ [E] $\dfrac{2}{219}$

41 If the slope of line ℓ_1 is $x+1$, and the slope of line ℓ_2 is $x-1$, and ℓ_1 is perpendicular to ℓ_2, x equals

[A] 0 [B] -1 [C] 1
[D] ± 1 [E] none of these

42 The graph of $P(x) = 3x^5 - 2x^4 + 5x^3 + 7x^2 - 8x + 5$ must cross the x-axis at least r times. The value of r is

[A] 0 [B] 1 [C] 2 [D] 3 [E] 5

43 What is the mean of this set of data: 1, 1, 2, 2, 3, 3, 3, 3?

[A] 2 [B] 3 [C] 2.5
[D] 2.75 [E] 2.25

44 Which of the following functions has an inverse which is also a function?
 I. $y = x^2 - 2x + 4$
 II. $y = |x+1|$
 III. $y = \sqrt{4-9x^2}$

[A] I only [B] II only
[C] III only [D] I, II, & III
[E] none of these

45 If the parabola $ay^2 + by + c = x$ passes through the points $(-4, 17)$, $(5, 11)$, and $(8, 1)$, the value of $a+b+c$ equals

[A] $\dfrac{17}{43}$ [B] $\dfrac{5}{2}$ [C] $\dfrac{14}{27}$
[D] 8 [E] -11

46 The sum of all the numerical coefficients of $(x-y)^{17}$ is

[A] $\dbinom{17}{8}$ [B] $2 \cdot \dbinom{17}{9}$ [C] 1
[D] 0 [E] 17

47 Given that $\sqrt{10} \approx 3.16$, which does $\sqrt{.4}$ equal approximately?

[A] .63 [B] .13 [C] 1.26
[D] .20 [E] .87

48 $2x + 3y - 4 - |2x + 3y - 4| = 0$, for all values of x and y belonging to the set $\{(x,y) : p \text{ is true}\}$. p must be the statement

[A] $2x + 3y > 4$
[B] $2x + 3y = 4$
[C] $2x + 3y < 4$
[D] $2x + 3y \geq 4$
[E] $x = 0$ and $y = 0$

49 The sum of the reciprocals of the roots of $x^3 + ax^2 + bx + c = 0$ is

[A] $\dfrac{a}{b}$ [B] $\dfrac{a}{c}$ [C] $-\dfrac{a}{c}$
[D] $-\dfrac{b}{c}$ [E] $\dfrac{1}{a+c}$

50 If $x^2 - 2px + p^2 - 1 = 0$ were solved for x, the absolute value of the difference between the two roots would be

[A] 0 [B] 1 [C] 2 [D] p [E] p^2

MODEL TEST 4

50 questions
1 hour

Directions: For each of the following problems, decide which is the best of the choices given. Then blacken the corresponding space on the answer sheet on page 139. Answers may be found on page 143, and solutions begin on page 151.

Notes: (1) Figures that accompany problems in this test are intended to provide information useful in solving the problems. They are drawn as accurately as possible EXCEPT when it is stated in a specific problem that its figure is not drawn to scale. All figures lie in a plane unless otherwise indicated.

(2) Unless otherwise specified, the domain of a function f is assumed to be the set of all real numbers x for which $f(x)$ is a real number.

1. The plane whose equation is $3x+4y-5z=60$ intersects the xy-plane in the line whose equation is

  [A] $3x+4y=60$ [B] $x=20$ [C] $y=15$
  [D] $3x-4y=0$ [E] $z=-12$

2. If 5 and $3+\sqrt{2}$ are zeros of the integral polynomial $P(x)=ax^4+bx^3-cx+d$, which of the following must also be a zero?

  I. -5
  II. $3-\sqrt{2}$
  III. 0

  [A] I only   [B] II only
  [C] III only   [D] I and II only
  [E] II and III only

3. If $3x+4=2(y+2)$, the ratio of $y:x$ is

  [A] 2:3 [B] 3:2 [C] 1:1 [D] 7:4
  [E] 4:7

4. The value of $\log_8 16$ is

  [A] $\dfrac{1}{2}$ [B] $\dfrac{3}{4}$ [C] $\dfrac{4}{3}$ [D] 2 [E] 8

5. If $f(x)=3$ and $g(x)=2$, then $f(g(x))=$

  [A] 2 [B] 3 [C] 6 [D] 5 [E] 0

6. In the figure, $r \sin \theta$ equals

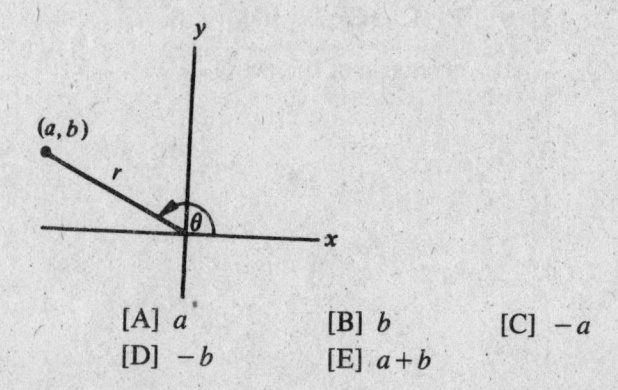

  [A] a [B] b [C] $-a$
  [D] $-b$ [E] $a+b$

7. If the zeros of the function $f(x)$ are 3, -2, and 1, what are the zeros of $f(x-3)$?

   [A] $0, -5, -2$ [B] $6, 1, 4$
   [C] $9, -6, 3$ [D] $-9, 6, -3$
   [E] $1, -\frac{2}{3}, \frac{1}{3}$

8. What is the radius of the sphere whose equation is $x^2+y^2+z^2-4x-6y+2z=0$?

   [A] 0   [B] $\sqrt{14}$ [C] 14   [D] $\sqrt{8}$
   [E] 8

9. If $f(x)=x^2+1$ and $f(g(x))=4x-3$, then $g(x)=$

   [A] $2\sqrt{x}-1$ [B] $2\sqrt{x-1}$ [C] $\sqrt{x-4}$
   [D] $\sqrt{4x+4}$ [E] $\frac{\sqrt{x-1}}{4}$

10. Which of the following is the equation of the perpendicular bisector of the segment with endpoints $(3,7)$ and $(5,-5)$?

    [A] $6x+y=25$ [B] $x-6y=2$
    [C] $x-6y=-2$ [D] $x-y=-3$
    [E] $x-y=3$

11. There are n integers in the solution set of $x(x-2)(x+3)(x+5)<0$. n equals

    [A] 2   [B] 6   [C] 4   [D] 3
    [E] more than 6

12. Which of the following functions is an odd function?

    [A] $f(x)=x^3+1$ [B] $f(x)=\frac{x}{x-1}$
    [C] $f(x)=x^3+x$ [D] $f(x)=2x^4$
    [E] $f(x)=\cos x$

13. If x^2-3x+2 is a factor of $P(x)=3x^4-2x^3+Ax^2+Bx-8$, then $A+B=$

    [A] 3   [B] -19 [C] 26   [D] 7
    [E] -73

14. If $A=e^{kt}$, then $k=$

    [A] $\log_{10} A^{1/t}$ [B] $A^{1/t}$ [C] $\frac{A}{e^t}$
    [D] $\frac{1}{t}\log_e A$ [E] $e\sqrt{A}$

15. If a square prism is inscribed in a right circular cylinder of radius 4 and height 10, what is the total surface area of the prism?

    [A] 192
    [B] 88
    [C] $160\sqrt{2}$
    [D] $160\sqrt{2}+64$
    [E] 320

16. The domain of $f(x)=\frac{x^2-1}{x^2-x}$ is

    [A] all real numbers
    [B] all reals except $x=1$
    [C] all reals except $x=0$
    [D] all reals except $x=-1$
    [E] all reals except $x=0$ or $x=1$

17. If the ratio of $\sin x$ to $\cos x$ is 1 to 2, then the ratio of $\tan x$ to $\cot x$ is

    [A] $1:4$ [B] $1:2$ [C] $1:1$
    [D] $2:1$ [E] $4:1$

18. Given the statement "Only if you work hard will you succeed," if it is known that you "always work hard," then which of the following must be true?

    [A] You always succeed.
    [B] You never succeed.
    [C] You sometimes succeed.
    [D] No conclusion can be drawn.
    [E] None of the above.

19. If the numerical value of the arc length of a sector of a circle is the same as that of the area of the sector, what is the radius of the circle?

    [A] 1   [B] 2   [C] 4
    [D] 25   [E] cannot be determined

20. A point at which two branches of a curve meet and stop and have different tangents is called a *salient* point. Which of the following have salient points?
    I. $y=|x|$
    II. $y=x^{2/3}$
    III. $x^2-y^2=1$

    [A] I only   [B] II only
    [C] III only   [D] I & II only
    [E] II & III only

21. For all positive angles less than 360°, if $\csc(2x+30°)=\cos(3y-15°)$, the sum of x and y is

   [A] 185°   [B] 65°   [C] 35°
   [D] 215°   [E] 95°

22. The rate of growth of a certain organism varies jointly as the warmth of the sun and the square of the available food, and inversely as the number of enemies. If the growth rate remains constant when the warmth of the sun is cut in half and the number of enemies doubles, what can be said about the available food?

   [A] It is doubled.
   [B] It is four times as great.
   [C] It is cut in half.
   [D] It is divided by four.
   [E] It remains the same.

23. p_1 and p_2 are the points of intersection of two circles whose equations are $x^2+y^2=4$ and $(x-2)^2+(y-2)^2=4$. What is the slope of the line perpendicular to the line which passes through p_1 and p_2?

   [A] 0   [B] 1   [C] -1
   [D] 2   [E] undefined

24. $\lim\limits_{x\to a}\dfrac{2x^2-3ax+a^2}{x^2-a^2}$ equals

   [A] $\dfrac{1}{2}$ [B] 0   [C] $\dfrac{3}{2}$
   [D] a [E] undefined

25. If three non-colinear points determine a plane, how many planes are determined by ten points, no three of which are colinear?

   [A] 3   [B] 120   [C] 45
   [D] 90   [E] 720

26. In a race of several people, the probability that Bob will win is $\dfrac{1}{5}$ and the probability that Jim will win is $\dfrac{1}{4}$. What is the probability that either Bob or Jim will win?

   [A] $\dfrac{1}{20}$ [B] $\dfrac{9}{20}$ [C] $\dfrac{3}{5}$
   [D] $\dfrac{7}{20}$ [E] $\dfrac{2}{5}$

27. In a triangle with sides of 3, 5, and 7, the measure of the largest angle is

   [A] 75°   [B] 90°   [C] 120°
   [D] 135° [E] 150°

28. If $f(x)=|x|+2$ and $g(y)=3y-2$ and $h(z)=f(g(z))+g(z)$, the least value of $h(x)$ is

   [A] 0   [B] 2   [C] 4
   [D] -4 [E] a minimum value does not exist

29. The graph of the curve represented by $\begin{cases} x=3\sin\theta \\ y=3\sin\theta \end{cases}$ is

   [A] a line
   [B] a horizontal line segment
   [C] a circle
   [D] a line segment three units long
   [E] a line segment with slope 1

30. In the figure, S is the set of points in the shaded region. Which of the following represents the set T consisting of all points $(x-y,y)$, where (x,y) is a point in S?

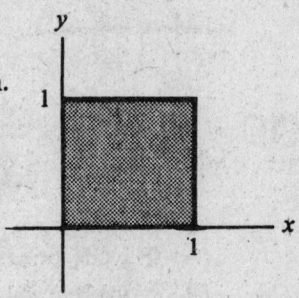

   [A]

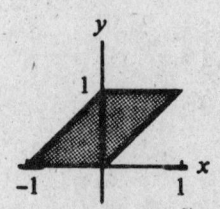

   [B]

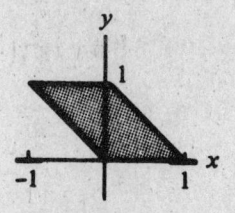

[C]

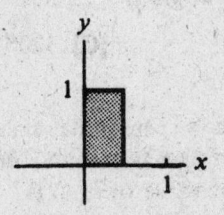

[D]

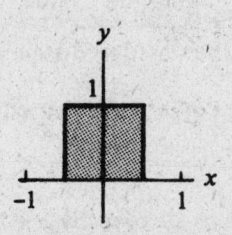

[E]

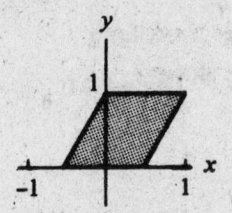

31 If the following steps are followed in order, what sequence of numbers will be printed?
  1. Let $x = A = 2$.
  2. If $x > 5$, stop.
     If $x < 5$, print the value of x.
  3. Replace x by $x + 1$.
  4. Replace A by $2A - 1$.
  5. If $A < x$, print the value of A.
     If $A > x$, go back to step 2.
  6. Stop.

[A] 2,3,4   [B] 2,3,4,5
[C] 2,3,3,4 [D] 2,3
[E] 2,3,3,4,5

32 If $f(x,y) = \tan x + \tan y$ and $g(x,y) = 1 - \tan x \cdot \tan y$, then $\dfrac{f\left(\frac{\pi}{12}, \frac{\pi}{6}\right)}{g\left(\frac{\pi}{12}, \frac{\pi}{6}\right)}$ equals

[A] 0   [B] 1   [C] $\dfrac{\sqrt{3}}{3}$
[D] $\sqrt{3}$ [E] ∞

33 If the graph at right represents the function $f(x)$, which of the following could represent the graph of $y = \dfrac{1}{f(x)}$?

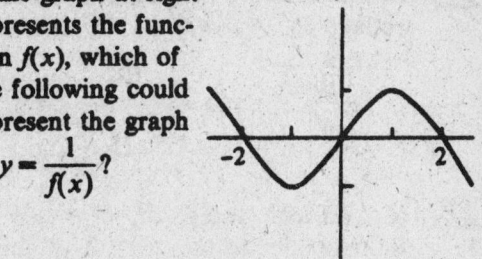

[A]

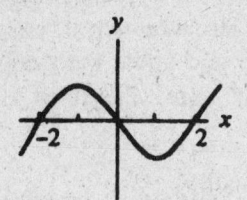

[B]

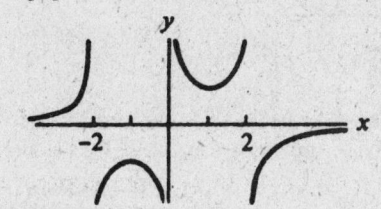

[C]

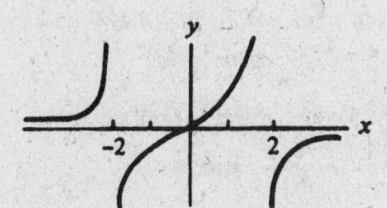

[D]

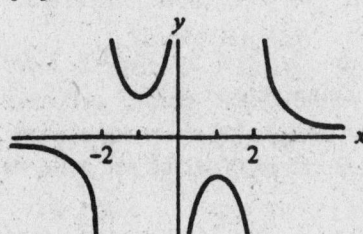

[E]

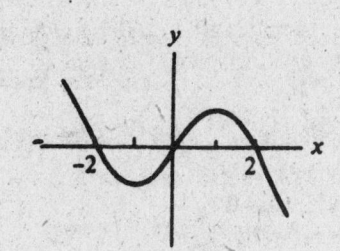

34. If $p(x)=3x^2+4x+1$ and $p(a)=0$, then a equals

[A] 1 only [B] $\frac{1}{3}$ only

[C] -1 only [D] $\frac{1}{3}$ and -1

[E] -1 and $-\frac{1}{3}$

35. What positive values of $x \leq 2\pi$ does $\sin^2 x \cdot \cos^2 x + \sin^2 x + \cos^4 x = 1$?

[A] $\frac{\pi}{2}, \frac{3\pi}{2}$ only [B] $\pi, 2\pi$ only

[C] $\frac{\pi}{4}, \frac{3\pi}{4}, \frac{5\pi}{4}, \frac{7\pi}{4}$ only [D] all values of x

[E] no values of x

36. The shaded area in the figure is represented by which of the following?

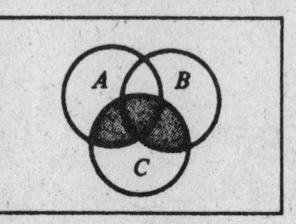

[A] $(A \cap B) \cap C$ [B] $A \cup (B \cup C)$
[C] $(A \cap B) \cup C$ [D] $A \cap (B \cup C)$
[E] $(A \cup B) \cap C$

37. If $\frac{x+3}{x} < 5$, then the solution set is

[A] $x > \frac{3}{4}$ [B] $x \neq \frac{3}{4}$ [C] $x > \frac{4}{3}$

[D] $x < 0$ or $x > \frac{3}{4}$ [E] none of these

38. The function defined by $f(x) = \sqrt{3} \cos x + 3 \sin x$ has an amplitude of

[A] $3 - \sqrt{3}$ [B] $\sqrt{3}$ [C] $2\sqrt{3}$
[D] $3 + \sqrt{3}$ [E] $3\sqrt{3}$

39. What is the equation of the set of points situated in a plane such that the distance between any point and (0,0) is twice the distance between that point and the x-axis?

[A] $3x^2 - y^2 = 0$
[B] $x^2 - 3y^2 = 0$
[C] $x^2 + y^2 - 2y = 0$
[D] $x^2 + y^2 - 2x = 0$
[E] $4x^2 + 3y^2 = 0$

40. Given the four equations, where $a \neq b \neq 0$:
I. $ax + by = c$
II. $ax - by = c$
III. $-ax + by = c$
IV. $bx - ay = c$.
Which pair represents perpendicular lines?

[A] I & IV [B] II & IV
[C] II & III [D] I & II
[E] III & IV

41. The lines $4x - 3y - 7 = 0$ and $8x - 6y + 11 = 0$ are parallel. The perpendicular distance between them is

[A] 2.5 [B] 2.4 [C] 4
[D] 2.7 [E] 3

42. When $5x^{13} + 3x^{10} - K$ is divided by $x+1$, the remainder is 20. The value of K is

[A] 8 [B] 14 [C] -12
[D] -22 [E] 28

43. If the equation $x^3 + 9x^2 - ax - b = 0$ has three equal roots, then

[A] $a = 27$
[B] $ab = 0$
[C] $b = 27$
[D] each root equals 3
[E] each root equals -3

44. Which of the following is the largest interval over which the graph of $f(x) = \sin 2x$ is always above the graph of $g(x) = \tan x$?

[A] $\left(0, \frac{\pi}{6}\right)$ [B] $\left(0, \frac{\pi}{4}\right)$
[C] $\left(0, \frac{\pi}{3}\right)$ [D] $\left(0, \frac{\pi}{2}\right)$

[E] the graph of $f(x)$ is not above the graph of $g(x)$ on any of these intervals

45. $\tan \frac{A}{2} + \cot \frac{A}{2}$ equals

[A] $2 \sin A$ [B] $2 \cos A$
[C] $2 \sec A$ [D] $2 \csc A$
[E] none of these

46. If $f(x) = \frac{x^2 - 1}{x + 1}$, what does $f(i)$ equal, where $i = \sqrt{-1}$?

[A] 0 [B] $\frac{2}{i+1}$ [C] $i - 1$
[D] -2 [E] $i + 1$

47 If $x, 3x+3, 5x+5$ are three consecutive terms of a geometric sequence, the sum of these three terms is

[A] $-\dfrac{9}{4}$ [B] $-\dfrac{49}{4}$ [C] $-\dfrac{59}{4}$

[D] $-\dfrac{19}{4}$ [E] -10

48 In order for the inverse of $f(x) = 2\sin x$ to be a function, the domain of f must be limited to

[A] $0° \leq x \leq 180°$
[B] $90° \leq x \leq 270°$
[C] $135° \leq x \leq 315°$
[D] $45° \leq x \leq 135°$
[E] $180° \leq x \leq 360°$

49 $(2 \operatorname{cis} 50°)^3$ written in rectangular form is

[A] $4\sqrt{3} + 4i$ [B] $4 - 4\sqrt{3}\,i$
[C] $4\sqrt{3} - 4i$ [D] $-4\sqrt{3} + 4i$
[E] $4 + 4\sqrt{3}\,i$

50 Mary, John, and Fran are on a trip. Mary drives during the first hour at an average speed of 50 mi/h. John drives during the next 2 hours at an average speed of 48 mi/h. Fran drives for the next 3 hours at an average speed of 52 mi/h. They reach their destination after exactly 6 hours. Their mean speed (in miles per hour) was

[A] 52   [B] 50   [C] $51\dfrac{1}{3}$ [D] $50\dfrac{1}{3}$ [E] $50\dfrac{2}{3}$

MODEL TEST 5

50 questions
1 hour

<u>Directions:</u> For each of the following problems, decide which is the best of the choices given. Then blacken the corresponding space on the answer sheet on page 139. Answers may be found on page 143, and solutions begin on page 153.

<u>Notes:</u> (1) Figures that accompany problems in this test are intended to provide information useful in solving the problems. They are drawn as accurately as possible EXCEPT when it is stated in a specific problem that its figure is not drawn to scale. All figures lie in a plane unless otherwise indicated.

(2) Unless otherwise specified, the domain of a function f is assumed to be the set of all real numbers x for which $f(x)$ is a real number.

1. The amplitude of the function $f(x) = -\sin x \cdot \cos x$ is

   [A] 1 [B] 2 [C] $\frac{1}{2}$ [D] $-\frac{1}{2}$ [E] -1

2. The graph of $y^4 - 3x^2 + 7 = 0$ is symmetric with respect to which of the following:

   I. the x-axis
   II. the y-axis
   III. the origin

   [A] I only   [B] II only
   [C] III only   [D] I and II only
   [E] I, II, and III

3. The negation of the statement "For all sets, there is one subset" is

   [A] for all sets, there is not one subset
   [B] for no sets, there is one subset
   [C] for some sets, there is not one subset
   [D] for some sets, there is one subset
   [E] for no sets, there is not one subset

4. In a group of 30 students, 20 take French, 15 take Spanish, and 5 take neither. How many students take both French and Spanish?

   [A] 0 [B] 5 [C] 10 [D] 15 [E] 20

5. The domain of $f(x) = \log_{10}(\sin x)$ contains which of the following intervals?

   [A] $0 \leq x \leq \pi$ [B] $-\frac{\pi}{2} \leq x \leq \frac{\pi}{2}$
   [C] $0 < x < \pi$ [D] $-\frac{\pi}{2} < x < \frac{\pi}{2}$
   [E] $\frac{\pi}{2} < x < \frac{3\pi}{2}$

6. Which of the following is the ratio of the surface area of a sphere with radius r to its volume?

[A] $\dfrac{4}{\pi}$ [B] $\dfrac{r}{\pi}$ [C] $\dfrac{3}{r}$ [D] $\dfrac{r}{4}$ [E] $\dfrac{4}{r}$

7. The point (3,2) lies on the graph of the inverse of $f(x) = 2x^3 + x + A$. The value of A is

[A] 15 [B] −15 [C] 18 [D] 54 [E] −54

8. If $\tan x = 3$, the numerical value of $\dfrac{\sin^2 x}{\cos^2 x}$ is

[A] 3 [B] 9 [C] $\sqrt{3}$ [D] 6 [E] 1

9. If $f(x) = \dfrac{x+1}{x^2+1}$ and $g(x) = \dfrac{x^2+1}{x+1}$, find the equation of the line that passes through the points $(0, g(0))$ and $(a, f(a))$ where $f(a) = 0$.

[A] $x = 0$ [B] $y = -1$
[C] $x + y = -1$ [D] $x + y = 1$
[E] $x - y = -1$

10. If $f(x) = ax^2 + bx + c$ and $f(1) = 3$ and $f(-1) = 3$, then $a + c$ equals

[A] −3 [B] 0 [C] 3 [D] 6 [E] 2

11. As x increases from $-\dfrac{\pi}{4}$ to $\dfrac{3\pi}{4}$, the value of $\cos x$

[A] always increases
[B] always decreases
[C] increases then decreases
[D] decreases then increases
[E] none of these

12. The value of $\cos 10° \cdot \cos 20° - \sin 10° \cdot \sin 20°$ is

[A] $\dfrac{\sqrt{3}-1}{2}$ [B] $\cos 10°$
[C] $\dfrac{1}{2}$ [D] $\dfrac{\sqrt{3}}{2}$
[E] cannot be determined without tables

13. Solve the equation $\sin x + \cos x = 0$ for all positive values of $x < 360°$.

[A] 45° and 135°
[B] 135° and 315°
[C] 45° and 225°
[D] 225° and 315°
[E] 45°, 135°, 225°, and 315°

14. If $f(x) = x^3 + x + k$ has exactly one real zero, then which of the following is true?

[A] k must be negative
[B] k must be zero
[C] k must be positive
[D] k must be 0
[E] k can be any real number

15. If $f(x) = x + 3$ and $g(x) = x^2$, for what value(s) of x does $f(g(x)) = g(f(x))$?

[A] 0 [B] −1 [C] 1 [D] 3 [E] ±1

16. The primary period of the function $f(x) = 2 \cdot \cos^2 3x$ is

[A] 2π [B] $\dfrac{\pi}{3}$ [C] π [D] $\dfrac{\pi}{2}$ [E] 3π

17. If five coins are flipped and all the different ways they could fall are listed, how many elements of this list contain more than three heads?

[A] 16 [B] 10 [C] 5
[D] 6 [E] 32

18. What is the remainder when $3x^4 - 2x^3 - 20x^2 - 12$ is divided by $x + 2$?

[A] −4 [B] −28 [C] −6
[D] −36 [E] −60

19. The graph of the curve represented by $\begin{cases} x = \sec\theta \\ y = \cos\theta \end{cases}$ is

[A] a line
[B] a hyperbola
[C] an ellipse
[D] a line segment
[E] a portion of a hyperbola

20. $\text{Log}_2 \left(\cos \dfrac{\pi}{3} \right)$ equals

[A] $\dfrac{\sqrt{3}}{2}$ [B] $\dfrac{1}{2}$ [C] −1
[D] 1 [E] $\dfrac{\pi}{3}$

21. The lines $3x - 4y + 8 = 0$ and $8x + 6y - 4 = 0$ intersect at point P. One angle formed at point P contains

[A] 30° [B] 45°
[C] 60° [D] 90°
[E] point P does not exist because the lines are parallel

22 In $\triangle ABC$, $\angle B = 42°$, $\angle C = 30°$, and $AB = 100$. The length of BC is

[A] $50 \sin 72°$
[B] $100 \sin 42°$
[C] $200 \sin 72°$
[D] $200 \sin 42°$
[E] none of these

23 If the two solutions of $x^2 - 9x + c = 0$ are complex conjugates, which of the following describes all possible values of c?

[A] $c = 0$ [B] $c \neq 0$ [C] $c > \dfrac{81}{4}$
[D] $c > 9$ [E] $c < 9$

24 In the figure, the graph of $f(x)$ has two transformations performed on it. First it is rotated $180°$ about the origin, then it is reflected about the x-axis. Which of the following is the equation of the resulting curve?

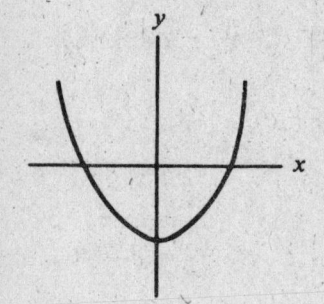

[A] $y = -f(x)$
[B] $y = f(x+2)$
[C] $x = f(y)$
[D] $y = f(x)$
[E] none of these

25 If (x,y) represents a point on the graph of $y = 2x + 1$, which of the following could be a portion of the graph of the set of points (x^2, y)?

[A]

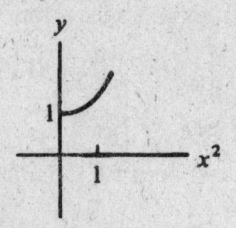

[B]

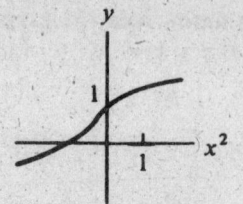

[C]

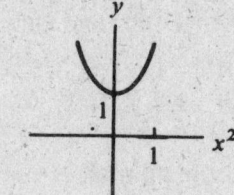

[D]

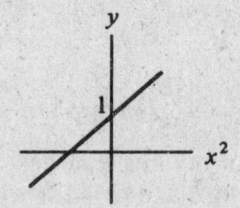

[E]

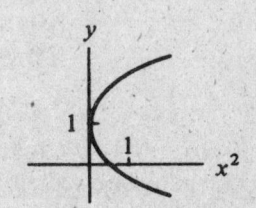

26 A central angle of two concentric circles is $\dfrac{3\pi}{4}$. The area of the large sector is twice the area of the small sector. What is the ratio of the radii of the two circles?

[A] $\dfrac{1}{2}$ [B] $\dfrac{\sqrt{2}}{2}$ [C] $\dfrac{1}{4}$
[D] 1 [E] cannot be determined

27 $\displaystyle\lim_{x \to \infty} \dfrac{3x^3 - 7x^2 + 2}{4x^2 - 3x - 1}$ equals

[A] $\dfrac{3}{4}$ [B] 0 [C] 3 [D] 1 [E] ∞

28 In the figure, the bases, ABC and DEF, of the right prism are equilateral triangles of side s. The altitude of the prism BE is h. If a plane cuts the figure through the points A, C, and E, two solids, $E\text{-}ABC$ and $E\text{-}ACFD$, are formed. What is the ratio of the volume of $E\text{-}ABC$ to the volume of $E\text{-}ACFD$?

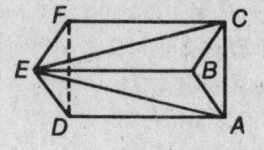

[A] $\dfrac{1}{2}$ [B] $\dfrac{1}{3}$ [C] $\dfrac{\sqrt{3}}{3}$ [D] $\dfrac{\sqrt{3}}{4}$ [E] $\dfrac{1}{4}$

29. How many times does the graph of x^3+x^2+x+1 cross the x-axis?

   [A] 0   [B] 1   [C] 2
   [D] 3   [E] cannot be determined

30. The vertical distance between the minimum and maximum value of the function $y=|2\sin 3x|$ is

   [A] 2   [B] 4   [C] 3   [D] 6   [E] $\frac{2\pi}{3}$

31. Two roots of $x^3+3x^2+Kx-12=0$ are unequal but have the same absolute value. The value of K is

   [A] 4   [B] -4 [C] 6
   [D] -6 [E] -9

32. If n is an integer, what is the remainder when $3x^{2n+3}-4x^{2n+2}+5x^{2n+1}-8$ is divided by $x+1$?

   [A] -4 [B] -10 [C] 0
   [D] -20 [E] cannot be determined

33. Four men, A, B, C, and D, line up in a row. What is the probability that man A is at either end of the row?

   [A] $\frac{1}{2}$ [B] $\frac{1}{3}$ [C] $\frac{1}{4}$ [D] $\frac{1}{6}$ [E] $\frac{1}{12}$

34. $\sum_{n=1}^{\infty} \log\left(\frac{1}{n}\right)$ equals

   [A] 0   [B] 1   [C] $-\infty$
   [D] ∞ [E] cannot be determined

35. In $a+bi$ form, the reciprocal of $2+6i$ is

   [A] $-\frac{1}{16}+\frac{3}{16}i$
   [B] $\frac{1}{16}+\frac{3}{16}i$
   [C] $\frac{1}{20}-\frac{3}{20}i$
   [D] $\frac{1}{20}+\frac{3}{20}i$
   [E] none of these

36. If the domain of $f(x)=-|x|+2$ is $\{x: -1 < x < 3\}$, $f(x)$ has a minimum value when x equals

   [A] 0   [B] -1 [C] 1
   [D] 3   [E] there is no minimum value

37. If the region bounded by the lines $y=-\frac{4}{3}x+4$, $x=0$, and $y=0$ is rotated about the y-axis, the volume of the figure formed is

   [A] 6π [B] 12π [C] 18π
   [D] 27π [E] 36π

38. The plane whose equation is $2x+3y+5z=30$ forms a pyramid in the first octant with the coordinate planes. Its volume is

   [A] 120   [B] 150   [C] 180
   [D] 240   [E] 900

39. If there are known to be 4 broken transistors in a box of 12, and 3 transistors are drawn at random, what is the probability that all 3 are not broken?

   [A] $\frac{3}{8}$ [B] $\frac{14}{55}$ [C] $\frac{1}{4}$
   [D] $\frac{3}{4}$ [E] $\frac{5}{9}$

40. In order for the inverse of $f(x)=\sin 2x$ to be a function, the domain of f can be limited to

   [A] $-\frac{\pi}{2} \leq x \leq \frac{\pi}{2}$
   [B] $0 \leq x \leq \frac{\pi}{2}$
   [C] $\frac{\pi}{4} \leq x \leq \frac{3\pi}{4}$
   [D] $\frac{\pi}{2} \leq x \leq \pi$
   [E] $0 \leq x \leq \pi$

41. If $f(x,y)=\dfrac{\log x}{\log y}$, $f(4,2)$ equals

   [A] 0   [B] $\frac{1}{2}$ [C] 1   [D] 2   [E] $\log 2$

42. If $3x-x^2 \geq 2$ and $y^2+y \leq 2$, then

   [A] $-1 \leq xy \leq 2$
   [B] $-2 \leq xy \leq 2$
   [C] $-4 \leq xy \leq 4$
   [D] $-4 \leq xy \leq 2$
   [E] 1, 2, and 4 only

43. In the $\triangle ABC$, if $\sin A=\frac{1}{3}$ and $\sin B=\frac{1}{4}$, $\sin C$ equals

   [A] $\frac{1}{7}$ [B] $\frac{7}{12}$
   [C] $\dfrac{2\sqrt{2}+\sqrt{15}}{12}$ [D] $\dfrac{2\sqrt{2}-\sqrt{15}}{12}$
   [E] $\frac{5}{12}$

44 The solution set of $\frac{|x-1|}{x} > 2$ is

[A] $0 < x < \frac{1}{3}$ [B] $x < \frac{1}{3}$
[C] $x > \frac{1}{3}$ [D] $\frac{1}{3} < x < 1$
[E] $x > 0$

45 A positive rational root of the equation $4x^3 - x^2 + 16x - 4 = 0$ is

[A] $\frac{1}{2}$ [B] 1 [C] $\frac{1}{4}$ [D] 2 [E] $\frac{3}{4}$

46 R varies as the square of z and inversely as the cube of T. If z is tripled and T is doubled, the value of R is

[A] multiplied by three
[B] multiplied by $\frac{9}{8}$
[C] multiplied by 8
[D] divided by 3
[E] divided by $\frac{2}{3}$

47 If $x + 3y = 6$ and $2x - y = 3$, then $\frac{x}{y}$ equals

[A] $\frac{5}{3}$ [B] $\frac{3}{7}$ [C] $\frac{3}{5}$ [D] $\frac{15}{7}$ [E] $\frac{9}{7}$

48 Given the set of data 1, 1, 2, 2, 2, 3, 3, x, y, where x and y represent two different integers. If the mode is 2, which of the following statements must be true?

[A] if $x = 1$ or 3, then y must $= 2$.
[B] Both x and y must be > 3. [C] x or y must be 2.
[D] It does not matter what values x and y have.
[E] x or y must $= 3$ and the other must $= 1$.

49 If a coordinate system is devised so that the positive y-axis makes an angle of 60° with the positive x-axis, what is the distance between the points with coordinates $(4, -3)$ and $(5, 1)$?

[A] $\sqrt{17}$ [B] $\sqrt{21}$ [C] $\sqrt{15}$
[D] $\sqrt{13}$ [E] $\sqrt{51}$

50 The norm of the vector $\vec{V} = 3\vec{i} - \sqrt{2}\,\vec{j}$ is

[A] $3\sqrt{2}$ [B] $\sqrt{6}$ [C] $\sqrt{13}$
[D] $\sqrt{11}$ [E] $3 - \sqrt{2}$

MODEL TEST 6

50 questions
1 hour

Directions: For each of the following problems, decide which is the best of the choices given. Then blacken the corresponding space on the answer sheet on page 140. Answers may be found on page 143, and solutions begin on page 156.

Notes: (1) Figures that accompany problems in this test are intended to provide information useful in solving the problems. They are drawn as accurately as possible EXCEPT when it is stated in a specific problem that its figure is not drawn to scale. All figures lie in a plane unless otherwise indicated.

(2) Unless otherwise specified, the domain of a function f is assumed to be the set of all real numbers x for which $f(x)$ is a real number.

1. If the point (a,b) lies on the graph of the function, f, which of the following points must lie on the graph of the inverse of f?

   [A] (a,b) [B] $(-a,b)$ [C] $(a,-b)$
   [D] (b,a) [E] $(-b,-a)$

2. Harry had grades of 70, 80, 85, and 80 on his quizzes. If all quizzes have the same weight, what grade must he get on his next quiz so that his average will be 80?

   [A] 85   [B] 90   [C] 95
   [D] 100   [E] more than 100

3. The line $4x+3y=12$ forms a triangle with the x- and y-axes. If this triangle is rotated about the y-axis, a right circular cone is formed. The volume of this cone is

   [A] 36π [B] 12π [C] 48π [D] 16π
   [E] 64π

4. A trace of the plane $5x-2y+3z=10$ is

   [A] $5x+2y=10$ [B] $3z=2y$
   [C] $2y+3z=10$ [D] $5x+3z=10$
   [E] $2y=5x+10$

5. The sum of the roots of $3x^3+4x^2-4x=0$ is

   [A] $\frac{4}{3}$ [B] 0   [C] $-\frac{4}{3}$ [D] 4   [E] $-\frac{3}{4}$

6. The graph of $|y-1|=|x+1|$ forms an X. The two branches of the X intersect at a point whose coordinates are

   [A] $(1,1)$ [B] $(-1,1)$ [C] $(1,-1)$
   [D] $(-1,-1)$ [E] $(0,0)$

7. The graph of $y=(x+2)(2x-3)$ can be expressed as a set of parametric equations. If $x=2t-2$ and $y=f(t)$, then $f(t)$ equals

   [A] $2t(4t-5)$ [B] $(2t-2)(4t-7)$
   [C] $2t(4t-7)$ [D] $(2t-2)(4t-5)$
   [E] $2t(4t+1)$

121

8. For what values of x is $\dfrac{1-\cos x}{\sin x} = \dfrac{\sin x}{1+\cos x}$?

   [A] all values of x
   [B] no value of x
   [C] multiples of $\dfrac{\pi}{3}$ only
   [D] multiples of $\dfrac{\pi}{6}, \dfrac{\pi}{4}, \dfrac{\pi}{3}$ only
   [E] all values of x except multiples of π

9. If $\sin x \cdot \cos \dfrac{7\pi}{4} \cdot \sec 135° \cdot \tan \dfrac{2\pi}{3} \cdot \sin 330° = \dfrac{\sqrt{3}}{2}$, x equals

   [A] 60° or 120°
   [B] 240° or 300°
   [C] $\dfrac{\pi}{2}$
   [D] 270°
   [E] $\dfrac{2\pi}{3}$ or $\dfrac{4\pi}{3}$

10. If $x-1$ is a factor of x^2+ax-4, then a has the value

    [A] 4
    [B] 3
    [C] 2
    [D] 1
    [E] none of these

11. The negation of the statement "If a dog is hungry, then he will howl" is

    [A] if a dog is not hungry, then he will not howl
    [B] if a dog does not howl, then he is not hungry
    [C] the dog is hungry and he does not howl
    [D] the dog is not hungry and he howls
    [E] none of these is the negation

12. If $i = \sqrt{-1}$, and n is a positive integer, which of the following statements is FALSE?

    [A] $i^{4n} = 1$
    [B] $i^{4n+1} = -i$
    [C] $i^{4n+2} = -1$
    [D] $i^{n+4} = i^n$
    [E] $i^{4n+3} = -i$

13. $\log_x 2 = \log_2 x$ is satisfied by two values of x. Their sum equals

    [A] 0 [B] 1 [C] 2 [D] $\dfrac{5}{2}$ [E] 4

14. In the expansion of $(2b^2 - 3b^{-3})^n$, if the fifth term does not contain a factor of b, what is the value of n?

    [A] 6 [B] 10 [C] $\dfrac{15}{2}$ [D] $\dfrac{25}{2}$ [E] 9

15. If $\sin x = \tan x$, then $|2 \cos x|$ equals

    [A] -2 [B] -1 [C] 2
    [D] 1 [E] none of these

16. The period of the graph of the function $y = |2 \sin 3x|$ is

    [A] π [B] $\dfrac{\pi}{3}$ [C] $\dfrac{2\pi}{3}$ [D] $\dfrac{\pi}{6}$ [E] 3π

17. If $f(x,y) = 2x^2 - y^2$ and $g(x) = 2^x$, which one of the following is equal to 2^{2x}?

    [A] $f(x, g(x))$
    [B] $f(g(x), x)$
    [C] $f(g(x), g(x))$
    [D] $f(g(x), 0)$
    [E] $g(f(x,x))$

18. The angle between the lines $\sqrt{3}y = 3x + 4$ and $2x + 2y = 2$ could be

    [A] 30° [B] 45° [C] 90° [D] 60°
    [E] 75°

19. Two positive numbers, a and b, are in the sequence 4, a, b, 12. The first three numbers form a geometric sequence, and the last three numbers form an arithmetic sequence. The difference $b - a$ equals

    [A] 1 [B] $1\dfrac{1}{2}$ [C] 2 [D] $2\dfrac{1}{2}$ [E] 3

20. If $\dfrac{\cos x + 2 \sin \dfrac{\pi}{6}}{\sin \dfrac{5\pi}{3} - \cos 180°} = 1$ and $0 \leq x \leq \pi$, then x equals

    [A] $\dfrac{\pi}{6}$ [B] $\dfrac{\pi}{3}$ [C] $\dfrac{2\pi}{3}$ [D] $\dfrac{5\pi}{6}$ [E] 1

21. If the points $(1, y_1)$ and $(-1, y_2)$ lie on the graph of $y = x^3 + ax^2 + bx + c$, and $y_1 - y_2 = 3$, then b equals

    [A] 1
    [B] $\dfrac{1}{2}$
    [C] $\dfrac{a+c}{2}$
    [D] $\dfrac{a-c}{2}$
    [E] $-\dfrac{1}{2}$

22. Which one of the following is NOT a fifth root of 1?

    [A] $1(\cos 0 + i \cdot \sin 0)$
    [B] $1(\cos 72° + i \cdot \sin 72°)$
    [C] $1(\cos 154° + i \cdot \sin 154°)$
    [D] $1(\cos 216° + i \cdot \sin 216°)$
    [E] $1(\cos 288° + i \cdot \sin 288°)$

23. If a and b are real numbers, $a > b$, and $|a| < |b|$ then

   [A] $a > 0$ [B] $a < 0$ [C] $b > 0$
   [D] $b < 0$ [E] none of these

24. If $[x]$ is defined to represent the greatest integer less than or equal to x, and $f(x) = |x - [x] - \frac{1}{2}|$, the maximum value of $f(x)$ is

   [A] 1   [B] 2   [C] 0   [D] $\frac{1}{2}$ [E] -1

25. $\lim_{x \to 2} \dfrac{x^3 - 8}{x^2 - 4}$ equals

   [A] 0   [B] 2   [C] 3   [D] 1   [E] ∞

26. A right circular cone whose base radius is 4 is inscribed in a sphere of radius 5. What is the ratio of the volume of the cone to the volume of the sphere?

   [A] $\dfrac{1}{3}$ [B] $\dfrac{108}{125}$ [C] $\dfrac{24}{125}$
   [D] $\dfrac{36}{125}$ [E] $\dfrac{32}{125}$

27. Which of the following is an asymptote of $f(x) = \dfrac{x^2 + 3x + 2}{x + 2} \cdot \tan \pi x$?

   [A] $x = 2$ [B] $x = 1$ [C] $x = -2$
   [D] $x = -1$ [E] $x = \dfrac{1}{2}$

28. $\sum_{k=1}^{250} \left(\dfrac{1}{k+1} - \dfrac{1}{k} \right)$ equals

   [A] $\dfrac{1}{251}$ [B] $\dfrac{250}{251}$ [C] $-\dfrac{250}{251}$
   [D] $\dfrac{252}{251}$ [E] $-\dfrac{252}{251}$

29. If $f(x) = x - \dfrac{1}{x}$, then $f(a) + f\left(\dfrac{1}{a}\right)$ equals

   [A] 0   [B] $2a - \dfrac{2}{a}$ [C] $a - \dfrac{1}{a}$
   [D] $\dfrac{a^4 - a^2 + 1}{a(a^2 - 1)}$ [E] 1

30. For what positive values of $x \leq 2\pi$ does $\sin^2 x + \tan^2 x + \cos^2 x = \sec^2 x$?

   [A] $\dfrac{\pi}{2}, \dfrac{3\pi}{2}$ only
   [B] $\pi, 2\pi$ only
   [C] $\dfrac{\pi}{4}, \dfrac{\pi}{2}, \dfrac{3\pi}{4}, \pi, \dfrac{5\pi}{4}, \dfrac{3\pi}{2}, \dfrac{7\pi}{4}, 2\pi$ only
   [D] all values of x for which the functions are defined
   [E] no values of x

31. If i is a root of $x^4 + 2x^3 - 3x^2 + 2x - 4 = 0$, the product of the real roots is

   [A] 0   [B] -2 [C] 2   [D] 4   [E] -4

32. The graph of $P(x) = x^4 - 4x^3 + 6x^2 - 4x + 2$ crosses the x-axis in how many points?

   [A] 0   [B] 1   [C] 2   [D] 4
   [E] cannot be determined

33. If (x, y) represents a point on the graph of $y = x^2 + 1$, which of the following could be a portion of the graph of the set of points (x^2, y)?

[A]

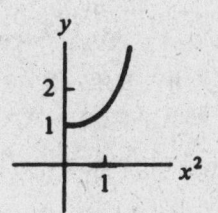

[B]

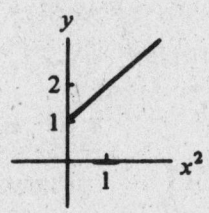

[C]

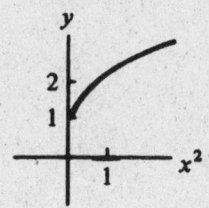

[D]

(graph showing curve opening right with vertex near (1,1), y-axis labeled y, x-axis labeled x^2)

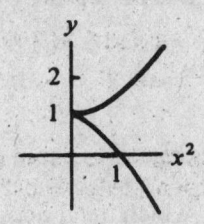

[E]

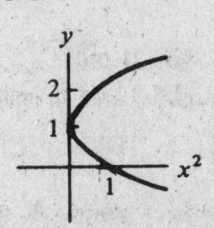

(graph showing curve opening right with vertex near (1,1), y-axis labeled y, x-axis labeled x^2)

34 The binary operation * is defined over the set of non-negative real numbers to be $a*b = \sqrt{ab}$. Which of the following is true?

[A] 1 is the identity element.
[B] Every number is its own inverse.
[C] The associative property holds.
[D] Zero is the identity element.
[E] None of these is true.

35 If the center of the circle $x^2 + y^2 + ax + by + 2 = 0$ is the point $(4, -8)$, $a + b$ equals

[A] -4 [B] 4 [C] 8 [D] -8 [E] 24

36 If $p(x) = 3x^2 + 4x + 1$ and $p(a) = 0$, then a equals

[A] 1 only
[B] $\frac{1}{3}$ only
[C] -1 only
[D] $\frac{1}{3}$ and -1
[E] -1 and $-\frac{1}{3}$

37 What positive values of $x \leq 2\pi$ does $\sin^2 x \cdot \cos^2 x + \sin^2 x + \cos^4 x = 1$?

[A] $\frac{\pi}{2}, \frac{3\pi}{2}$ only
[B] $\pi, 2\pi$ only
[C] $\frac{\pi}{4}, \frac{3\pi}{4}, \frac{5\pi}{4}, \frac{7\pi}{4}$ only
[D] all values of x
[E] no values of x

38 For each positive integer n, let $S_n =$ the sum of all positive integers less than or equal to n. S_{51} equals

[A] 50   [B] 1326   [C] 1275
[D] 1250   [E] 51

39 If the graphs of $3x^2 + 4y^2 - 6x + 8y - 5 = 0$ and $(x-2)^2 = 4(y+2)$ are drawn on the same coordinate system, in how many points do they intersect?

[A] 0   [B] 1   [C] 2   [D] 3   [E] 4

40 If $f(x) \geq 0$ for all x, then $f(2-x)$ is

[A] ≥ 0 [B] ≥ 2 [C] ≥ -2
[D] ≤ 2 [E] ≤ 0

41 If x varies directly as t and t varies inversely as the square of y, what is the relationship between x and y?

[A] x varies as y^2
[B] x varies inversely as y^2
[C] y varies as x^2
[D] y varies inversely as x^2
[E] no variation between x and y can be determined

42 Sin (2 Arctan 3) equals

[A] $\frac{\sqrt{10}}{6}$ [B] $\frac{3}{5}$ [C] $\frac{4}{5}$
[D] $\frac{\sqrt{10}}{5}$ [E] $\frac{\sqrt{3}}{2}$

43 How many four-digit numbers can be formed from the numbers 0, 2, 4, 8, if no digit is repeated?

[A] 24   [B] 18   [C] 64
[D] 36   [E] 27

44 If 10 coins are to be flipped, and the first five all come up heads, what is the probability that exactly 3 more heads will be flipped?

[A] $\frac{5}{16}$ [B] $\frac{45}{1024}$ [C] $\frac{3}{5}$ [D] $\frac{15}{128}$
[E] $\frac{1}{8}$

45 If $f(x)=x^2+3x-4$ and $g(x)=x-4$, then $f(g(x))=$

[A] x^2+3x [B] x^2+3x-8
[C] $x^2-8x+16$ [D] $x^2-11x+32$
[E] x^2-5x

46 If $f(x)=4x^2$ and $g(x)=f(\sin x)+f(\cos x)$, then g is

[A] the zero function
[B] a non-zero constant function
[C] the inverse of f
[D] an odd function
[E] a periodic function

47 $ABCD$ is a square. M is the point one-third of the way from B to C. N is the point one-third of the way from D to C. $\cos\theta=$

[A] $\dfrac{4}{5}$ [B] $\dfrac{3}{5}$ [C] $\dfrac{\sqrt{3}}{2}$
[D] $\dfrac{\sqrt{10}}{5}$ [E] $\dfrac{1}{2}$

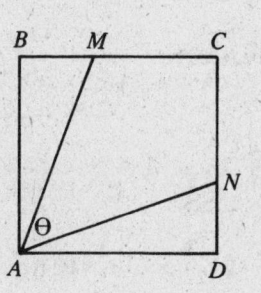

48 A real zero of the polynomial $y=2x^3-3x^2-22x-5$ lies between

[A] -2 and -1 [B] -1 and 0 [C] 0 and 1
[D] 1 and 2 [E] 2 and 3

49 Which of the following equations have graphs consisting of two perpendicular lines?

I. $xy=0$
II. $|y|=|x|$
III. $|xy|=1$

[A] I only [B] II only
[C] III only [D] I and II only
[E] I, II, and III

50 A line, m, is parallel to a plane, X, and 6 inches from X. The set of points which are 6 inches from m and 1 inch from X form

[A] a line parallel to m
[B] 2 lines parallel to m
[C] 4 lines parallel to m
[D] one point
[E] the empty set

MODEL TEST 7

50 questions
1 hour

Directions: For each of the following problems, decide which is the best of the choices given. Then blacken the corresponding space on the answer sheet on page 140. Answers may be found on page 143, and solutions begin on page 159.

Notes: (1) Figures that accompany problems in this test are intended to provide information useful in solving the problems. They are drawn as accurately as possible EXCEPT when it is stated in a specific problem that its figure is not drawn to scale. All figures lie in a plane unless otherwise indicated.

(2) Unless otherwise specified, the domain of a function f is assumed to be the set of all real numbers x for which $f(x)$ is a real number.

1. A cylindrical bar of metal has a base radius of 2 and a height of 9. It is melted down and reformed into a sphere. The radius of the sphere is

[A] 3   [B] 4.5   [C] 6   [D] $\sqrt[3]{36}$
[E] $\sqrt[3]{9}$

2. In 3-dimensions, what is the set of all points for which $x=0$?

[A] the origin
[B] a line parallel to the x-axis
[C] the yz-plane
[D] a plane containing the x-axis
[E] the x-axis

3. Expressed with positive exponents only $\dfrac{ab^{-1}}{a^{-1}-b^{-1}}$ is equivalent to

[A] $\dfrac{a^2}{a-b}$ [B] $\dfrac{a^2}{a-1}$ [C] $\dfrac{b-a}{ab}$
[D] $\dfrac{a^2}{b-a}$ [E] $\dfrac{1}{a-b}$

4. If $f_n = \begin{cases} \dfrac{f_{n-1}}{2} & \text{if } f_{n-1} \text{ is an even number} \\ 3 \cdot f_{n-1}+1 & \text{if } f_{n-1} \text{ is an odd number} \end{cases}$

and $f_1 = 3$
then $f_5 =$
[A] 1   [B] 2   [C] 4   [D] 8   [E] 16

5. $x > \sin x$ for

[A] all $x > 0$
[B] all $x < 0$
[C] all x for which $x \neq 0$
[D] all x
[E] all x for which $-\dfrac{\pi}{2} < x < 0$

6. The range of the function $f(x) = x^2 - 2x + 3$ is the set of

[A] all real numbers
[B] all real numbers ≥ 0
[C] all real numbers ≥ 1
[D] all real numbers ≥ 3
[E] all real numbers ≥ 2

7. If $f(x)=2x+5$ and $|f(x)-f(2)|<1$, then $|x-2|<L$. The smallest value of L would be

[A] 1 [B] 7 [C] $\frac{3}{2}$ [D] $\frac{1}{2}$

[E] There is no smallest value of L.

8. For what value(s) of k is $x^2 - kx + k$ divisible by $x - k$?

[A] 0 only
[B] 0 or $-\frac{1}{2}$ only
[C] 1 only
[D] any value of k
[E] no value of k

9. The domain of the function $f(x) = \tan x$ consists of all real numbers except those of the form (where k is an integer)

[A] $\frac{\pi}{2}k$ [B] $(2k-1)\frac{\pi}{2}$ [C] $2k\pi$

[D] $(2k+1)\pi$ [E] $k\pi$

10. $P(x) = x^5 + x^4 - 2x^3 - x - 1$ has at most n positive zeros. $n =$

[A] 0 [B] 1 [C] 2 [D] 3 [E] 5

11. Which of the following equations have roots which are $\frac{2 \pm i\sqrt{2}}{3}$?

[A] $9x^2 - 12x + 2 = 0$ [B] $9x^2 + 12x + 2 = 0$
[C] $9x^2 + 12x + 6 = 0$ [D] $9x^2 - 12x + 6 = 0$
[E] $9x^2 + 12x - 6 = 0$

12. $P(x) = x^3 + 18x - 30$ has a zero in the interval

[A] $(-2, -1)$ [B] $(-1, 0)$ [C] $(0, 1)$
[D] $(1, 2)$ [E] $(2, 3)$

13. The radius of a circle is one-half the length of an arc of the circle. How large is the central angle that intercepts that arc?

[A] 60° [B] 120° [C] π^R
[D] 1^R [E] 2^R

14. In a triangle with sides of 7, 8, and 13, the measure of the largest angle is

[A] 60° [B] 90° [C] 120° [D] 135°
[E] 150°

15. If the graphs of $x^2 = 4(y+9)$ and $x + ky = 6$ intersect on the x-axis, then k equals

[A] 0 [B] 6
[C] -6 [D] no real number
[E] any real number

16. The 71st term of $30, 27, 24, 21, \ldots$ is

[A] 180 [B] -183 [C] -180
[D] 240 [E] 5325

17. What positive values of $x \leq 2\pi$ makes $\sin 2x - 2 \cdot \cos 2x + 4 \cdot \cos^4 x = 2$?

[A] $\frac{\pi}{2}, \frac{3\pi}{2}$ only

[B] $\pi, 2\pi$ only

[C] $\frac{\pi}{4}, \frac{5\pi}{4}$ only

[D] $\frac{\pi}{4}, \frac{\pi}{2}, \frac{5\pi}{4}, \frac{3\pi}{2}$ only

[E] $\frac{\pi}{4}, \frac{\pi}{2}, \pi, \frac{5\pi}{4}, \frac{3\pi}{2}, 2\pi$ only

18. The value of $\cos\left(2 \cdot \text{Arcsin}\left(-\frac{3}{5}\right)\right)$ is

[A] $\frac{4}{5}$ [B] $\frac{9}{25}$ [C] $\frac{7}{25}$

[D] $-\frac{7}{25}$ [E] $-\frac{4}{5}$

19. The length of the latus rectum of the hyperbola whose equation is $x^2 - 4y^2 = 16$ is

[A] 1 [B] $\sqrt{20}$ [C] 16
[D] $2\sqrt{20}$ [E] 2

20. 80% of all articles in a box are satisfactory, while 20% are not. The probability of obtaining exactly five good items out of eight randomly selected articles is

[A] $\binom{5}{4} \cdot (80) \cdot (20)$ [B] $\binom{8}{5} \cdot (.8)^8 \cdot (.2)^5$

[C] $\binom{8}{5} \cdot (.8)^5 \cdot (.2)^3$ [D] $\binom{5}{4} \cdot (.8)^5 \cdot (.2)^3$

[E] .132

21. If the operation $*$ is defined on the set of ordered triples as follows: $(a,b,c)*(x,y,z) =$

$(ax, b+y, cz)$, which of the following represents the identity element for this operation?

[A] (0,0,0)   [B] (1,1,1)   [C] (1,0,1)
[D] (0,1,0)   [E] (1,1,0)

22 The approximate value of $(1.001)^9$ is closest to

[A] 1.008   [B] 1.009   [C] 1.010
[D] 1.011   [E] 1.012

23 The statement "If it is green, then it is a turkey" is a true statement. Which of the following is (are) true?

I. It is not green, therefore it is not a turkey.
II. It is a turkey, therefore it is green.
III. It is not a turkey, therefore it is not green.

[A] I only   [B] II only
[C] III only   [D] I and III only
[E] II and III only

24 If the probability that the Giants will win the NFC championship is p and if the probability that the Raiders will win the AFC championship is q, what is the probability that only one of them will win a championship?

[A] pq [B] $1-pq$ [C] $|p-q|$
[D] $p+q-2pq$ [E] $2pq-p-q$

25 $(1-i)^8$ equals

[A] 0   [B] $8i$
[C] $\sqrt{2}-\sqrt{2}i$ [D] $1-i$
[E] 16

26 If $f(x,y) = 2x - y$ and $g(z) = \sin z$, then $g\left(f\left(\pi, \frac{\pi}{2}\right)\right)$ equals

[A] $\frac{3\pi}{2}$ [B] -1 [C] 0

[D] $\frac{5\pi}{2}$ [E] 1

27 $|x-y| \leq |y-x|$ is true for

[A] $x < y$
[B] $y < x$
[C] $x > 0$ and $y < 0$
[D] no values of x and y
[E] all values of x and y

28 Let S be the sum of the first n terms of the arithmetic sequence 3, 7, 11,... and let T be the sum of the first n terms of the arithmetic sequence 8, 10, 12,... For $n > 1$, $S = T$ for

[A] no values of n [B] one value of n
[C] two values of n [D] three values of n
[E] four values of n

29 What is the period of the graph of the function $y = \cos^4 x - \sin^4 x$?

[A] 2π [B] π [C] $\frac{\pi}{2}$ [D] $\frac{\pi}{4}$ [E] 4π

30 In the figure, c equals

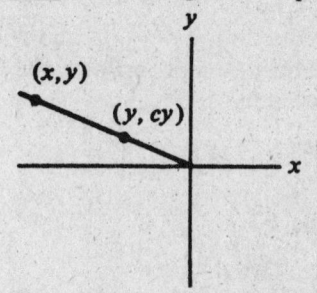

[A] 1   [B] xy [C] $\frac{x}{y}$ [D] $\frac{y}{x}$ [E] -1

31 Each of a group of 50 students studies either French or Spanish but not both, and either math or physics but not both. If 16 students study French and math, 26 study Spanish, and 12 study physics, how many study both Spanish and physics?

[A] 5   [B] 6   [C] 8   [D] 4   [E] 10

32 What is the equation of the set of points situated in a plane such that the distance between any point and (0,0) is twice the distance between that point and the x-axis?

[A] $3x^2 - y^2 = 0$ [B] $x^2 - 3y^2 = 0$
[C] $x^2 + y^2 - 2y = 0$ [D] $x^2 + y^2 - 2x = 0$
[E] $4x^2 + 3y^2 = 0$

33 $\text{Arctan } \frac{1}{5} + \text{Arctan } \frac{1}{2} + \text{Arctan } \frac{1}{8} =$

[A] 0   [B] $\text{Arctan } \frac{33}{40}$

[C] $\text{Arcsin } \frac{33}{\sqrt{2689}}$ [D] $\frac{\pi}{4}$

[E] $\text{Arccos}\left(\frac{5}{\sqrt{26}} + \frac{2}{\sqrt{5}} + \frac{8}{\sqrt{65}}\right)$

34 If the parameter is removed from $\begin{cases} x=1-e^t \\ y=1+e^{-t} \end{cases}$ the resulting equation is

[A] $y=\dfrac{x}{1-x}$ [B] $y=\dfrac{2x}{1+x}$

[C] $y=\dfrac{x+2}{x+1}$ [D] $y=\dfrac{x+2}{1-x}$

[E] $y=\dfrac{2-x}{1-x}$

35 (p,q) is called a *lattice point* if p and q are both integers. How many lattice points lie in the area between the two curves $x^2+y^2=9$ and $x^2+y^2-6x+5=0$?

[A] 0 [B] 1 [C] 2 [D] 3 [E] 4

36 If $\sin A=\dfrac{3}{5}$, $90°<A<180°$, $\cos B=\dfrac{1}{3}$, and $270°<B<360°$, the value of $\sin(A+B)$ is

[A] $-\dfrac{1}{3}$ [B] $\dfrac{11}{15}$

[C] $\dfrac{-6\sqrt{2}-4}{15}$ [D] $\dfrac{3-8\sqrt{2}}{15}$

[E] $\dfrac{3+8\sqrt{2}}{15}$

37 The plane $ax+by+cz=12$ intersects the x-axis at $(2,0,0)$, the y-axis at $(0,-3,0)$, and the z-axis at $(0,0,-4)$. $a+b+c=$

[A] 13 [B] 7 [C] 6 [D] 0 [E] -1

38 A point, p, is 5 inches from a plane, X. The set of points 13 inches from p and 2 inches from X forms

[A] a circle parallel to X with a radius <13
[B] two circles parallel to X with radii >10
[C] a sphere with radius <13
[D] two lines parallel to X
[E] two planes parallel to X

39 The operation # is defined by the equation $a\#b=\dfrac{a}{b}-\dfrac{b}{a}$. If $2\#k=3\#3$, find the value of k.

[A] 0 [B] ± 2 [C] ± 3 [D] $\pm\sqrt{3}$
[E] $\pm\sqrt{2}$

40 The graph of which of the following is a portion of an ellipse?

[A] $xy=1$ [B] $|y|=x^2$
[C] $y=\sqrt{1+4x^2}$ [D] $y=\sqrt{4-x^2}$
[E] $y=\sqrt{1-4x^2}$

41 For what values of x is $\log_2(-x)$ a negative number?

[A] $x<1$ [B] $x>-1$ [C] $0<x<1$
[D] $-1<x<0$ [E] $x<-1$

42 The amount of heat received by a body varies inversely as the square of its distance from the heat source. In comparison, how much heat will be received by a body which is 3 times as far from the heat source?

[A] 3 times as much [B] 9 times as much
[C] $\dfrac{1}{3}$ as much [D] $\dfrac{1}{9}$ as much
[E] $\dfrac{1}{6}$ as much

43 For what value of k are the roots of $x^2-2kx^2=6-8x$ equal?

[A] $\dfrac{6}{11}$ [B] $\dfrac{11}{6}$ [C] $\dfrac{5}{6}$ [D] $\dfrac{6}{5}$ [E] $\dfrac{4}{3}$

44 How many positive integers are there in the solution set of $\dfrac{x}{x-2}>5$?

[A] 0 [B] 2 [C] 4
[D] 5 [E] an infinite number

45 The sum of the roots of $3x-7x^{-1}+3=0$ is

[A] $\dfrac{7}{3}$ [B] 1 [C] $-\dfrac{7}{3}$ [D] -1 [E] $\dfrac{3}{7}$

46 If $f(x)=x^2-4$, for what real number values of x will $f(f(x))=0$?

[A] $\sqrt{6}$ [B] $\pm\sqrt{6}$
[C] 2 or 6 [D] $\pm\sqrt{2}$ or $\pm\sqrt{6}$
[E] no values

47. Given that $f(x)$ is an even function and $g(x)$ is an odd function, which of the following functions is an odd function for all values for which they are defined?

 I. $h(x) = \dfrac{f(x)}{g(x)}$
 II. $k(x) = f(x) + g(x)$
 III. $m(x) = f(g(x))$

[A] I only [B] II only
[C] III only [D] I and II only
[E] I and III only

48. If $f(g(x)) = x^2 - 1$ and $g(x) = x + 3$, then $f(x) =$

[A] $x^2 + 6x + 8$ [B] $x^2 + 2$
[C] $x^2 - 6x + 8$ [D] $x^2 + 8$
[E] $x^2 - 4$

49. If $f(x)$ is a linear function and $f(2) = 1$ and $f(4) = -2$, then $f(x) =$

[A] $-\dfrac{3}{2}x + 4$ [B] $\dfrac{3}{2}x - 2$

[C] $-\dfrac{3}{2}x + 2$ [D] $\dfrac{3}{2}x - 4$

[E] $-\dfrac{2}{3}x + \dfrac{7}{3}$

50. The domain of the function $f(x) = \dfrac{(x-1)(x+2)}{(x-1)(x+2)}$ is the set of

[A] all real numbers
[B] all real numbers except 1 or -2
[C] all real numbers except -1 or 2
[D] all real numbers greater than -2 and less than 1
[E] all real numbers greater than -1 and less than 2

MODEL TEST 8

50 questions
1 hour

<u>Directions:</u> For each of the following problems, decide which is the best of the choices given. Then blacken the corresponding space on the answer sheet on page 141. Answers may be found on page 144, and solutions begin on page 161.

<u>Notes:</u> (1) Figures that accompany problems in this test are intended to provide information useful in solving the problems. They are drawn as accurately as possible EXCEPT when it is stated in a specific problem that its figure is not drawn to scale. All figures lie in a plane unless otherwise indicated.

(2) Unless otherwise specified, the domain of a function f is assumed to be the set of all real numbers x for which $f(x)$ is a real number.

1. The equation $\sec^2 x - \tan x - 1 = 0$ has n solutions between 10° and 350°. $n =$

   [A] 0 [B] 1 [C] 2 [D] 3 [E] 4

2. A sphere has a surface area of 36π. Its volume is

   [A] 36π [B] 108π [C] 64π
   [D] 27π [E] 288π

3. If (a,b) is a solution of the system of equations $\begin{cases} 2x - y = 7 \\ x + y = 8 \end{cases}$ then the difference, $a - b =$

   [A] 0 [B] 2 [C] 4 [D] -12
   [E] -10

4. If $f(x) = x - 1$, $g(x) = 3x$, and $h(x) = \dfrac{5}{x}$, $f^{-1}(g(h(5))) =$

   [A] 4 [B] 2 [C] $\dfrac{5}{12}$ [D] $\dfrac{5}{6}$ [E] $\dfrac{1}{2}$

5. In three dimensions, the set of points described by the equations $x = 3$ and $y = 2$ is

   [A] a line perpendicular to the z-axis
   [B] the point (3,2,0)
   [C] a line perpendicular to the xy-plane
   [D] a line in the xy-plane
   [E] a line passing through (3,2,0) and (0,0,0)

6. The period of the function $f(x) = k \cos kx$ is $\dfrac{\pi}{2}$. The amplitude of f is

   [A] 2 [B] $\dfrac{1}{2}$ [C] 1 [D] $\dfrac{1}{4}$ [E] 4

7. The nature of the roots of the equation $3x^4 + 4x^3 + x - 1 = 0$ is

   [A] 3 positive real roots and 1 negative real root
   [B] 3 negative real roots and 1 positive real root

[C] 1 negative real root and 3 complex roots
[D] 1 positive real root, 1 negative real root, and 2 complex roots
[E] 2 positive real roots, 1 negative real root, and 1 complex root

8. For what value(s) of k is x^2+3x+k divisible by $x+k$?

[A] 0 only [B] 0 or 2 only
[C] 0 or -4 only [D] no value of k
[E] any value of k

9. The hour hand of a clock moves k radians in 48 minutes. $k=$

[A] 24 [B] $\dfrac{8\pi}{5}$ [C] 288
[D] $\dfrac{2\pi}{15}$ [E] $\dfrac{\pi}{6}$

10. An expression equivalent to $\cos^4 x - \sin^4 x$ is

[A] $\tan^2 x - \sec^2 x$ [B] $\sec^2 x - \tan^2 x$
[C] $\cot^2 x - \csc^2 x$ [D] $\cos 2x$
[E] $\sin 2x$

11. If x^4+x^2+x+2 is divided by $x+2$, the remainder is

[A] 30 [B] 24 [C] 20 [D] 16
[E] -4

12. If $f(x)=x^3-4$, then the inverse of $f=$

[A] $-x^3+4$ [B] $\sqrt[3]{x+4}$ [C] $\sqrt[3]{x-4}$
[D] $\dfrac{1}{x^3-4}$ [E] $\dfrac{4}{\sqrt[3]{x}}$

13. If f is an odd function and $f(a)=b$, which of the following must also be true?

I. $f(a)=-b$
II. $f(-a)=b$
III. $f(-a)=-b$

[A] I only [B] II only
[C] III only [D] I and II only
[E] II and III only

14. If the following instructions are followed, what number will be printed in line 6?
1. Let $A=1$.
2. Let $x=4$.
3. Let A be replaced by the sum of A and x.
4. Increase the value of x by 3.
5. If $x<9$ go back to step 3. If $x\geq 9$ go to step 6.
6. Print the value of A.

[A] 7 [B] 10 [C] 12 [D] 0 [E] 9

15. Given $A=\{a,b\}$ and $C=\{a,b,c,d\}$, if $A\subseteq B\subseteq C$, how many different sets, B, are there which satisfy this condition? ($A\subseteq B$ means all the elements of set A are elements of set B.)

[A] 2 [B] 4 [C] 6 [D] 8 [E] 10

16. Given the statement, "All vacationers are tourists," which of the conclusions follows logically?

[A] If the Browns are tourists, then they are vacationers.
[B] If the Smiths are not tourists, then they are not vacationers.
[C] If the Polks are not vacationers, then they are not tourists.
[D] All tourists are vacationers.
[E] Some tourists are not vacationers.

17. In trigonometric form, the complex number $2-2\sqrt{3}i$ is equivalent to

[A] $2(\cos 60° - i\cdot\sin 60°)$
[B] $4(\cos 60° - i\cdot\sin 60°)$
[C] $2(\cos 30° - i\cdot\sin 30°)$
[D] $4(\cos 30° - i\cdot\sin 30°)$
[E] none of the above

18. In how many ways can four couples form a circle for a dance if each woman's escort is on her left?

[A] 6 [B] 24 [C] 2520
[D] 5040 [E] 40320

19. The sum of the first 21 terms of the sequence $\left\{(-1)^n\cdot\cos\dfrac{3\pi}{2}\right\}$ is

[A] 0 [B] 1 [C] -1
[D] 21 [E] -21

20. The probability of getting at least three heads when flipping five coins is

[A] $\dfrac{2}{5}$ [B] $\dfrac{5}{16}$ [C] $\dfrac{3}{5}$ [D] $\dfrac{5}{8}$ [E] $\dfrac{1}{2}$

21. If Arcsin $x = 2$ Arccos x, then x equals

[A] $\dfrac{1}{2}$ [B] $\dfrac{\sqrt{3}}{2}$ [C] 0

[D] $\pm\dfrac{\sqrt{3}}{2}$ [E] $\pm\dfrac{1}{2}$

22. What is the period of the graph of the function $y = \dfrac{\sin x}{1 + \cos x}$?

[A] 2π [B] π [C] $\dfrac{\pi}{2}$ [D] $\dfrac{\pi}{4}$ [E] 4π

23. A man piles 150 toothpicks in layers so that each layer has 1 less toothpick than the layer below. If the top layer has 3 toothpicks, how many layers are there?

[A] 15 [B] 17 [C] 20
[D] 148 [E] 11322

24. A point moves in a plane so that its distance from the origin is always twice its distance from the point (1,1). All such points would form

[A] a line [B] a circle
[C] a parabola [D] an ellipse
[E] a hyperbola

25. If $f(x,y) = x+y$ and $g(x,y) = x-y$, then $f(g(2,1), f(2,-1))$ equals

[A] 0 [B] 1 [C] 2 [D] 4 [E] 6

26. If the circle $x^2 + y^2 - 2x - 6y = r^2 - 10$ is tangent to the line $5x + 12y = 60$, the value of r is

[A] $\sqrt{10}$ [B] $\dfrac{19}{13}$ [C] $\dfrac{13}{12}$

[D] $\dfrac{41}{13}$ [E] $\dfrac{41\sqrt{10}}{10}$

27. $|x+2| > x-1$ is true for

[A] $-2 \leq x \leq -\dfrac{1}{2}$ [B] $x < -\dfrac{1}{2}$
[C] $x \leq -2$ [D] $x \geq -2$
[E] all values of x

28. A red box contains 8 items, of which three are defective, and a blue box contains 5 items; of which 2 are defective. An item is drawn at random from each box. What is the probability that one is defective and one is not?

[A] $\dfrac{5}{8}$ [B] $\dfrac{19}{40}$ [C] $\dfrac{17}{32}$ [D] $\dfrac{17}{20}$ [E] $\dfrac{9}{40}$

29. The distance between the center of the ellipse $5x^2 + 8y^2 + 10x - 32y - 3 = 0$ and the center of the hyperbola $8x^2 - 5y^2 - 16x + 20y - 3 = 0$ is

[A] 0 [B] 2 [C] 4 [D] $2\sqrt{5}$ [E] 3

30. If $f(x) = x + g(x)$ and $g(x) = x \cdot f(x)$, then $\dfrac{f(x)}{g(x)} =$

[A] x [B] $\dfrac{1}{x}$ [C] $\dfrac{x}{1-x}$

[D] $\dfrac{1-x}{x^2}$ [E] 1

31. What is the equation of the set of points which are 5 units from the point (2,3,4)?

[A] $2x + 3y + 4z = 5$
[B] $x^2 + y^2 + z^2 - 4x - 6y - 8z = 25$
[C] $(x-2)^2 + (y-3)^2 + (z-4)^2 = 25$
[D] $x^2 + y^2 + z^2 = 5$
[E] $\dfrac{x}{2} + \dfrac{y}{3} + \dfrac{z}{4} = 5$

32. The force of the wind on a sail varies jointly as the area of the sail and the square of the wind velocity. On a sail of area 50 sq yds, the force of a 15 mi/hr wind is 45 lbs. Find the force on the sail if the wind increases to 45 mi/hr.

[A] 135 lbs [B] 405 lbs [C] 450 lbs
[D] 225 lbs [E] 675 lbs

33. If $g(x-1) = x^2 + 2$, then $g(x) =$

[A] $x^2 - 2x + 3$ [B] $x^2 + 2x + 3$
[C] $x^2 + 2$ [D] $x^2 - 2$
[E] $x^2 - 3x + 2$

134   Section Three: Model Examinations

34 In the diagram the circle has a radius of 1 and center at the origin. If the point F represents a complex number, $a+bi$, which of the other points could represent the conjugate of F?

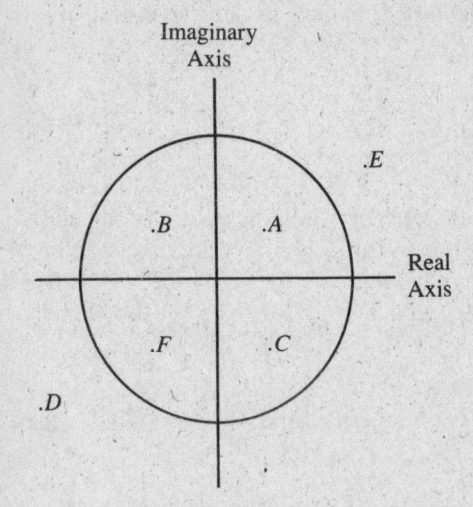

[Diagram: unit circle centered at origin with Imaginary Axis (vertical) and Real Axis (horizontal). Points: E in upper right outside/on circle, B in upper left inside, A in upper right inside, F in lower left inside, C in lower right inside, D outside lower left.]

[A] A [B] B [C] C [D] D [E] E

35 If $\cos(5n-30)° = \sin 50°$ and $0° < n < 90°$, then $n =$

[A] $2°$ [B] $14°$ [C] $88°$ [D] $16°$
[E] $76°$

36 For what values of k will the roots of the equation $kx^2 + 4x + k = 0$ be real and unequal?

[A] $0 < k < 2$ [B] $|k| < 2$
[C] $|k| > 2$ [D] $k > 2$
[E] $-2 < k < 0$ or $0 < k < 2$

37 Given that $f(x) = 1 - 2\cos x$. For how may values of x (where $0 \leq x < 2\pi$) does the graph of f cross the horizontal axis?

[A] 0 [B] 1 [C] 2 [D] 3 [E] 4

38 The operation # is defined by the equation $a \# b = \dfrac{a}{b} - \dfrac{b}{a}$. What is the value of k if $3 \# k = k \# 2$?

[A] ± 2 [B] ± 3 [C] ± 6
[D] $\pm\sqrt{6}$ [E] $\pm\sqrt{30}$

39 For what values of x is $\log_{1/3}(-x)$ a positive number?

[A] $x < 1$ [B] $x > -1$ [C] $0 < x < 1$
[D] $-1 < x < 0$ [E] $x < -1$

40 If $f(g(x)) = x^2 + 1$ and $f(x) = x + 3$, then $g(x) =$

[A] $x^2 + 4$ [B] $x^2 - 2$
[C] $x^2 + 6x + 10$ [D] $x^2 + 2$
[E] $x^2 - 6x + 10$

41 Expressed as a function of an acute angle, $\cos 310° + \cos 190° =$

[A] $-\cos 40°$ [B] $\cos 70°$
[C] $-\cos 50°$ [D] $\sin 20°$
[E] $-\cos 70°$

42 If the following equations are graphed, which will have more than one asymptote?

I. $x^2 - y^2 = 4$
II. $y = \dfrac{x}{x^2 + 1}$
III. $y = \dfrac{x}{x - 1}$

[A] III only [B] I and II only
[C] I and III only [D] II and III only
[E] I, II, and III

43 By how much does the sum of the roots of the equation $6x^4 - 3x^2 + 7x = 0$ exceed the product of the roots?

[A] $\sqrt{3}$ [B] $\sqrt[3]{3} - 1$ [C] 0
[D] $\dfrac{1}{2}$ [E] $\dfrac{7}{6}$

44 If $f(x) = \dfrac{x+2}{(x-2)(x^2-4)}$, its graph will have

[A] 1 horizontal and 3 vertical asymptotes
[B] 1 horizontal and 2 vertical asymptotes
[C] 1 horizontal and 1 vertical asymptote
[D] 0 horizontal and 1 vertical asymptote
[E] 0 horizontal and 2 vertical asymptotes

45 The line passing through $(1,4,-2)$ and $(2,1,4)$ could be represented by the set of equations

[A] $x = -1 + 2d$ [B] $x = 1 + 2d$
    $y = 3 + d$ $y = 4 + d$
    $z = -2 + 4d$ $z = -2 + 4d$

[C] $x = 2 + d$ [D] $x = 1 + d$
    $y = 1 + 4d$ $y = 4 - 3d$
    $z = 4 - 2d$ $z = -2 + 6d$

[E] $x = 1 - d$
    $y = -3 - 4d$
    $z = 2 + 2d$

46. The coefficient of the middle term of the expansion of $(2a^2 - a^{1/4})^6$ is

[A] 240 [B] 160 [C] 60
[D] −30 [E] −160

47. The range of $f(x) = \dfrac{x^2}{x^2 - 4}$ is the set of

[A] all real numbers
[B] all real numbers >1
[C] all real numbers except ± 2
[D] all real numbers except 0 and ± 2
[E] all real numbers greater than 1 or less than or equal to 0

48. In triangle ABC, $\angle A = 45°$, $\angle B = 30°$, and $b = 8$. Side $a =$

[A] 12 [B] 16 [C] $\dfrac{8\sqrt{6}}{3}$

[D] $8\sqrt{2}$ [E] $8\sqrt{3}$

49. The equations of the asymptotes of the graph of $4x^2 - 9y^2 = 36$ are

[A] $y = x$ and $y = -x$
[B] $y = 0$ and $x = 0$
[C] $y = \dfrac{2}{3}x$ and $y = -\dfrac{2}{3}x$
[D] $y = \dfrac{3}{2}x$ and $y = -\dfrac{3}{2}x$
[E] $y = \dfrac{4}{9}x$ and $y = -\dfrac{4}{9}x$

50. If $f(x) = 3x^3 - 2x^2 + x - 2$, then $f(i) =$

[A] $-2i - 4$ [B] $4i$ [C] $4i - 4$
[D] $-2i$ [E] 0

Model Test 4

1. Ⓐ Ⓑ Ⓒ Ⓓ Ⓔ
2. Ⓐ Ⓑ Ⓒ Ⓓ Ⓔ
3. Ⓐ Ⓑ Ⓒ Ⓓ Ⓔ
4. Ⓐ Ⓑ Ⓒ Ⓓ Ⓔ
5. Ⓐ Ⓑ Ⓒ Ⓓ Ⓔ
6. Ⓐ Ⓑ Ⓒ Ⓓ Ⓔ
7. Ⓐ Ⓑ Ⓒ Ⓓ Ⓔ
8. Ⓐ Ⓑ Ⓒ Ⓓ Ⓔ
9. Ⓐ Ⓑ Ⓒ Ⓓ Ⓔ
10. Ⓐ Ⓑ Ⓒ Ⓓ Ⓔ
11. Ⓐ Ⓑ Ⓒ Ⓓ Ⓔ
12. Ⓐ Ⓑ Ⓒ Ⓓ Ⓔ
13. Ⓐ Ⓑ Ⓒ Ⓓ Ⓔ
14. Ⓐ Ⓑ Ⓒ Ⓓ Ⓔ
15. Ⓐ Ⓑ Ⓒ Ⓓ Ⓔ
16. Ⓐ Ⓑ Ⓒ Ⓓ Ⓔ
17. Ⓐ Ⓑ Ⓒ Ⓓ Ⓔ
18. Ⓐ Ⓑ Ⓒ Ⓓ Ⓔ
19. Ⓐ Ⓑ Ⓒ Ⓓ Ⓔ
20. Ⓐ Ⓑ Ⓒ Ⓓ Ⓔ
21. Ⓐ Ⓑ Ⓒ Ⓓ Ⓔ
22. Ⓐ Ⓑ Ⓒ Ⓓ Ⓔ
23. Ⓐ Ⓑ Ⓒ Ⓓ Ⓔ
24. Ⓐ Ⓑ Ⓒ Ⓓ Ⓔ
25. Ⓐ Ⓑ Ⓒ Ⓓ Ⓔ
26. Ⓐ Ⓑ Ⓒ Ⓓ Ⓔ
27. Ⓐ Ⓑ Ⓒ Ⓓ Ⓔ
28. Ⓐ Ⓑ Ⓒ Ⓓ Ⓔ
29. Ⓐ Ⓑ Ⓒ Ⓓ Ⓔ
30. Ⓐ Ⓑ Ⓒ Ⓓ Ⓔ
31. Ⓐ Ⓑ Ⓒ Ⓓ Ⓔ
32. Ⓐ Ⓑ Ⓒ Ⓓ Ⓔ
33. Ⓐ Ⓑ Ⓒ Ⓓ Ⓔ
34. Ⓐ Ⓑ Ⓒ Ⓓ Ⓔ
35. Ⓐ Ⓑ Ⓒ Ⓓ Ⓔ
36. Ⓐ Ⓑ Ⓒ Ⓓ Ⓔ
37. Ⓐ Ⓑ Ⓒ Ⓓ Ⓔ
38. Ⓐ Ⓑ Ⓒ Ⓓ Ⓔ
39. Ⓐ Ⓑ Ⓒ Ⓓ Ⓔ
40. Ⓐ Ⓑ Ⓒ Ⓓ Ⓔ
41. Ⓐ Ⓑ Ⓒ Ⓓ Ⓔ
42. Ⓐ Ⓑ Ⓒ Ⓓ Ⓔ
43. Ⓐ Ⓑ Ⓒ Ⓓ Ⓔ
44. Ⓐ Ⓑ Ⓒ Ⓓ Ⓔ
45. Ⓐ Ⓑ Ⓒ Ⓓ Ⓔ
46. Ⓐ Ⓑ Ⓒ Ⓓ Ⓔ
47. Ⓐ Ⓑ Ⓒ Ⓓ Ⓔ
48. Ⓐ Ⓑ Ⓒ Ⓓ Ⓔ
49. Ⓐ Ⓑ Ⓒ Ⓓ Ⓔ
50. Ⓐ Ⓑ Ⓒ Ⓓ Ⓔ

Model Test 5

1. Ⓐ Ⓑ Ⓒ Ⓓ Ⓔ
2. Ⓐ Ⓑ Ⓒ Ⓓ Ⓔ
3. Ⓐ Ⓑ Ⓒ Ⓓ Ⓔ
4. Ⓐ Ⓑ Ⓒ Ⓓ Ⓔ
5. Ⓐ Ⓑ Ⓒ Ⓓ Ⓔ
6. Ⓐ Ⓑ Ⓒ Ⓓ Ⓔ
7. Ⓐ Ⓑ Ⓒ Ⓓ Ⓔ
8. Ⓐ Ⓑ Ⓒ Ⓓ Ⓔ
9. Ⓐ Ⓑ Ⓒ Ⓓ Ⓔ
10. Ⓐ Ⓑ Ⓒ Ⓓ Ⓔ
11. Ⓐ Ⓑ Ⓒ Ⓓ Ⓔ
12. Ⓐ Ⓑ Ⓒ Ⓓ Ⓔ
13. Ⓐ Ⓑ Ⓒ Ⓓ Ⓔ
14. Ⓐ Ⓑ Ⓒ Ⓓ Ⓔ
15. Ⓐ Ⓑ Ⓒ Ⓓ Ⓔ
16. Ⓐ Ⓑ Ⓒ Ⓓ Ⓔ
17. Ⓐ Ⓑ Ⓒ Ⓓ Ⓔ
18. Ⓐ Ⓑ Ⓒ Ⓓ Ⓔ
19. Ⓐ Ⓑ Ⓒ Ⓓ Ⓔ
20. Ⓐ Ⓑ Ⓒ Ⓓ Ⓔ
21. Ⓐ Ⓑ Ⓒ Ⓓ Ⓔ
22. Ⓐ Ⓑ Ⓒ Ⓓ Ⓔ
23. Ⓐ Ⓑ Ⓒ Ⓓ Ⓔ
24. Ⓐ Ⓑ Ⓒ Ⓓ Ⓔ
25. Ⓐ Ⓑ Ⓒ Ⓓ Ⓔ
26. Ⓐ Ⓑ Ⓒ Ⓓ Ⓔ
27. Ⓐ Ⓑ Ⓒ Ⓓ Ⓔ
28. Ⓐ Ⓑ Ⓒ Ⓓ Ⓔ
29. Ⓐ Ⓑ Ⓒ Ⓓ Ⓔ
30. Ⓐ Ⓑ Ⓒ Ⓓ Ⓔ
31. Ⓐ Ⓑ Ⓒ Ⓓ Ⓔ
32. Ⓐ Ⓑ Ⓒ Ⓓ Ⓔ
33. Ⓐ Ⓑ Ⓒ Ⓓ Ⓔ
34. Ⓐ Ⓑ Ⓒ Ⓓ Ⓔ
35. Ⓐ Ⓑ Ⓒ Ⓓ Ⓔ
36. Ⓐ Ⓑ Ⓒ Ⓓ Ⓔ
37. Ⓐ Ⓑ Ⓒ Ⓓ Ⓔ
38. Ⓐ Ⓑ Ⓒ Ⓓ Ⓔ
39. Ⓐ Ⓑ Ⓒ Ⓓ Ⓔ
40. Ⓐ Ⓑ Ⓒ Ⓓ Ⓔ
41. Ⓐ Ⓑ Ⓒ Ⓓ Ⓔ
42. Ⓐ Ⓑ Ⓒ Ⓓ Ⓔ
43. Ⓐ Ⓑ Ⓒ Ⓓ Ⓔ
44. Ⓐ Ⓑ Ⓒ Ⓓ Ⓔ
45. Ⓐ Ⓑ Ⓒ Ⓓ Ⓔ
46. Ⓐ Ⓑ Ⓒ Ⓓ Ⓔ
47. Ⓐ Ⓑ Ⓒ Ⓓ Ⓔ
48. Ⓐ Ⓑ Ⓒ Ⓓ Ⓔ
49. Ⓐ Ⓑ Ⓒ Ⓓ Ⓔ
50. Ⓐ Ⓑ Ⓒ Ⓓ Ⓔ

Model Test 6

1. Ⓐ Ⓑ Ⓒ Ⓓ Ⓔ
2. Ⓐ Ⓑ Ⓒ Ⓓ Ⓔ
3. Ⓐ Ⓑ Ⓒ Ⓓ Ⓔ
4. Ⓐ Ⓑ Ⓒ Ⓓ Ⓔ
5. Ⓐ Ⓑ Ⓒ Ⓓ Ⓔ
6. Ⓐ Ⓑ Ⓒ Ⓓ Ⓔ
7. Ⓐ Ⓑ Ⓒ Ⓓ Ⓔ
8. Ⓐ Ⓑ Ⓒ Ⓓ Ⓔ
9. Ⓐ Ⓑ Ⓒ Ⓓ Ⓔ
10. Ⓐ Ⓑ Ⓒ Ⓓ Ⓔ
11. Ⓐ Ⓑ Ⓒ Ⓓ Ⓔ
12. Ⓐ Ⓑ Ⓒ Ⓓ Ⓔ
13. Ⓐ Ⓑ Ⓒ Ⓓ Ⓔ
14. Ⓐ Ⓑ Ⓒ Ⓓ Ⓔ
15. Ⓐ Ⓑ Ⓒ Ⓓ Ⓔ
16. Ⓐ Ⓑ Ⓒ Ⓓ Ⓔ
17. Ⓐ Ⓑ Ⓒ Ⓓ Ⓔ
18. Ⓐ Ⓑ Ⓒ Ⓓ Ⓔ
19. Ⓐ Ⓑ Ⓒ Ⓓ Ⓔ
20. Ⓐ Ⓑ Ⓒ Ⓓ Ⓔ
21. Ⓐ Ⓑ Ⓒ Ⓓ Ⓔ
22. Ⓐ Ⓑ Ⓒ Ⓓ Ⓔ
23. Ⓐ Ⓑ Ⓒ Ⓓ Ⓔ
24. Ⓐ Ⓑ Ⓒ Ⓓ Ⓔ
25. Ⓐ Ⓑ Ⓒ Ⓓ Ⓔ
26. Ⓐ Ⓑ Ⓒ Ⓓ Ⓔ
27. Ⓐ Ⓑ Ⓒ Ⓓ Ⓔ
28. Ⓐ Ⓑ Ⓒ Ⓓ Ⓔ
29. Ⓐ Ⓑ Ⓒ Ⓓ Ⓔ
30. Ⓐ Ⓑ Ⓒ Ⓓ Ⓔ
31. Ⓐ Ⓑ Ⓒ Ⓓ Ⓔ
32. Ⓐ Ⓑ Ⓒ Ⓓ Ⓔ
33. Ⓐ Ⓑ Ⓒ Ⓓ Ⓔ
34. Ⓐ Ⓑ Ⓒ Ⓓ Ⓔ
35. Ⓐ Ⓑ Ⓒ Ⓓ Ⓔ
36. Ⓐ Ⓑ Ⓒ Ⓓ Ⓔ
37. Ⓐ Ⓑ Ⓒ Ⓓ Ⓔ
38. Ⓐ Ⓑ Ⓒ Ⓓ Ⓔ
39. Ⓐ Ⓑ Ⓒ Ⓓ Ⓔ
40. Ⓐ Ⓑ Ⓒ Ⓓ Ⓔ
41. Ⓐ Ⓑ Ⓒ Ⓓ Ⓔ
42. Ⓐ Ⓑ Ⓒ Ⓓ Ⓔ
43. Ⓐ Ⓑ Ⓒ Ⓓ Ⓔ
44. Ⓐ Ⓑ Ⓒ Ⓓ Ⓔ
45. Ⓐ Ⓑ Ⓒ Ⓓ Ⓔ
46. Ⓐ Ⓑ Ⓒ Ⓓ Ⓔ
47. Ⓐ Ⓑ Ⓒ Ⓓ Ⓔ
48. Ⓐ Ⓑ Ⓒ Ⓓ Ⓔ
49. Ⓐ Ⓑ Ⓒ Ⓓ Ⓔ
50. Ⓐ Ⓑ Ⓒ Ⓓ Ⓔ

Model Test 7

1. Ⓐ Ⓑ Ⓒ Ⓓ Ⓔ
2. Ⓐ Ⓑ Ⓒ Ⓓ Ⓔ
3. Ⓐ Ⓑ Ⓒ Ⓓ Ⓔ
4. Ⓐ Ⓑ Ⓒ Ⓓ Ⓔ
5. Ⓐ Ⓑ Ⓒ Ⓓ Ⓔ
6. Ⓐ Ⓑ Ⓒ Ⓓ Ⓔ
7. Ⓐ Ⓑ Ⓒ Ⓓ Ⓔ
8. Ⓐ Ⓑ Ⓒ Ⓓ Ⓔ
9. Ⓐ Ⓑ Ⓒ Ⓓ Ⓔ
10. Ⓐ Ⓑ Ⓒ Ⓓ Ⓔ
11. Ⓐ Ⓑ Ⓒ Ⓓ Ⓔ
12. Ⓐ Ⓑ Ⓒ Ⓓ Ⓔ
13. Ⓐ Ⓑ Ⓒ Ⓓ Ⓔ
14. Ⓐ Ⓑ Ⓒ Ⓓ Ⓔ
15. Ⓐ Ⓑ Ⓒ Ⓓ Ⓔ
16. Ⓐ Ⓑ Ⓒ Ⓓ Ⓔ
17. Ⓐ Ⓑ Ⓒ Ⓓ Ⓔ
18. Ⓐ Ⓑ Ⓒ Ⓓ Ⓔ
19. Ⓐ Ⓑ Ⓒ Ⓓ Ⓔ
20. Ⓐ Ⓑ Ⓒ Ⓓ Ⓔ
21. Ⓐ Ⓑ Ⓒ Ⓓ Ⓔ
22. Ⓐ Ⓑ Ⓒ Ⓓ Ⓔ
23. Ⓐ Ⓑ Ⓒ Ⓓ Ⓔ
24. Ⓐ Ⓑ Ⓒ Ⓓ Ⓔ
25. Ⓐ Ⓑ Ⓒ Ⓓ Ⓔ
26. Ⓐ Ⓑ Ⓒ Ⓓ Ⓔ
27. Ⓐ Ⓑ Ⓒ Ⓓ Ⓔ
28. Ⓐ Ⓑ Ⓒ Ⓓ Ⓔ
29. Ⓐ Ⓑ Ⓒ Ⓓ Ⓔ
30. Ⓐ Ⓑ Ⓒ Ⓓ Ⓔ
31. Ⓐ Ⓑ Ⓒ Ⓓ Ⓔ
32. Ⓐ Ⓑ Ⓒ Ⓓ Ⓔ
33. Ⓐ Ⓑ Ⓒ Ⓓ Ⓔ
34. Ⓐ Ⓑ Ⓒ Ⓓ Ⓔ
35. Ⓐ Ⓑ Ⓒ Ⓓ Ⓔ
36. Ⓐ Ⓑ Ⓒ Ⓓ Ⓔ
37. Ⓐ Ⓑ Ⓒ Ⓓ Ⓔ
38. Ⓐ Ⓑ Ⓒ Ⓓ Ⓔ
39. Ⓐ Ⓑ Ⓒ Ⓓ Ⓔ
40. Ⓐ Ⓑ Ⓒ Ⓓ Ⓔ
41. Ⓐ Ⓑ Ⓒ Ⓓ Ⓔ
42. Ⓐ Ⓑ Ⓒ Ⓓ Ⓔ
43. Ⓐ Ⓑ Ⓒ Ⓓ Ⓔ
44. Ⓐ Ⓑ Ⓒ Ⓓ Ⓔ
45. Ⓐ Ⓑ Ⓒ Ⓓ Ⓔ
46. Ⓐ Ⓑ Ⓒ Ⓓ Ⓔ
47. Ⓐ Ⓑ Ⓒ Ⓓ Ⓔ
48. Ⓐ Ⓑ Ⓒ Ⓓ Ⓔ
49. Ⓐ Ⓑ Ⓒ Ⓓ Ⓔ
50. Ⓐ Ⓑ Ⓒ Ⓓ Ⓔ

Model Test 8

1. Ⓐ Ⓑ Ⓒ Ⓓ Ⓔ
2. Ⓐ Ⓑ Ⓒ Ⓓ Ⓔ
3. Ⓐ Ⓑ Ⓒ Ⓓ Ⓔ
4. Ⓐ Ⓑ Ⓒ Ⓓ Ⓔ
5. Ⓐ Ⓑ Ⓒ Ⓓ Ⓔ
6. Ⓐ Ⓑ Ⓒ Ⓓ Ⓔ
7. Ⓐ Ⓑ Ⓒ Ⓓ Ⓔ
8. Ⓐ Ⓑ Ⓒ Ⓓ Ⓔ
9. Ⓐ Ⓑ Ⓒ Ⓓ Ⓔ
10. Ⓐ Ⓑ Ⓒ Ⓓ Ⓔ
11. Ⓐ Ⓑ Ⓒ Ⓓ Ⓔ
12. Ⓐ Ⓑ Ⓒ Ⓓ Ⓔ
13. Ⓐ Ⓑ Ⓒ Ⓓ Ⓔ
14. Ⓐ Ⓑ Ⓒ Ⓓ Ⓔ
15. Ⓐ Ⓑ Ⓒ Ⓓ Ⓔ
16. Ⓐ Ⓑ Ⓒ Ⓓ Ⓔ
17. Ⓐ Ⓑ Ⓒ Ⓓ Ⓔ
18. Ⓐ Ⓑ Ⓒ Ⓓ Ⓔ
19. Ⓐ Ⓑ Ⓒ Ⓓ Ⓔ
20. Ⓐ Ⓑ Ⓒ Ⓓ Ⓔ
21. Ⓐ Ⓑ Ⓒ Ⓓ Ⓔ
22. Ⓐ Ⓑ Ⓒ Ⓓ Ⓔ
23. Ⓐ Ⓑ Ⓒ Ⓓ Ⓔ
24. Ⓐ Ⓑ Ⓒ Ⓓ Ⓔ
25. Ⓐ Ⓑ Ⓒ Ⓓ Ⓔ
26. Ⓐ Ⓑ Ⓒ Ⓓ Ⓔ
27. Ⓐ Ⓑ Ⓒ Ⓓ Ⓔ
28. Ⓐ Ⓑ Ⓒ Ⓓ Ⓔ
29. Ⓐ Ⓑ Ⓒ Ⓓ Ⓔ
30. Ⓐ Ⓑ Ⓒ Ⓓ Ⓔ
31. Ⓐ Ⓑ Ⓒ Ⓓ Ⓔ
32. Ⓐ Ⓑ Ⓒ Ⓓ Ⓔ
33. Ⓐ Ⓑ Ⓒ Ⓓ Ⓔ
34. Ⓐ Ⓑ Ⓒ Ⓓ Ⓔ
35. Ⓐ Ⓑ Ⓒ Ⓓ Ⓔ
36. Ⓐ Ⓑ Ⓒ Ⓓ Ⓔ
37. Ⓐ Ⓑ Ⓒ Ⓓ Ⓔ
38. Ⓐ Ⓑ Ⓒ Ⓓ Ⓔ
39. Ⓐ Ⓑ Ⓒ Ⓓ Ⓔ
40. Ⓐ Ⓑ Ⓒ Ⓓ Ⓔ
41. Ⓐ Ⓑ Ⓒ Ⓓ Ⓔ
42. Ⓐ Ⓑ Ⓒ Ⓓ Ⓔ
43. Ⓐ Ⓑ Ⓒ Ⓓ Ⓔ
44. Ⓐ Ⓑ Ⓒ Ⓓ Ⓔ
45. Ⓐ Ⓑ Ⓒ Ⓓ Ⓔ
46. Ⓐ Ⓑ Ⓒ Ⓓ Ⓔ
47. Ⓐ Ⓑ Ⓒ Ⓓ Ⓔ
48. Ⓐ Ⓑ Ⓒ Ⓓ Ⓔ
49. Ⓐ Ⓑ Ⓒ Ⓓ Ⓔ
50. Ⓐ Ⓑ Ⓒ Ⓓ Ⓔ

ANSWER KEY

MODEL TEST 1

1. E	6. C	11. E	16. B	21. B	26. B	31. A	36. A	41. B	46. D
2. A	7. D	12. D	17. C	22. B	27. C	32. E	37. C	42. B	47. E
3. C	8. D	13. B	18. C	23. B	28. A	33. A	38. C	43. A	48. E
4. B	9. B	14. C	19. D	24. E	29. D	34. A	39. E	44. E	49. D
5. A	10. B	15. E	20. E	25. D	30. E	35. D	40. A	45. B	50. D

MODEL TEST 2

1. E	6. C	11. B	16. E	21. D	26. D	31. B	36. B	41. C	46. B
2. B	7. B	12. C	17. A	22. D	27. E	32. A	37. D	42. C	47. A
3. C	8. A	13. C	18. E	23. A	28. D	33. D	38. B	43. C	48. E
4. A	9. D	14. B	19. E	24. D	29. C	34. D	39. B	44. A	49. C
5. C	10. D	15. C	20. B	25. B	30. D	35. B	40. D	45. B	50. C

MODEL TEST 3

1. E	6. C	11. D	16. B	21. D	26. D	31. D	36. C	41. A	46. D
2. A	7. D	12. A	17. D	22. B	27. E	32. A	37. E	42. B	47. A
3. E	8. D	13. D	18. B	23. A	28. E	33. E	38. C	43. E	48. D
4. B	9. B	14. B	19. A	24. E	29. C	34. D	39. C	44. E	49. D
5. C	10. B	15. D	20. B	25. B	30. A	35. E	40. B	45. D	50. C

MODEL TEST 4

1. A	6. B	11. A	16. E	21. A	26. D	31. D	36. E	41. A	46. C
2. B	7. B	12. C	17. A	22. A	27. C	32. B	37. D	42. D	47. B
3. B	8. B	13. D	18. D	23. B	28. B	33. B	38. C	43. E	48. B
4. C	9. B	14. D	19. B	24. A	29. E	34. E	39. B	44. B	49. D
5. B	10. C	15. D	20. D	25. B	30. B	35. D	40. A	45. D	50. D

MODEL TEST 5

1. C	6. C	11. C	16. B	21. D	26. B	31. B	36. D	41. D	46. B
2. E	7. B	12. D	17. D	22. C	27. E	32. D	37. B	42. D	47. A
3. C	8. B	13. B	18. B	23. C	28. A	33. A	38. C	43. C	48. A
4. C	9. E	14. E	19. E	24. D	29. B	34. C	39. B	44. A	49. B
5. C	10. C	15. B	20. C	25. E	30. A	35. C	40. C	45. C	50. D

MODEL TEST 6

1. D	6. B	11. C	16. B	21. B	26. E	31. E	36. E	41. B	46. B
2. A	7. C	12. B	17. C	22. C	27. E	32. A	37. D	42. B	47. B
3. B	8. E	13. D	18. E	23. D	28. C	33. B	38. B	43. B	48. B
4. D	9. D	14. B	19. E	24. D	29. A	34. E	39. C	44. A	49. D
5. C	10. B	15. C	20. D	25. C	30. D	35. C	40. A	45. E	50. B

MODEL TEST 7

1. A	6. E	11. D	16. C	21. C	26. B	31. D	36. E	41. D	46. D
2. C	7. D	12. D	17. E	22. B	27. E	32. B	37. E	42. D	47. A
3. D	8. A	13. E	18. C	23. C	28. B	33. D	38. B	43. B	48. C
4. D	9. B	14. C	19. E	24. D	29. B	34. E	39. B	44. A	49. A
5. A	10. B	15. E	20. C	25. E	30. D	35. D	40. E	45. D	50. B

MODEL TEST 8

1. D	6. E	11. C	16. B	21. B		26. B	31. C	36. E	41. E	46. E
2. A	7. D	12. B	17. B	22. A		27. E	32. B	37. C	42. C	47. E
3. B	8. B	13. C	18. A	23. A		28. B	33. B	38. D	43. C	48. D
4. A	9. D	14. C	19. A	24. B		29. B	34. B	39. D	44. D	49. C
5. C	10. D	15. B	20. E	25. C		30. B	35. B	40. B	45. D	50. D

ANSWER EXPLANATIONS

MODEL TEST 1

1. **E** Complete the square to get $x^2+(y-5)^2=61$. Center at (0,5). [4.1].

2. **A** $1 \div \frac{1}{3} = 3 = r$. Third term $= 1 \cdot 3 = 3$. [5.4].

3. **C** $\frac{453!}{450! \cdot 3!} = \frac{453 \cdot 452 \cdot 451 \cdot 450!}{3 \cdot 2 \cdot 1 \cdot 450!}$
   $\approx 2 \times 10^2 \cdot 2 \times 10^2 \cdot 5 \times 10^2 = 2 \times 10^7$. [5.1].

4. **B** $g(2)=3$. $f(g(2))=f(3)=9$. [1.2].

5. **A** $\left[\left(-\frac{1}{64}\right)^{1/3}\right]^2 = \left(-\frac{1}{4}\right)^2 = \frac{1}{16}$. [4.2].

6. **C** Solve for y. $y=\frac{4}{3}x-4$. Slope $=\frac{4}{3}$. Slope of perpendicular $= -\frac{3}{4}$. [2.2].

7. **D** Volume of a sphere is $\frac{4}{3}\pi r^3$. Volume of region
   $= \frac{4}{3} \cdot \pi \cdot 5^3 - \frac{4}{3} \cdot \pi \cdot 2^3 = \frac{4}{3} \cdot \pi \cdot (125-8) = 156\pi$. [5.5].

8. **D** $(4^2)^x = 4^1$ so $x = \frac{1}{2}$. $5^{x+y}=5^4$, so $x+y=4$. $y=\frac{7}{2}$. [4.2].

9. **B** Slope of line $=\frac{3}{2}$, so $\tan A = \frac{3}{2}$. [2.2].

10. **B** $\frac{(x-4)^2}{16}+\frac{(y-3)^2}{4}=1$. An ellipse with $a^2=16$. Sum of distances $=2a=8$. [4.1].

11. **E** $t=\frac{y}{2}$. Eliminate parameter and get $x=\frac{y^2}{4}+1$ or $y^2=4x-4$. [4.6].

12. **D** $f(1)=0$ and $f(2)=0$ imply that $x-1$ and $x-2$ are factors of $f(x)$. Their product x^2-3x+2 is also a factor. [2.4].

13. **B** It does not matter what is substituted for x, $f(x)$ still equals 2. [1.2].

14. **C** $\sin A = \frac{8}{10} = \frac{4}{5}$. [3.1].

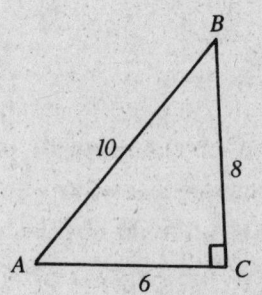

15. **E** $f(-2.5)=2.5-3=-.5$ and $f(1.5)=1.5+1=2.5$. Therefore, $f(-2.5)+f(1.5)=2$. [4.3, 4.4].

16. **B** Number of terms is always one more than the exponent of the binomial. Number of terms $= 8$. [5.2].

17. **C** If it passes through the origin $x=0$ and $y=0$, which gives $\frac{4k^2}{1}-\frac{9k^2}{3}=1$. $k^2=1$, so $k=\pm 1$. [4.1].

18. **C** Law of sines: $\frac{\sin 120°}{\sqrt{6}} = \frac{\sin B}{2}$. $\sin 120° = \frac{\sqrt{3}}{2}$, so $\sin B = \frac{\sqrt{2}}{2}$. $B=45°$, so $C=15°$. [3.7].

19. **D** Substituting -1 for x gives 3. [2.4].

20. **E** Substituting for x and solving for y gives $4(7-y^2)+8y^2=64$. $4y^2=36$, so $y^2=9$ and $y=\pm 3$. BUT if $y^2=9$, from the second equation $x^2=-7$, which is not a real value for x. Therefore, they don't intersect. [4.1].

21. **B** $t_n \cdot t_{n+1} = K$. $2 \cdot 6 = K = 12$. Therefore, $6 \cdot t_3 = 12$ so $t_3 = 2$. Continuing this process gives all odd terms to be 2 and all even terms to be 6. [5.6, 5.4].

22. **B** As x increases from negative to positive the slope increases from very negative to zero at (0,0) to very positive. The only graph that has these characteristics is [B]. [5.5, 5.9].

Answer Explanations: Model Test 1   145

23. **B** Converting from polar to rectangular coordinates: $r = \dfrac{1}{\cos\theta}$. $1 = \dfrac{1}{r\cos\theta}$. $1 = \dfrac{1}{x}$. $x = 1$, which is a straight line. [4.7].

24. **E** When x is negative the function takes on the values $f(-x)$, which makes the graph symmetric about the y-axis. The range: $3 \le y \le 11$. [2.2, 4.3].

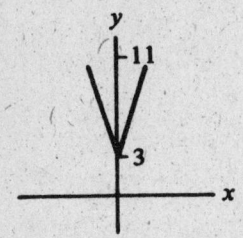

25. **D** Circumference of base of cone = length of arc $AB = 5 \cdot \dfrac{8\pi}{5} = 8\pi$. Circumference $= 2\pi r = 8\pi$. Therefore, radius of base is 4. Height of cone is 3. Therefore, volume $= \dfrac{1}{3}\pi r^2 h = \dfrac{1}{3}\pi \cdot 16 \cdot 3 = 16\pi$. [5.5, 3.2].

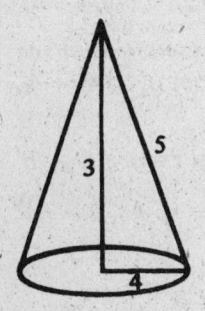

26. **B** $\tan 15° = \cot 75°$. [3.1].

27. **C** $\dfrac{\cot\theta}{\csc\theta} = \dfrac{\frac{\cos\theta}{\sin\theta}}{\frac{1}{\sin\theta}} = \cos\theta$. [3.1].

28. **A** Law of cosines: $12 = x^2 + (3x+2)^2 - 2x(3x+2)\cos 60°$. $12 = x^2 + 9x^2 + 12x + 4 - 3x^2 - 2x$. $7x^2 + 10x - 8 = 0$. $(7x-4)(x+2) = 0$, so $x = \dfrac{4}{7}$ or -2. [3.7].

29. **D** If the roots are r and $2r$, $3r = -\dfrac{k}{3}$ and $2r^2 = \dfrac{54}{3} = 18$. Therefore, $r = \pm 3$ and $k = \pm 27$. [2.3].

30. **E** The solution set of $x^2 + 3x + 2 < 0$ is $-2 < x < -1$. $f(-2) = 12$ and $f(-1) = 6$. Intermediate values of x indicate that $f(x)$ is always between 6 and 12 when $-2 < x < -1$. [4.3, 2.3].

31. **A** For each value of x in the domain of g and h, there is more than one value of y, so these are not functions. [1.1].

32. **E** $\dfrac{|5(3a) - 12(a) - 3|}{\sqrt{5^2 + (-12)^2}} = 6$. $|3a - 3| = 6 \cdot 13$. If $3a - 3 > 0$, then $3a - 3 = 78$ and $a = 27$. If $3a - 3 < 0$, then $-3a + 3 = 78$ and $a = -25$. [1.1].

33. **A** Set of points equidistant from two vertices is the perpendicular bisecting plane of the segment joining them. The set of points 2 inches from the third vertex is a sphere with center at the third vertex and radius 2. The plane and sphere intersect in a circle. [5.5].

34. **A** Complete the square: $f(x) = 3\left(x^2 + \dfrac{4}{3}x + \dfrac{4}{9}\right) + 5 - \dfrac{4}{3} = 3\left(x + \dfrac{2}{3}\right)^2 + \dfrac{11}{3}$. Substitute $x - k$ for x in this equation in an attempt to get an equation of the form $3x^2 + \dfrac{11}{3}$ which is symmetric about the y-axis. This form is obtained if $k = \dfrac{2}{3}$. [2.3].

35. **D** $\dfrac{3}{2}$ is the only possibility for a root because 3 is a factor of 6 and 2 is a factor of 4. [2.4].

36. **A** This is of the form $A^3 - 3A^2 + 3A - 1$, which $= (A-1)^3$. Substituting $x + 2$ for A gives [A]. [5.2].

37. **C** f is an even function, g is not; therefore $f \cdot g$ is not even. $f(g(x)) = \cos(2x + 1)$, which is a cosine curve shifted less than π to the left. Thus it is not even. $g(f(x)) = 2\cos x + 1$ is a cosine curve with period 2π, amplitude 2, shifted one unit up. Thus it is even. [1.4].

38. **C** $\binom{x \text{ people}}{2} = 28$. $\dfrac{x(x-1)}{2 \cdot 1} = 28$. $x^2 - x = 56$. $x = 8$. [5.1].

39. **E** Sum of pairs of roots multiplied $= -\dfrac{-4}{3} = \dfrac{4}{3}$. [2.4].

40. **A** Reduce $f(x)$ to 2 for all $x \ne 1$. Therefore, $f^n(f(x)) = 2$ for all values of n. [1.2, 5.9].

41. **B** Let $\theta = \text{Arctan}\dfrac{1}{3}$. Thus, $\tan\theta = \dfrac{1}{3}$. $\sin\theta = \dfrac{\sqrt{10}}{10}$. [3.6].

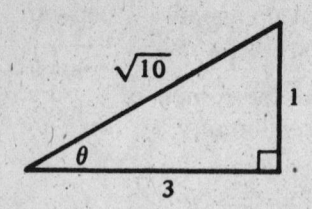

42. **B** Height of cylinder is 8.
Volume of sphere = $\frac{4}{3}\pi r^3 = \frac{4}{3}\pi(125) = \frac{500\pi}{3}$.
Volume of cylinder = $\pi r^2 h = \pi(9)8$.
Difference = $166\frac{2}{3}\pi - 72\pi = 94\frac{2}{3}\pi$. [5.5].

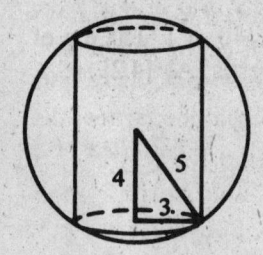

43. **A** Denominator determinant is $\begin{vmatrix} 3 & 2 \\ 1 & -3 \end{vmatrix}$. [B] and [D] are ruled out. The numerator determinant is the same with the y-coefficients replaced by 1 and 2. Thus, $\begin{vmatrix} 3 & 1 \\ 1 & 2 \end{vmatrix}$. [5.9].

44. **E** The graph never repeats itself. Each piece of the graph extends lower than the preceding one. [4.4].

45. **B** The period is $\frac{\pi}{2}$, so [A] and [E] are ruled out. Choice [C] is the same as $\frac{1}{2}\sin 4x$ and is ruled out because amplitude is $\frac{1}{2}$. Choice [D] is ruled out because $y < 0$ when $x < \frac{\pi}{4}$. [3.4].

46. **D** This will be a maximum when $\sin(x + \frac{\pi}{6}) = -1$. $x + \frac{\pi}{6} = \frac{3\pi}{2}$. Therefore, $x = \frac{4\pi}{3}$. [3.4].

47. **E** Since there is no way to reduce the fraction to eliminate the $x-2$ in the denominator, $f(x)$ has an asymptote when x approaches 2. Therefore, there is no way to define k to make f continuous. [4.5].

48. **E** Range = largest value − smallest value = 18 − 8 = 10. [4.7].

49. **D** The graph of f must be symmetric about the line $y = x$. In I, f becomes $y = -1x + b$, which is symmetric about $y = x$. In II, f becomes $y = x$, which is symmetric about $y = x$ since it is $y = x$. In III, f becomes $y = ax$, which is not necessarily symmetric about $y = x$. [1.3].

50. **D** $\tan\frac{\theta}{2} = \frac{\sqrt{3}}{3}$. Therefore, $\frac{\theta}{2} = 30°$, or 210°, so $\theta = 60°$ or 420°. [3.5].

MODEL TEST 2

1. **E** $f(x) = \frac{x-2}{(x-2)(x+2)} = \frac{1}{x+2}$. $f(x)$ is undefined at $x = 2$ and at $x = -2$. Since the $x - 2$ divides out, there is a hole in the graph at $x = 2$. The only vertical asymptote occurs when $x + 2 = 0$. [4.5].

2. **B** The only pairs of parallel edges are the opposite sides of the square base. [5.5].

3. **C** Complete the square: $(x^2 + 2x + 1) + (y^2 - 2y + 1) + z^2 = 10 + 1 + 1$. $r = \sqrt{12}$. [5.5].

4. **A** Since a polynomial of even degree does not have to cross the x-axis, the answer is [A]. [2.4].

5. **C** In any quadratic equation the sum of the roots $= -\frac{b}{a}$ and the product of the roots $= \frac{c}{a}$. $r + s = -b$. $(r+s)^2 = b^2$. $rs = c$. Therefore, $\frac{(r+s)^2}{rs} = \frac{b^2}{c}$. [2.3].

6. **C** $6 \cdot \sin x \cdot \cos x = 3 \cdot \sin 2x$ whose amplitude is 3. [3.5].

7. **B** $f\left(2, \frac{\pi}{3}\right) = 2 \cdot \cos\frac{\pi}{3} = 2 \cdot \frac{1}{2} = 1$. [4.7].

8. **A** Sum of zeros $= -\frac{b}{a} = 3$. [2.4].

9. **D** $i^{14}(1 + i + i^2 + i^3)$. $i^2 = -1$ and $i^3 = -i$. $i^{14}(1 + i - 1 - i) = 0$. [4.7].

10. **D** $y = \sin 2x$ has a period of $\frac{2\pi}{2} = \pi$. $\sin 2x = 0$ when $2x = 0°, 180°, 360°, 540°, 720°, \ldots$ so $x = 0°, 90°, 180°, 270°, 360°, \ldots$ Three values lie between 10° and 350°. [3.4].

11. **B** $x^{-2/3} = \frac{1}{x^{2/3}}$. Since all values of x are squared and $x \neq 0$, $y > 0$. [1.1, 4.2].

12. **C** Surface area = $4\pi r^2$. Volume = $\frac{4}{3}\pi r^3$. $4\pi r^2 = \frac{4}{3}\pi r^3$. $r = 3$. [5.5].

13. **C** A unit vector parallel to $\vec{V} = \frac{\vec{V}}{|\vec{V}|}$. $|\vec{V}| = \sqrt{2^2 + (-3)^2 + 6^2} = 7$. A unit vector would either be $\left(\frac{2}{7}, -\frac{3}{7}, \frac{6}{7}\right)$ which is parallel to and in the same direction as $\vec{V}$ or $\left(-\frac{2}{7}, \frac{3}{7}, -\frac{6}{7}\right)$ which is parallel to and in the opposite direction as $\vec{V}$. [5.5].

14. **B** $s = r\theta$. $12 = r$. [3.2].

15. **C** $g(x)$ is the inverse function of $f(x)$ so $g(f(x)) = f(g(x)) = x$. $f(g(x)) = \frac{1}{3}(g(x)) + 2 = x$. $\frac{1}{3}(g(x)) = x - 2$. $g(x) = 3x - 6$. [1.3].

16. **E** Consider the associated equation $x(x-3)(x+2) = 0$ which is solved when $x = 0, 3$, or -2. Check numbers in the intervals in the original inequality. Numbers between -2 and 0 or greater than 3 satisfy the inequality. [2.5].

17. **A** The slope of any tangent line is a positive number. Therefore, every point on the graph of g must be positive. [5.5, 5.9].

18. **E** Sketching the graphs indicates 2 points of intersection, one of which appears to be $\left(1, \frac{\pi}{2}\right)$. This is not one of the choices, but the point is also named by $\left(-1, \frac{3\pi}{2}\right)$. Checking $\left(\frac{\sqrt{2}}{2}, \frac{\pi}{4}\right)$ in both equations shows the answer is [E]. [4.8].

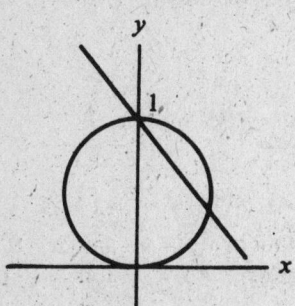

19. **E** $\log(a^2 - b^2) = \log(a+b)(a-b) = \log(a+b) + \log(a-b)$. [4.2].

20. **B** Substituting for y gives
$$x\left(\frac{4t}{t-3}\right) - 4x - 2\left(\frac{4t}{t-3}\right) - 4 = 0$$
and simplifies to
$$12x - 12t + 12 = 0. \; x = t - 1.$$
[4.7].

21. **D** Completing the square: $a\left(x^2 + \frac{b}{a}x + \frac{b^2}{4a^2}\right) + c - \frac{b^2}{4a} = a\left(x + \frac{b}{2a}\right)^2 - \frac{b^2 - 4ac}{4a}$. To have symmetry about the y-axis, $\left(x - 3 + \frac{b}{2a}\right)^2 = x^2$. Therefore, $b = 6a$. [2.3, 5.5].

22. **D** Slope of the first line is $-\frac{1}{2}$. Slope of the second line is $-\frac{a}{3}$. To be perpendicular $-\frac{1}{2} = \frac{3}{a}$. $a = -6$. [2.2].

23. **A** The first equation converts into $b^y = x^2$; the second $b^y = 2x$. $x^2 = 2x$, so $x = 0$ or 1. The log of 0 does not exist so the answer is [A]. [4.2].

24. **D** Since one man must be on a committee, the problem changes to be "form a committee of 3 from 8 men." $\binom{8}{3} = \frac{8 \cdot 7 \cdot 6}{3 \cdot 2 \cdot 1} = 56$. [5.1].

25. **B** $\sin(A + 30°) = \sin A \cdot \cos 30° + \cos A \cdot \sin 30°$
$= \frac{\sqrt{3}}{2} \sin A + \frac{1}{2} \cos A$.
$\cos(A + 60°) = \cos A \cdot \cos 60° - \sin A \cdot \sin 60°$
$= \frac{1}{2} \cos A - \frac{\sqrt{3}}{2} \sin A$.
Therefore, $\sin(A + 30°) + \cos(A + 60°) = \cos A$. [3.3].

26. **D** Use half angle formula: $\sin 15° = \sin \frac{1}{2}(30°)$
$= \sqrt{\frac{1 - \cos 30°}{2}} = \sqrt{\frac{1 - \frac{\sqrt{3}}{2}}{2}} = \frac{\sqrt{2 - \sqrt{3}}}{2}$.
[3.5].

27. **E** When the exponent is not a positive integer the expansion has an infinite number of terms. [5.2].

28. **D** $LD^2 = K$. If D is divided by 2 and then squared, the 4 in the denominator will cancel out the 4 times L. [5.6].

29. **C** Contrapositive is $q' \rightarrow (p \wedge q)'$, which is equivalent to $q' \rightarrow (p' \wedge q')$. [5.7].

30. **D** $f(a,b,0) = 2a + 3b = f(0,a,b) = 3a - b$. $4b = a$. $\frac{b}{a} = \frac{1}{4}$. [1.5].

31. **B** $f^2(x) = f(f(x)) = f\left(\frac{x}{x-1}\right)$

$$= \frac{\frac{x}{x-1}}{\frac{x}{x-1}-1} = x.$$

$f^3(x) = f(f^2(x)) = f(x) = \frac{x}{x-1}.$

[1.2, 5.9].

32. **A** See the sketch. [4.3].

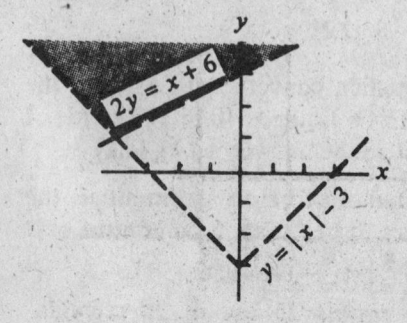

33. **D** Fold the graph about the line $y = x$ and the resulting graph will be [D]. [1.3].

34. **D** Since these points are on a line, all slopes must be equal.

$$\text{Slope} = \frac{-2-1}{1-(-2)} = -1 = \frac{1.5-(-2)}{x-1}.$$

Solving for x gives -2.5. [2.2].

35. **B** If it is an even term it must be negative, and if it is an odd term it must be positive. Therefore, the answer must be [B] or [C]. The coefficient of [B] comes from $\binom{18}{3} = 816$. [5.2].

36. **B** This is a parabola that opens up with vertex at $(0, 2)$. Minimum value of 2 occurs when $x = 0$. [2.3].

37. **D** Multiply through to get rid of the fraction and obtain $\sin x + \cos\frac{\pi}{3} = 0$. $\sin x = -\frac{1}{2}$. $x = 210°$ or $330°$. [3.3].

38. **B** There are 8 elements in the sample space of a coin being flipped 3 times. 7 of these ways contain one head. Of these 7, three (HHT, HTH, THH) contain two heads. Probability $= \frac{3}{7}$. [5.3].

39. **B** Distance from origin to the line equals the radius of the circle.

$$\text{Distance} = \frac{|3 \cdot 0 - 4 \cdot 0 - 10|}{\sqrt{3^2 + 4^2}} = \frac{10}{5} = 2.$$

Equation is $x^2 + y^2 = 4$. [4.1].

40. **D**

$$\frac{f(x+t) - f(x)}{t}$$

$$= \frac{(3 - 2(x+t) + (x+t)^2) - (3 - 2x + x^2)}{t}$$

$$= \frac{3 - 2x - 2t + x^2 + 2xt + t^2 - 3 + 2x - x^2}{t}$$

$$= 2x - 2.$$

[1.2].

41. **C** Cofunctions of complementary angles are equal. x and y are complementary. Tan and cot are cofunctions. [3.1].

42. **C** $\vec{V} + \vec{U} = (-4, -3)$.
$|\vec{V} + \vec{U}| = \sqrt{(-4)^2 + (-3)^2} = 5$. [5.5].

43. **C** $2x + 3 = (x + 2) + d$ and $5x - 6 = (2x + 3) + d$. Eliminate the d. $x + 1 = 3x - 9$. $2x = 10$. $x = 5$. [5.4].

44. **A** $f(x) \cdot g(x) = x^5 + x^3$. An odd function because $-F(x) = F(-x)$. $f(g(x)) = (x^2 + 1)^3$. Not an odd function because $-F(x) \neq F(-x)$. $g(f(x)) = x^6 + 1$. Not an odd function because $-F(x) \neq F(-x)$. [1.4].

45. **B** Does not intersect positive x-axis by Descartes' Rule of Signs but does intersect the negative x-axis once. [2.4].

46. **B** A sketch of $y = \sin x$ and $y = \cos x$ on the same axes gives [B]. [3.4].

47. **A** Mean $= \frac{1 + 2 + 3 + 1 + 2 + 5 + x}{7} = \frac{14 + x}{7} = 3$. Therefore, $x = 7$. [5.8].

48. **E** Law of sines: $\frac{\frac{1}{3}}{5} = \frac{\frac{3}{5}}{KL}$. $KL = 9$. [3.7].

49. **C** Law of cosines: $c^2 = 16 + 1 - 8 \cdot \frac{1}{2} = 13$. [3.7].

50. **C** A table of values indicates the graph is [C]. [5.5].

x	$-.5$	0	1	2	3
y	0	1	3	5	7
y^2	0	1	9	25	49

MODEL TEST 3

1. **E** $2^{(\log_2 5)} = 5$ because the base of the exponent and the base of the log are both 2. [4.2].

Answer Explanations: Model Test 3

2. **A** The set of points equidistant from 2 parallel planes is a plane parallel to both planes half way between them. This plane cuts the sphere through its center. This intersection is a circle. [5.5].

3. **E** The domain of f^{-1} is the same as the range of f which contains all numbers greater than or equal to zero. [1.3].

4. **B** Letting $A = 2x$ in the double angle formula, $\cos 2A = 1 - 2\sin^2 A$.
$\dfrac{1-\cos 4x}{2} = \dfrac{1-\cos 2A}{2} = \dfrac{2\sin^2 A}{2} = \sin^2 A$
$= \sin^2 2x$. [3.5].

5. **C** $g(-1) = (-1)^3 - (-1)^2 - (-1) - 1 = -1 - 1 + 1 - 1 = -2$. [1.2].

6. **C** $\tan 5x = -1$ implies that $5x = 135°$. So $x = 27°$. [3.5].

7. **D** $\log(\sqrt{A} \cdot B^2) = \dfrac{1}{2}\log A + 2\cdot\log B$
$= \dfrac{1}{2}(.2222) + 2(.3333) = .7777$
[4.2].

8. **D** Using the half-angle formula for sin, the equation $\sin\dfrac{1}{2}x = \cos x$ becomes $\dfrac{1-\cos x}{2} = \cos^2 x$. $2\cos^2 x + \cos x - 1 = 0$. $(2\cos x - 1)(\cos x + 1) = 0$. $\cos x = \dfrac{1}{2}, -1$. $x = 60°, 300°,$ or $180°$. [3.5].

9. **B** $\sin(90° + x) = \sin 90°\cdot\cos x + \cos 90°\cdot\sin x = \cos x$. [3.1].

10. **B** -3 and 1 are zeros. $-b = -3 + 1 = -2$ so $b = 2$. $c = (-3)(1) = -3$. $b + c = -1$. [2.3].

11. **D** Since the coefficients are all real numbers, the complex factors must come in conjugate pairs: $x - 2i$ and $x + 2i$. [2.4].

12. **A** The point is -3 from the x-axis in the y direction and 4 in the z direction. Therefore, the distance from the point to the x-axis is the hypotenuse of a right triangle with legs 3 and 4. Distance $= 5$. [5.5].

13. **D** By the remainder theorem, the remainder when $P(x)$ is divided by $x - 1$ is $P(1) = 3$. [2.4].

14. **B** Since f is an even function, $f(-x) = f(x)$. Since the inverse of an even function is not an even function, $h(-x) = f(-x) + 1 = f(x) + 1$. $h(x)$ is the only even function. [1.4].

15. **D** $f^{-1}\left(\dfrac{1}{2}\right) = $ the value of x for which $f(x) = \dfrac{1}{2}$. Therefore, $4x + 2 = \dfrac{1}{2}$ and $x = -\dfrac{3}{8}$. [1.3].

16. **B** By the law of cosines, $9 = 4 + 16 - 2\cdot 2\cdot 4\cdot\cos x$. $\cos x = \dfrac{-11}{-16} = \dfrac{11}{16}$. [3.7].

17. **D** Sketch the graph and check points in the different regions. [2.5].

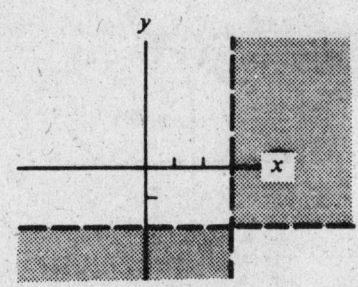

18. **B** The particle travels 25 cm in 30 seconds. $s = r\theta$. $\theta = 30° = \dfrac{\pi}{6}$. $25 = r\dfrac{\pi}{6}$. $r = \dfrac{150}{\pi}$. [3.2].

19. **A** The line must cut the axes at two points $(c, 0)$ and $(0, c)$. $\dfrac{b-0}{a-c} = \dfrac{b-c}{a-0} = $ slope. Therefore $c = a + b$. Area $= \dfrac{1}{2}(a+b)^2$. [2.2].

20. **B** A rough sketch shows intersection only when $x = 0$. [3.4].

21. **D** The diameter of the small sphere equals the side of the cube, say s. The diameter of the large sphere equals the diagonal of the cube, which is $s\sqrt{3}$. $\dfrac{\text{Small Volume}}{\text{Large Volume}} = \dfrac{s^3}{(s\sqrt{3})^3} = \dfrac{1}{3\sqrt{3}} = \dfrac{\sqrt{3}}{9}$ (i.e., volumes of similar figures are to one another as the cubes of linear corresponding parts). [5.5].

22. **B** $y \geq 0$ since it equals a square root. [4.1].

23. **A** $\text{Log}(\sec\theta) = \log\left(\dfrac{1}{\cos\theta}\right)$
$= \log 1 - \log(\cos\theta) = 0 - p$.
[4.2].

24. **E** Using 1 coin, $\binom{5}{1}$; using 2 coins, $\binom{5}{2}$; using 3 coins, $\binom{5}{3}$; using 4 coins, $\binom{5}{4}$; using 5 coins, $\binom{5}{5}$. Total $= 5 + 10 + 10 + 5 + 1 = 31$. [5.1].

25. **B** Substituting $2x$ in place of x leads to graph [B]. [5.5].

26. **D** $x^2 + y^2 = r^2$ and $x = r\cdot\cos\theta$ gives an equation of $r^2 - 4r\cos\theta + 2 = 0$. [4.7].

27. **E** Eliminating t^3 gives
$$y = \frac{3}{4}(x-9) + 7 = \frac{3}{4}x - \frac{27}{4} + 7.$$
[4.7].

28. **E** By dropping a vertical line from Q and a horizontal line through P to meet forms a right triangle with sides 5, 12, and 13. $\sin\theta = \frac{12}{13}$. [3.1].

29. **B** $f(g(z)) = f(z^2 - 1) = 3(z^2-1) - 5 = 3z^2 - 8$. [1.2].

30. **A** Sketch a portion of the graph to see that it is continuous. [4.5, 4.6, 4.4].

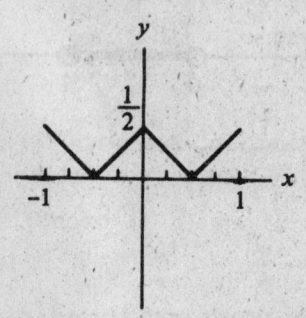

31. **D** Volume of cone $= \frac{1}{3}\pi r^2 h = \frac{1}{3}\pi(r^2)r = \frac{1}{3}r^3\pi$. Volume of hemisphere $= \frac{2}{3}\pi r^3$. $V_c : V_h = 1:2$. [5.5].

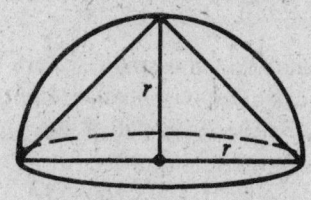

32. **A** A table of values indicates the graph is [A]. [5.5].

x	0	1	2	3	4
$\sqrt{x}$	0	1	$\sqrt{2}$	$\sqrt{3}$	2
y	2	3	4	5	6

33. **E** By synthetic division. [2.4].

```
1  -97  -199   99  0  0  -2   190  |99
     99   198  -99  0  0   0  -198
_____
1    2   -1    0   0  0  -2   |-8
```

34. **D** Listing the first few terms indicates that the series is an arithmetic series with first term $\frac{3}{2}$ and common difference of $\frac{1}{2}$.

$$S_n = \frac{n}{2}\left(3 + (n-1)\frac{1}{2}\right) = \frac{n}{2}\left(\frac{5}{2} + \frac{n}{2}\right) = \frac{n^2 + 5n}{4}.$$
[5.4].

35. **E** Substituting -1 for x gives $-a + b - c + 3 = 0$. Substituting 1 for x gives $a - b + c + 3 = k$. Adding the two equations gives $k = 6$. [2.4].

36. **C** $(a*b) + (a*c) = \log_a b + \log_a c = \log_a bc = (a*bc)$. [5.8].

37. **E** To get 4 good apples he must make one of $\binom{5}{4} = 5$ selections of good apples and $\binom{5}{0} = 1$ selection of bad apples. There can be $\binom{10}{4}$ selections of four apples from the total of ten. Probability of 4 good apples $= \dfrac{\binom{5}{4}\binom{5}{0}}{\binom{10}{4}} = \dfrac{1}{42}$. [5.2].

38. **C** $\dfrac{hr^2}{V} = K$. [5.6].

39. **C** Let Arcsin $.8 = A$ and Arccos $.8 = B$ so $\sin A = \frac{4}{5}$ and $\cos B = \frac{4}{5}$. $A + B = 90°$ since the cofunctions are equal. Since $\sin 90° = 1$, Arcsin 1. [3.6].

40. **B** There are $\binom{4}{1}$ ways to draw one of the four aces and there are $\binom{4}{1}$ ways to draw one of the four tens. There are $\binom{52}{2}$ ways to draw any two cards from the deck. Probability of getting one ace and one ten $= \dfrac{\binom{4}{1}\binom{4}{1}}{\binom{52}{2}} = \dfrac{8}{663}$. [5.3].

41. **A** $(x+1)(x-1) = -1$. $x^2 - 1 = -1$. $x = 0$. [2.2].

42. **B** Since this is an odd power polynomial it must cross the x-axis at least once. [2.4].

43. **E** The mean $= \dfrac{\text{sum of the numbers}}{\text{number of numbers}} = \dfrac{18}{8} = 2.25$. [5.8].

44. **E** The graph of I is a parabola. The graph of II is a "V." The graph of III is the top half of an ellipse. If any is reflected about the line $y = x$, the graph formed will not be a function. [1.3].

Answer Explanations: Model Test 4　　151

45. **D** Substitute the point (8,1) into the equation to get $a+b+c=8$. [4.1].

46. **D** Since the signs alternate, the coefficients are symmetric about the middle terms, there are an even number of terms, and the sum of all the coefficients is zero. [5.2].

47. **A** $\sqrt{\dfrac{2}{5}} = \dfrac{\sqrt{10}}{5} \approx \dfrac{3.16}{5} = .63$. [5.9].

48. **D** If $2x+3y-4>0$, $|2x+3y-4|=2x+3y-4$ and everything on the left drops out. [4.4].

49. **D** If $p, q,$ and r are the 3 roots, the sum of the reciprocals is $\dfrac{1}{p}+\dfrac{1}{q}+\dfrac{1}{r}=\dfrac{qr+pr+pq}{pqr}$. $qr+pr+pq=b$ and $pqr=-c$. [2.4].

50. **C** Using the General Quadratic Formula $x=p\pm 1$. Difference $=2$. [2.3].

MODEL TEST 4

1. **A** The xy-plane is intersected when $z=0$ so the equation of the line is $3x+4y=60$. [5.5]

2. **B** Irrational roots come in conjugate pairs but no relationship between the other roots is necessary. [2.4].

3. **B** $3x+4=2y+4$. $\dfrac{3x}{2x}=\dfrac{2y}{2x}\cdot\dfrac{y}{x}=\dfrac{3}{2}$. [5.6].

4. **C** $\log_8 16=x$ implies $8^x=16$. $2^{3x}=2^4$. $3x=4$. $x=\dfrac{4}{3}$. [4.2].

5. **B** $f(x)=3$ for all values of x including $g(x)$. Therefore, $f(g(x))=3$. [1.2].

6. **B** $\sin\theta=\dfrac{b}{r}$. [3.1].

7. **B** $f(x-3)=0$ when $x-3=3, -2,$ or 1. Therefore, $x=6, 1,$ or 4. [2.4].

8. **B** Complete the square on $x^2-4x+y^2-6y+z^2+2z=0$.
$x^2-4x+4+y^2-6y+9+z^2+2z+1=4+9+1$.
$(x-2)^2+(y-3)^2+(z+1)^2=14$. Therefore, radius $=\sqrt{14}$. [5.5].

9. **B** $f(x)=x^2+1$. $f(g(x))=(g(x))^2+1=4x-3$. $(g(x))^2=4x-4$. $g(x)=\pm\sqrt{4x-4}=\pm 2\sqrt{x-1}$. Therefore, $2\sqrt{x-1}$ is an expression for $g(x)$. [1.2].

10. **C** Slope of segment $=\dfrac{-5-7}{5-3}=\dfrac{-12}{2}=-6$.
Slope of perpendicular $=\dfrac{1}{6}$. Midpoint of segment is $\left(\dfrac{3+5}{2},\dfrac{7-5}{2}\right)=(4,1)$. Equation of perpendicular bisector is $y-1=\dfrac{1}{6}(x-4)$. $6y-6=x-4$. $x-6y=-2$. [2.2].

11. **A** Consider the equality to find four points of separation of the inequalities $(0,2,-3,-5)$. The regions satisfying the inequality contain only 2 integers. [2.5].

```
  |---o---|---|---o---|---o---|---o---|
 -5  -4  -3  -2  -1   0   1   2
```

12. **C** With an odd function $f(-x)=-f(x)$. $f(x)=x^3+x$ is the only possible odd function since $f(-x)=(-x)^3+(-x)=-x^3-x=-(x^3+x)=-f(-x)$. [1.4].

13. **D** Since $x^2-3x+2=(x-2)(x-1)$, $x-2$ and $x-1$ are also factors of $P(x)$. By the factor theorem $P(1)=P(2)=0$. $P(1)=3-2+A+B-8=0$. So $A+B=7$. [2.4].

14. **D** $A=e^{kt}$ implies that $\log_e A=\log_e e^{kt}=kt(\log_e e)=kt(1)=kt$. $k=\dfrac{1}{t}\log_e A$. [4.2].

15. **D** A side of the square base is $4\sqrt{2}$. The area of the two square bases is 64. The area of each of the four sides is $40\sqrt{2}$. Total surface area $=160\sqrt{2}+64$. [5.5].

16. **E** $f(x)=\dfrac{x^2-1}{x(x-1)}$. Since division by zero is undefined, $x\neq 0$ or 1. Therefore, the domain is all real numbers except 0 or 1. [1.1].

17. **A** $\dfrac{\tan x}{\cot x}=\dfrac{\dfrac{\sin x}{\cos x}}{\dfrac{\cos x}{\sin x}}=\dfrac{\sin^2 x}{\cos^2 x}=\dfrac{1^2}{2^2}=\dfrac{1}{4}$. [3.1].

18. **D** This is equivalent to "If you succeed, then you worked hard." Since "you worked hard" is true, no conclusion can be drawn because "if you succeed" can be either true or false and the original statement remains true. [5.7].

19. **B** $r\theta=\dfrac{1}{2}r^2\theta$. $r=2$. [3.2].

20. **D** As x approaches zero from the left, the slope of I is -1 and the slope of II approaches ∞. As

x approaches zero from the right, the slope of I is 1 and the slope of II approaches $-\infty$. The branches of III do not meet. Therefore only I and II have salient points. [5.9].

21. **A** ± 1 are the only two numbers that are both within the range of csc and cos. If they each equal 1, $2x+30=90°$ and $3y-15=0°$. This is not allowed since $0°$ is not positive. If they each equal -1, $2x+30=270°$ and $3y-15=180°$. Thus, $x+y=185°$. [3.3].

22. **A** $\dfrac{RE}{WF^2}=K$. Let $R=5$, $E=2$, $W=4$, $F=1$ (arbitrary choices to make the arithmetic easy). $K=\dfrac{5\cdot 2}{4\cdot 1}=\dfrac{5}{2}$. Changing the variables as indicated: $\dfrac{5\cdot 4}{2F^2}=\dfrac{5}{2}$. $F^2=4$. Since F is the amount of food, $F=2$ only. The original value of F was doubled. [5.6].

23. **B** Substituting: $x^2+y^2=x^2-4x+4+y^2-4y+4$. Simplifying gives the equation of the line through p_1 and p_2: $x+y=2$. The slope of this line is -1, so the slope of the perpendicular is $+1$. [4.1].

24. **A** Factor and reduce, then take the limit: $\dfrac{(2x-a)(x-a)}{(x+a)(x-a)}$. Limit $=\dfrac{a}{2a}=\dfrac{1}{2}$. [4.5].

25. **B** From the ten points, 3 must be chosen each time. Therefore, $\binom{10}{3}=120$. [5.1].

26. **D** Probability that neither wins is $\dfrac{4}{5}\cdot\dfrac{3}{4}=\dfrac{3}{5}$. Probability that both win (which cannot happen) is $\dfrac{1}{5}\cdot\dfrac{1}{4}=\dfrac{1}{20}$. The probability that one of these happens is $\dfrac{13}{20}$ so the probability that one of them wins is $1-\dfrac{13}{20}=\dfrac{7}{20}$. [5.2].

27. **C** Law of cosines: $49=9+25-2\cdot 15\cos C$. Therefore, $\cos C=-\dfrac{1}{2}$ and $C=120°$. [3.7].

28. **B** $f(g(z))=f(3z-2)=3z-2+2$. Thus, $h(z)=3z-2+3z$. When $3z-2\geq 0$, $h(z)=6z-2$, which is ≥ 2. When $3z-2<0$, $h(z)=2$. [1.2, 4.3].

29. **E** Removing the parameter results in $y=x$. However, x and y are both limited between -3 and 3 by the $3\sin\theta$. [4.6].

30. **B** Setting up a table of a few representative values leads to [B]. [5.5].

x	0	0	0	$\frac{1}{2}$	$\frac{1}{2}$	$\frac{1}{2}$	1	1	1
y	0	$\frac{1}{2}$	1	0	$\frac{1}{2}$	1	0	$\frac{1}{2}$	1
$x-y$	0	$-\frac{1}{2}$	-1	$\frac{1}{2}$	0	$-\frac{1}{2}$	1	$\frac{1}{2}$	0

31. **D** Keeping track of the values of A and x:

Step	1	2	3	5	6
A	2		3		
x	2		3		
Printed		2		3	Stop.

[5.9].

32. **B**
$$\dfrac{f(x,y)}{g(x,y)}=\dfrac{\tan x+\tan y}{1-\tan x\cdot\tan y}$$
$$=\tan(x+y)=\tan\left(\dfrac{\pi}{12}+\dfrac{\pi}{6}\right)$$
$$=\tan\dfrac{\pi}{4}=1.$$
[1.5].

33. **B** Wherever $f(x)=0$, $\dfrac{1}{f(x)}$ is undefined so asymptotes occur. Choices [A], [C], and [E] are ruled out. The answer is [B] because $\dfrac{1}{f(x)}$ must be positive when $f(x)$ is positive and negative when $f(x)$ is negative. [5.5].

34. **E** $3a^2+4a+1=0$. $(3a+1)(a+1)=0$, so $a=-\dfrac{1}{3}$ or -1. [2.3].

35. **D** Factoring the first and third terms makes the equation $\cos^2 x(\sin^2 x+\cos^2 x)+\sin^2 x=1$, which becomes $\cos^2 x\cdot(1)+\sin^2 x=1$, which again simplifies to $1=1$. [3.5].

36. **E** The shaded region is the intersection of C and the union of A and B. [5.9].

37. **D** If $x>0$, $x+3<5x$ and $x>\frac{3}{4}$. If $x<0$, $x+3>5x$ and $x<\frac{3}{4}$. Therefore, $x>\frac{3}{4}$ or $x<0$. [2.5].

38. **C** Multiply and divide by $2\sqrt{3}$.
$$f(x) = 2\sqrt{3}\left(\frac{1}{2}\cos x + \frac{\sqrt{3}}{2}\sin x\right)$$
$$= 2\sqrt{3}(\sin 30° \cdot \cos x + \cos 30° \cdot \sin x)$$
$$= 2\sqrt{3}\sin(30° + x).$$
Therefore, the amplitude is $2\sqrt{3}$. [3.4].

39. **B** If (x,y) represents any such point, $|y|$ represents the distance between it and the x-axis. $d = \sqrt{(x-0)^2 + (y-0)^2} = \sqrt{x^2+y^2}$. Therefore, $\sqrt{x^2+y^2} = 2|y|$. Squaring both sides and simplifying: $x^2 - 3y^2 = 0$. [4.1].

40. **A** Slope of I is $-\frac{a}{b}$; slope of II is $\frac{a}{b}$; slope of III is $\frac{b}{a}$; slope of IV is $\frac{b}{a}$. I and IV are negative reciprocals. [2.2].

41. **A** Choose any point on one of the lines, such as $(1, -1)$ on the first line. The distance between this point and the other line equals $\frac{|8\cdot 1 - 6\cdot(-1) + 11|}{\sqrt{6^2+8^2}} = \frac{25}{10} = 2.5$. [2.2].

42. **D** Substituting -1 for x gives $-5+3-K=20$. $K = -22$. [2.4].

43. **E** $r+r+r = -9$, so $r = -3$. [2.4].

44. **B** Set $\sin 2x = \tan x$. $2\sin x \cdot \cos x = \frac{\sin x}{\cos x}$. $x = 0, \pm\frac{\pi}{4}$. A rough sketch of the graphs shows the region where $\sin 2x > \tan x$ is $\left(0, \frac{\pi}{4}\right)$. [3.4].

45. **D** Using $\tan\frac{\theta}{2}$ formulas,
$$\frac{1-\cos A}{\sin A} + \frac{1+\cos A}{\sin A} = \frac{2}{\sin A}.$$
[3.5].

46. **C** $f(x)$ reduces to $x-1$. Therefore, $f(i) = 1 - i$. [1.2].

47. **B** $3x+3 = rx$ and $\frac{5(x+1)}{3(x+1)} = r$. Therefore, $3x+3 = \frac{5}{3}x$ so $x = -\frac{9}{4}$. Sum $= 9x+8 = 9\left(-\frac{9}{4}\right)+8 = -\frac{49}{4}$. [5.4].

48. **B** In order for $f^{-1}(x)$ to be a function, the domain of f must be such that for each value of y there is only one value of x. This occurs when $90° < x < 270°$. [1.3, 3.6].

49. **D** $(2\text{cis }50°)^3 = 8\text{cis }150°$
$$= 8(\cos 150° + i \cdot \sin 150°)$$
$$= 8\left(-\frac{\sqrt{3}}{2} + \frac{1}{2}i\right)$$
$$= -4\sqrt{3} + 4i.$$
[4.7].

50. **D** Mary drove for $1 \times 50 = 50$ miles.
John drove for $2 \times 48 = 96$ miles.
Fran drove for $3 \times 52 = 156$ miles.
$$\text{Average (mean) speed} = \frac{\text{Total distance}}{\text{Total time}}$$
$$= \frac{302}{6} = 50\frac{1}{3}\text{ mph}. \text{ [5.8]}$$

MODEL TEST 5

1. **C** $f(x) = -\sin x \cos x = -\frac{1}{2}(2\sin x \cos x) = -\frac{1}{2}\sin 2x$. Amplitude $= \left|-\frac{1}{2}\right| = \frac{1}{2}$. [3.4].

2. **E** If x is replaced by $-x$, the original equation is unchanged, so it is symmetric with respect to the y-axis. If y is replaced by $-y$, the original equation is unchanged, so it is symmetric with respect to the x-axis. If x is replaced by $-x$ and y is replaced by $-y$, the equation is unchanged, so it is symmetric with respect to the origin. [1.4].

3. **C** The negation would be choice [C], by the definition of negation. [5.7].

4. **C** $a+b+c+d = 30$ (1)
$b+c = 20$ (2)
$c+d = 15$ (3)
$a = 5$

Subtract equation (2) from equation (1): $a+d = 10$. Since $a = 5$, $d = 5$. Substituting 5 for d in equation (3) leaves $c = 10$. [5.9].

All students

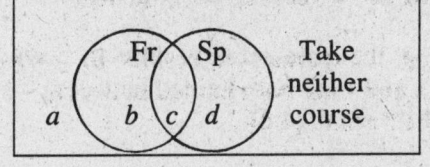

5. **C** The domain of $y = \log_{10} x$ consists of only positive numbers. Therefore, $\sin x > 0$. $\sin x > 0$ in quadrants I and II. Therefore, $0 < x < \pi$. [1.1, 3.4, 4.2].

6. **C** $\dfrac{\text{Surface area of sphere}}{\text{Volume of sphere}} = \dfrac{4\pi r^2}{\frac{4}{3}\pi r^3} = \dfrac{4}{\frac{4}{3}r \cdot 3} \cdot 3 = \dfrac{12}{4r} = \dfrac{3}{r}$. [5.5].

7. **B** If $(3,2)$ lies on the inverse of f, $(2,3)$ lies on f. Substituting in f gives $2 \cdot 2^3 + 2 + A = 3$. Therefore $A = -15$. [1.3].

8. **B** $\tan x = \dfrac{\sin x}{\cos x} = 3$. $\dfrac{\sin^2 x}{\cos^2 x} = 9$. [3.1].

9. **E** $g(0) = 1$. $f(a) = \dfrac{a+1}{a^2+1} = 0$ so $a = -1$. Slope $= \dfrac{g(0) - f(a)}{0 - a} = \dfrac{1 - 0}{0 - (-1)} = 1$. $y - 0 = 1(x - (-1))$. $y = x + 1$ or $x - y = -1$. [1.2, 2.2].

10. **C** Substitute 1 for x to get $a + b + c = 3$. Substitute -1 for x to get $a - b + c = 3$. Add these two equations to get $a + b = 3$. [2.3].

11. **C** Sketch the graph to see that the answer is [C]. [3.4].

12. **D** This is part of the formula for $\cos(A+B)$ so it is the same as $\cos(10° + 20°) = \cos 30° = \dfrac{\sqrt{3}}{2}$. [3.5].

13. **B** $\sin x + \cos x = 0$. $\sin x = -\cos x$. $\tan x = -1$. $x = 135°, 315°$. [3.5].

14. **E** If $f(x) = x^3 + x + k$ has only one real zero, there must be exactly one sign change in $f(x)$ or exactly one sign change in $f(-x)$—by Descartes' Rule of Signs. If $k > 0$, $f(x)$ has no sign change, but $f(-x) = -x^3 - x + k$ has one. If $k < 0$, $f(x)$ has one sign change, but $f(-x) = -x^3 - x + k$ has none (since k is negative). If $k = 0$, $f(x) = x(x^2 + 1)$ which has one real zero. Therefore, k can be any real number. [2.4].

15. **B** $f(g(x)) = g(x) + 3 = x^2 + 3$. $g(f(x)) = (f(x))^2 = (x+3)^2$. $x^2 + 3 = (x+3)^2$. $x^2 + 3 = x^2 + 6x + 9$. $x = -1$. [1.2].

16. **B** $f(x) = 2 \cdot \cos^2 3x = (2 \cdot \cos^2 3x - 1) + 1 = \cos 6x + 1$ (using double angle formula). Period $= \dfrac{2\pi}{6} = \dfrac{\pi}{3}$. [3.5].

17. **D** $\binom{5}{4} + \binom{5}{5} = 6$. [5.1].

18. **B** Substituting -2 for x gives $48 + 16 - 80 - 12 = -28$. [2.4].

19. **E** $x = \sec \theta = \dfrac{1}{\cos \theta} = \dfrac{1}{y}$. A portion of the hyperbola $xy = 1$ because $-1 \le y \le 1$. [4.6].

20. **C** $\log_2\left(\cos \dfrac{\pi}{3}\right) = \log_2\left(\dfrac{1}{2}\right) = n$. Therefore, $2^n = \dfrac{1}{2} = 2^{-1}$. [3.3, 4.2].

21. **D** Slope of first line is $\dfrac{3}{4}$ and second line is $-\dfrac{4}{3}$. Slopes are negative reciprocals so lines are perpendicular. [2.2].

22. **C** $\angle A = 108°$. Law of sines: $\dfrac{a}{\sin 108°} = \dfrac{100}{\sin 30°}$. Therefore, $a = 200 \sin 108° = 200 \sin 72°$. [3.7].

23. **C** $b^2 - 4ac = 81 - 4(1)(c) < 0$. Therefore, $c > \dfrac{81}{4}$. [2.3].

24. **D** The two transformations put the graph right back where it started. [5.5].

25. **E** A table of values indicates that the graph is choice [E]. [5.5].

x	-2	-1	0	1	2	3
x^2	4	1	0	1	4	9
y	-3	-1	1	3	5	7

26. **B** The central angle $\dfrac{3\pi}{4}$ is not necessary. $\dfrac{1}{2}R^2\theta = \dfrac{1}{2}r^2\theta \cdot 2$ so $\dfrac{1}{2} = \dfrac{r^2}{R^2}$. Therefore, $\dfrac{r}{R} = \dfrac{\sqrt{2}}{2}$. [3.2].

27. **E** Divide numerator and denominator through by x^3 and then let $x \to \infty$. The numerator approaches 3 and the denominator approaches 0, so the whole fraction approaches ∞. [4.5].

28. **A** The volume of the prism is area of base times height. The figure $E\text{-}ABC$ is a pyramid with base triangle ABC and height BE, the same as the base and height of the prism. The volume of the pyramid is $\dfrac{1}{3} \cdot$ base $\cdot$ height, $\dfrac{1}{3}$ the volume of the prism. Therefore, the other solid, $E\text{-}ACFD$, is $\dfrac{2}{3}$ the volume of the prism. The ratio of the volumes is $\dfrac{1}{2}$. [5.5].

29. **B** Factor: $x^2(x+1)+1(x+1)=(x^2+1)(x+1)$. Therefore, the graph crosses the x-axis only once, when $x=-1$. [2.4].

30. **A** The amplitude is 2, so maximum value is 2. Because of the absolute value the minimum value is 0. [3.4].

31. **B** The sum of the roots, $r+(-r)+p=-3$. Therefore, the third root, p, equals -3. Substituting -3 for x in the equation gives $-27+27+3K-12=0$. $K=-4$. [2.4].

32. **D** Substitute -1 for x. -1 to an even exponent $=1$, and to an odd exponent $=-1$. $3(-1)-4(1)+5(-1)-8=-20$. [2.4].

33. **A** In half of the arrangements man A is in one of the end seats. Therefore, probability that he is in the end seat is $\frac{1}{2}$. [5.4].

34. **C** Listing the first few terms:
$$\log 1 + \log\frac{1}{2} + \log\frac{1}{3} + \log\frac{1}{4} + \cdots$$
$$= \log\left(1 \cdot \frac{1}{2} \cdot \frac{1}{3} \cdot \frac{1}{4} \cdots\right).$$
As $n\to\infty$, this product approaches zero and log approaches $-\infty$. [4.2].

35. **C** $\frac{1}{2+6i}\cdot\frac{2-6i}{2-6i}=\frac{2-6i}{4-(-36)}=\frac{1}{20}-\frac{3}{20}i$. [4.7].

36. **D** Plot a few points. Minimum occurs when $x=3$. [4.3].

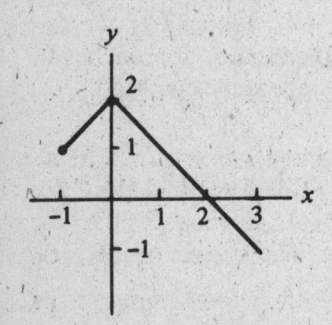

37. **B** The line cuts the x-axis at 3 and the y-axis at 4 to form a right triangle which, when rotated about the y-axis, forms a cone with radius 3 and altitude 4. Volume $=\frac{1}{3}\pi r^2 h = \frac{1}{3}\pi(9)(4) = 12\pi$. [5.5].

38. **B** The plane cuts the x-axis at 15, the y-axis at 10, and the z-axis at 6. The base is a right triangle with area $=75$. $V=\frac{Bh}{3}=\frac{1}{3}(75)(6)=150$. [5.5].

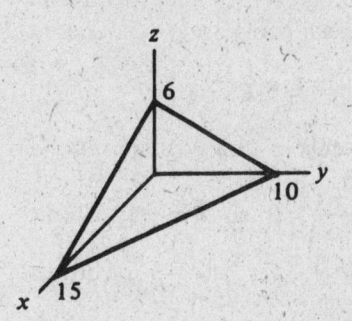

39. **B** There are $\binom{8}{3}=56$ ways to select 3 good transistors. There are $\binom{12}{3}=220$ ways to select any three transistors. $P(3 \text{ good ones})=\frac{56}{220}=\frac{14}{55}$. [5.3].

40. **C** In order for $f^{-1}(x)$ to be a function, the domain of f must be such that for each value of y there is only one value of x. This occurs when $\frac{\pi}{4} < x < \frac{3\pi}{4}$. [1.3, 3.6].

41. **D** $f(4,2)=\frac{\log 4}{\log 2}=\frac{\log 2^2}{\log 2}=\frac{2\cdot\log 2}{\log 2}=2$. [4.2].

42. **D** Solve the first inequality to get $1 \le x \le 2$. Solve the second inequality to get $-2 \le y \le 1$. The smallest product possible is -4 and the largest product possible is $+2$. [2.5].

43. **C** Sin $C=\sin(180°-(A+B))=\sin(A+B)=\sin A\cdot\cos B + \cos A\cdot\sin B = \frac{1}{3}\cdot\frac{\sqrt{15}}{4}+\frac{\sqrt{8}}{3}\cdot\frac{1}{4}=\frac{\sqrt{15}+2\sqrt{2}}{12}$. [3.8].

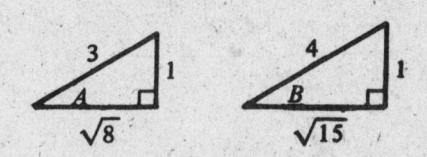

44. **A** If $x \geq 1$ the inequality becomes $x-1>2x$ and $x<-1$. Therefore, no points are possible. If $0<x<1$ the inequality becomes $-x+1>2x$ and $x<\frac{1}{3}$. Therefore, $0<x<\frac{1}{3}$. If $x<0$ the inequality becomes $-x+1<2x$ and $x>\frac{1}{3}$, which contains no points. [4.3, 2.5].

45. **C** The possible rational roots are
$$\pm 1, \pm 2, \pm 4, \pm \frac{1}{4}, \pm \frac{1}{2}.$$
Synthetic division yields $\frac{1}{4}$ as a positive rational root. [2.4].

46. **B** $\frac{RT^3}{Z^2} = K$. Let $R=1$, $T=2$, $Z=3$ (arbitrary choices) to get $K=\frac{8}{9}$. The new values are $Z=9$, $T=4$. $\frac{R(64)}{81} = \frac{8}{9}$. Therefore, $R=\frac{9}{8}$, and the original value of R has been multiplied by $\frac{9}{8}$. [4.5].

47. **A** Multiply the second equation by -2 and add to the first equation to get $-3x+5y=0$. $3x=5y$. $\frac{x}{y} = \frac{5}{3}$. [5.9].

48. **A** The number of 2's must exceed the number of other values. Eliminating some of the choices: [B] One of them could be 2. [C] x or y could be 2, but it is not necessary. [D] & [E] If $x=3$ and $y=1$, there would be no mode. Therefore, A is the answer. [5.9].

49. **B** From the sketch it can be seen that the lines drawn parallel to the axes and the line through the two points form a triangle with one angle of $120°$ and the adjacent sides of 4 and 1. Using the law of cosines, the distance between the two points equals: $d^2 = 1+16-2\cdot1\cdot4\cos 120°$. $d^2 = 17+4$. Thus, $d=\sqrt{21}$. [5.9].

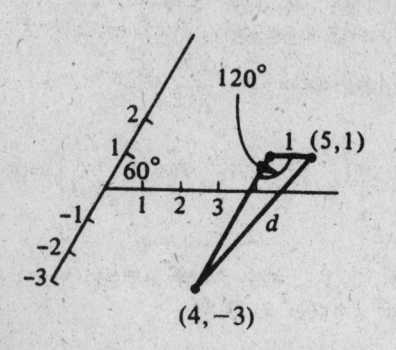

50. **D** $|\vec{V}| = \sqrt{3^2 + (-\sqrt{2})^2} = \sqrt{9+2} = \sqrt{11}$. [5.5].

MODEL TEST 6

1. **D** Since inverse functions are symmetric about the line $y=x$, if the point (a,b) lies on f, the point (b,a) must lie on f^{-1}. [1.3].

2. **A** Average $= \frac{70+80+85+80+x}{5} = 80$. Therefore, $x=85$. [5.8].

3. **B** The line intersects the y-axis at 4, so the altitude $=4$. The line intersects the x-axis at 3, so the radius $=3$. $V = \frac{1}{3}\pi r^2 h = 12\pi$. [5.5].

4. **D** The equation of a trace of a plane can be found by letting one of the variables $=0$. By inspection only [D] is a trace. [5.5].

5. **C** Factor out an x, giving 0 as one root. The sum of the roots of the remaining quadratic, $3x^2+4x-4=0$, equals $-\frac{b}{a} = -\frac{4}{3}$. $-\frac{4}{3}+0 = -\frac{4}{3}$. [2.3].

6. **B** The important point of this absolute value problem occurs when $y-1=0$ and $x+1=0$. Therefore, the pieces of the graph intersect at $(-1,1)$. [4.3].

7. **C** Substitute $2t-2$ for x. [4.6].

8. **E** Both expressions are formulas for $\tan \frac{x}{2}$. Therefore, they are equal for all values of x for which the denominators are not equal to zero. $\sin x=0$ for all multiples of π. $\cos x+1=0$ for all x which are odd multiples of π. Therefore, x can equal any number that is not a multiple of π. [3.5].

9. **D** $(\sin x)\left(\frac{\sqrt{2}}{2}\right)(-\sqrt{2})(-\sqrt{3})\left(-\frac{1}{2}\right) = \frac{\sqrt{3}}{2}$. $\sin x = -1$, so $x=270°$. [3.3].

10. **B** Substituting 1 for x gives $1+a-4=0$. $a=3$. [2.3].

11. **C** The negation of "If p, then q" is "p and $\sim q$." [5.7].

12. **B** $i^{4n}=1$; $i^{4n+1}=i$; $i^{4n+2}=-1$; $i^{4n+3}=-i$; $i^{4n+4}=(i^{4n})(i^4)=(1)(1)$. [4.7, 4.2].

13. **D** Let $y=\log_x 2 = \log_2 x$. $x^y=2$ and $2^y=x$. Substituting $(2^y)^y=2$. $2^{y^2}=2$. $y^2=1$. $y=\pm 1$. $\log_2 x=1$ or $\log_2 x=-1$. Therefore, $2^1=x$ or $2^{-1}=x$. Therefore, the sum of the two x's $=\frac{5}{2}$. [4.2].

Answer Explanations: Model Test 6 157

14. **B** The exponent on $(-3b^{-3})$ in the fifth term is 4. The exponent on $(2b^2)$ in the fifth term is $n-4$. $(b^2)^{n-4} \cdot (b^{-3})^4 = b^0$. $2(n-4) + (-3)4 = 0$. $2n - 8 - 12 = 0$. $2n = 20$. $n = 10$. [5.2].

15. **C** $\sin x = \frac{\sin x}{\cos x}$. $x = 0°, 180°, 360°$. $2 \cos x$ has values 2, 2, and 2 for these three angles. [3.1].

16. **B** The period of $y = 2 \sin 3x$ is $\frac{2\pi}{3}$. The absolute value cuts the period in half. [3.4].

17. **C** In order to get 2^{2x}, either one term must be zero in $f(x,y)$ or both must contain 2^x. [D] gives $2 \cdot 2^{2x}$, which is wrong. The only other possibility is [C], $f(g(x), g(x)) = 2(2^x)^2 - (2^x)^2 = 2^{2x}$. [1.2, 1.5, 4.2].

18. **E** Slope of first line is $\sqrt{3}$. Slope of second line is -1.
$\tan \theta = \frac{\sqrt{3}+1}{1-1\sqrt{3}} = \frac{1+\sqrt{3}}{1-\sqrt{3}} \cdot \frac{1+\sqrt{3}}{1+\sqrt{3}} = \frac{4+2\sqrt{3}}{-2}$,
which is not one of the familiar values of tangent such as $\tan 30°$, $\tan 45°$, $\tan 90°$, or $\tan 60°$. Therefore [E] is the only possible answer. [2.2, 3.1].

19. **E** From geometric sequence $b = a\left(\frac{a}{4}\right)$. From arithmetic sequence $2b - a = 12$, since $r = \frac{a}{4}$ and $d = b - a$. Substituting gives $2\left(\frac{a^2}{4}\right) - a = 12$. Solving, $a = 6$ or -4. Eliminate -4 since a is given to be positive. This gives $b = 9$. Therefore $b - a 3$. [5.4].

20. **D** $\cos x = \sin \frac{5\pi}{3} - \cos 180° - 2 \sin \frac{\pi}{6} =$
$-\frac{\sqrt{3}}{2} - (-1) - 2\left(\frac{1}{2}\right) = -\frac{\sqrt{3}}{2}$. [3.3].

21. **B** $y_1 = a + b + c + 1$ and $y_2 = a - b + c - 1$. Subtract to get $y_1 - y_2 = 2b + 2 = 3$, so $b = \frac{1}{2}$. [2.4].

22. **C** $1 = 1(\cos(0 + 360k)° + i \cdot \sin(0 + 360k)°)$.
$\sqrt[5]{1} = \sqrt[5]{1}(\cos \frac{1}{5}(0 + 360k)° + i \cdot \sin \frac{1}{5}(0 + 360k)°)$
as $k = 0, 1, 2, 3, 4$. $\sqrt[5]{1} = \cos 0° + i \cdot \sin 0°$ or $\cos 72° + i \cdot \sin 72°$ or $\cos 144° + i \cdot \sin 144°$ or $\cos 216° + i \cdot \sin 216°$ or $\cos 288° + i \cdot \sin 288°$. [4.7].

23. **D** a could be either positive or negative. b must be negative. [4.3, 2.5].

24. **D** Sketch a portion of the graph. Maximum is $\frac{1}{2}$. [4.3, 4.4].

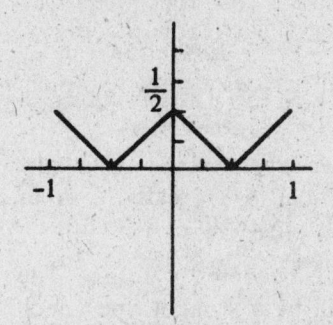

25. **C** $\lim_{x \to 2} \frac{x^3 - 8}{x^2 - 4} = \lim_{x \to 2} \frac{(x-2)(x^2 + 2x + 4)}{(x-2)(x+2)} = \frac{4+4+4}{2+2} = 3$. [4.5].

*26. **E** Height of cone is 8. Volume of cone $= \frac{1}{3}\pi r^2 h = \frac{1}{3}\pi(16)(8) = \frac{128\pi}{3}$. Volume of sphere $= \frac{4}{3}\pi r^3 = \frac{4}{3}\pi(125) = \frac{500\pi}{3}$. $V_c : V_s = 32 : 125$. [5.5].

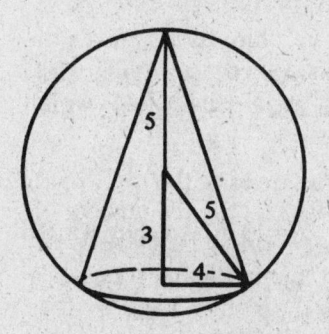

27. **E** $f(x) = \frac{(x+2)(x+1)}{(x+2)} \cdot \tan \pi x$. Since the $x+2$ divides out, the only asymptote occurs because of the $\tan \pi x$. Since $\tan$ has an asymptote at $\frac{\pi}{2}$, $x = \frac{1}{2}$. [4.5].

28. **C** List the first few terms: $\frac{1}{2} - \frac{1}{1} + \frac{1}{3} - \frac{1}{2} + \frac{1}{4} - \frac{1}{3} + \cdots + \frac{1}{251} - \frac{1}{250}$. Every positive term pairs with a negative term except -1 and $\frac{1}{251}$, leaving a total of $-\frac{250}{251}$. [5.4].

29. **A** $f(a) = a - \frac{1}{a}$. $f\left(\frac{1}{a}\right) = \frac{1}{a} - a$. $f(a) + f\left(\frac{1}{a}\right) = 0$. [1.2].

30. **D** $\sin^2 x + \cos^2 x = 1$ and $\tan^2 x = \sec^2 x - 1$. Therefore, $\sec^2 x = \sec^2 x$. [3.5].

31. **E** Product of roots $=(-1)^n \cdot \frac{\text{constant term}}{\text{leading coef.}}$ $= \frac{-4}{1}$. Since $-i$ is also a root (because i is a root) their product is $-i^2 = 1$. Descartes' rule of signs indicates the other roots are real, so their product $= -4$. [2.4].

32. **A** $P(x) = (x^4 - 4x^3 + 6x^2 - 4x + 1) + 1 = (x-1)^4 + 1$. Since $(x-1)^4$ is always greater than or equal to zero, $P(x) > 0$. [2.4].

33. **B** If $z = x^2$, $y = x^2 + 1$ becomes $y = z + 1$. The graph of $y = z + 1$ will be a straight line which rules out all choices but [B]. [5.5, 2.2, 2.3].

34. **E** $a*a = \sqrt{a \cdot a} = a$. $a*1 = \sqrt{a}$. $a*0 = 0$. Associative property does not hold. [5.9]

35. **C** $(x-4)^2 + (y+8)^2 = r^2$ is the equation of the circle. Multiplying these out indicates $a = -8$ and $b = 16$. $a + b = 8$. [4.1].

36. **E** $3a^2 + 4a + 1 = 0$. $(3a + 1)(a + 1) = 0$, so $a = -\frac{1}{3}$ or -1. [2.3].

37. **D** Factoring the first and third terms makes the equation $\cos^2 x(\sin^2 x + \cos^2 x) + \sin^2 x = 1$, which becomes $\cos^2 x \cdot (1) + \sin^2 x = 1$, which again simplifies to $1 = 1$. [3.5].

38. **B** This is an arithmetic series with $t_1 = 1$, $d = 1$. $S_{51} = \frac{51}{2}(2 + 50 \cdot 1) = 51 \cdot 26 = 1326$. An easy way to compute $51 \cdot 26$ is: $50(26) = 1300$; $1300 + 26 = 1326$. [5.4].

39. **C** Complete the square on the first equation: $3(x^2 - 2x + 1) + 4(y^2 + 2y + 1) = 5 + 3 + 4$. $3(x-1)^2 + 4(y+1)^2 = 12$. $\frac{(x-1)^2}{4} + \frac{(y+1)^2}{3} = 1$. From the sketch it is apparent that they intersect in two points. [4.1].

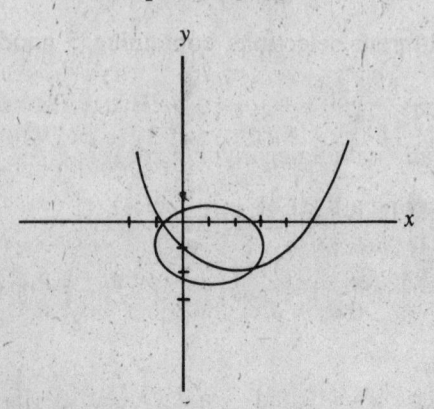

40. **A** Regardless of what is substituted for x, $f(x) \geq 0$. [1.2].

41. **B** $\frac{x}{t} = K$ and $ty^2 = C$. $t = \frac{x}{K}$. $\frac{xy^2}{K} = C$. $xy^2 = CK$. x varies inversely as y^2. [5.6].

42. **B** Let $\theta = \text{Arctan } 3$. Tan $\theta = 3$. Sin $2\theta = 2 \sin \theta \cdot \cos \theta = 2\left(\frac{3\sqrt{10}}{10}\right)\left(\frac{\sqrt{10}}{10}\right) = \frac{3}{5}$. [3.6].

43. **B** Only three of the numbers can be used in the thousands place, three are left to try in the hundreds place, two for the tens place, and only one for the units place. $3 \cdot 3 \cdot 2 \cdot 1 = 18$. [5.1].

44. **A** The first 5 flips have no effect on the last 5 flips, so the problem becomes, "What is the probability of getting exactly 3 heads in the flip of 5 coins?" $\binom{5}{3} = 10$ outcomes contain 3 heads out of a total of $2^5 = 32$ possible outcomes. $P(3H) = \frac{5}{16}$. [5.3].

45. **E** $f(g(x)) = f(x - 4) = (x-4)^2 + 3(x-4) - 4 = x^2 - 5x$. [1.2].

46. **B** $g(x) = 4(\sin x)^2 + 4(\cos x)^2 = 4(\sin^2 x + \cos^2 x) = 4$. [1.2, 3.5].

47. **B** Draw MN. Let $AB = AD = 3$. Thus $BM = DN = 1$ and $MC = CN = 2$. $\triangle CMN$ is an isosceles right triangle and $MN = 2\sqrt{2}$. $AM = AN = \sqrt{1^2 + 3^2} = \sqrt{10}$. By the law of cosines, $(2\sqrt{2})^2 = 10 + 10 - 2 \cdot \sqrt{10} \cdot \sqrt{10} \cdot \cos \theta$. $\cos \theta = \frac{-12}{-20} = \frac{3}{5}$. An alternate solution is possible. Let $x = \angle BAM = \angle NAD$. If $AB = AD = 3$, then $BM = DN = 1$. $\sin x = \frac{1}{\sqrt{10}}$. $\sin 2x = 2 \sin x \cos x = 2\left(\frac{1}{\sqrt{10}}\right)\left(\frac{3}{\sqrt{10}}\right) = \frac{6}{10} = \frac{3}{5}$. θ is the complement of $2x$, so $\cos \theta = \frac{3}{5}$. [3.7].

48. **B** Make a chart using consecutive integers for x until the sign of y changes. [2.4].

x	-2	-1	0
y	$+11$	$+12$	-5

49. **D** I. Graph consists of lines $x = 0$ and $y = 0$ (the x-axis and the y-axis), which are perpendicular. II. Graph consists of two lines $y = x$ and $y = -x$ which are perpendicular. III. Graph consists of two rectangular hyperbolas, one branch in each quadrant, with the axes as asymptotes, not lines. [4.3].

Answer Explanations: Model Test 7   159

50. **B** Points 6 inches from m form a cylinder, with m as axis, which is tangent to the plane X. Points 1 inch from X are two planes parallel to X, one above and one below X. The cylinder intersects only one of the planes in two lines parallel to m. [5.5].

MODEL TEST 7

1. **A** Volume of cylinder $=\pi r^2 h = 36\pi =$ volume of sphere $= \frac{4}{3}\pi R^3$. $r = 3$. [5.5].

2. **C** When $x = 0$, y and z can be any value. Therefore, any point in the yz-plane is a possible member of the set. [5.5].

3. **D** $\dfrac{\frac{a}{b}}{\frac{1}{a}-\frac{1}{b}} \cdot \dfrac{ab}{ab} = \dfrac{a^2}{b-a}$ [4.2].

4. **D**

n	1	2	3	4	5
f_n	3	10	5	16	8

[5.4].

5. **A** Sketch the graph of $y = x$ and $y = \sin x$ and the answer is obvious. [3.4].

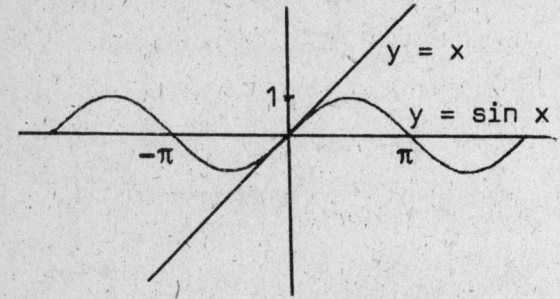

6. **E** $f(x) = x^2 - 2x + 3 = (x^2 - 2x + 1) + 2 = (x-1)^2 + 2$. This is the equation of a parabola with vertex at $(1,2)$ which opens up. Therefore, the range is $f(x) \geq 2$. [2.3].

7. **D** $|f(x) - f(2)| < 1$ implies $|(2x+5)-(4+5)| = |2x-4| = 2|x-2| < 1$. Therefore, $|x-2| < \frac{1}{2}$. [2.5, 4.3].

8. **A** Using the factor theorem, substitute k for x and set the result equal to zero. $k^2 - k^2 + k = 0$. $k = 0$. [2.4, 2.3].

9. **B** $\tan x$ is undefined for odd multiples of $\frac{\pi}{2}$. [3.4].

10. **B** By Descartes' rule of signs, the one sign change in $P(x)$ implies there will be exactly one positive real zero. [2.4].

11. **D** Sum of roots $= \frac{4}{3} = -\frac{b}{a}$. Product of roots $= \frac{4-2i^2}{9} = \frac{6}{9} = \frac{2}{3} = \frac{c}{a}$. [2.3].

12. **D** Make a chart using consecutive integers for x until the sign of y changes. [2.4].

x	-2	-1	0	1	2
y	-74	-49	-30	-11	14

13. **E** $s = r\theta$. $2r = r\theta$. $\theta = 2^R$. [3.2].

14. **C** By the law of cosines, $169 = 49 + 64 - 2\cdot 7 \cdot 8 \cdot \cos\theta$. $\cos\theta = \dfrac{56}{-112} = -\dfrac{1}{2}$. Therefore, $\theta = 120°$. [3.7].

15. **E** If they intersect on the x-axis, the value of y must be zero. Since the value of y is zero it does not matter what k is. [2.3].

16. **C** An arithmetic sequence with $t_1 = 30$, and $d = -3$. $t_{71} = 30 + 70(-3) = -180$. [5.4].

17. **E** $4\cos^4 x = (2\cos^2 x)^2 = (2\cos^2 x - 1 + 1)^2 = (\cos 2x + 1)^2$. Therefore the equation equals $\sin 2x - 2\cos 2x + \cos^2 2x + 2\cos 2x + 1 = 2$, which becomes $\sin 2x - \sin^2 2x = 0$. Therefore, $x = 0, \frac{\pi}{4}, \frac{\pi}{2}, \pi, \frac{5\pi}{4}, \frac{3\pi}{2}, 2\pi$. [3.5].

18. **C** Let $\theta = \text{Arcsin}\left(-\dfrac{3}{5}\right)$ so that $\sin\theta = -\dfrac{3}{5}$. The problem becomes $\cos 2\theta = 1 - 2\sin^2\theta = 1 - 2\left(-\dfrac{3}{5}\right)^2 = \dfrac{7}{25}$. [3.6].

19. **E** $a^2 = 16$. $b^2 = 4$. Latus rectum $= \dfrac{2b^2}{a} = 2$. [4.1].

20. **C** $\binom{8}{5}$ different selections containing 5 good items. In any one of these selections, 5 are good with a probability of .8 each, and 3 are bad with a probability of .2 each. Therefore, total probability $= \binom{8}{5}(.8)^5(.2)^3$. [5.3].

21. **C** The result $(ax, b+y, cz)$ must equal (a,b,c). Therefore the identity element must be $(1,0,1)$. [5.9].

22. **B** $(1.001)^9 = (1+.001)^9 \approx 1^9 + 9(1)^8(.001) + \binom{9}{2}(1)^7(.001)^2 = 1 + .009 + .000036 \approx 1.009$. [5.2].

23. **C** III is the contrapositive so it is true also. [5.7].

24. **D** Probability that they both win is pq. Probability that they both lose is $(1-p)(1-q)$. Probability that only one wins is $1-(pq+(1-p)(1-q))=1-(pq+1-p-q+pq)=p+q-2pq$. [5.3].

25. **E** $1-i=\sqrt{2}\left(\cos\left(-\frac{\pi}{4}\right)+i\cdot\sin\left(-\frac{\pi}{4}\right)\right)$. $(1-i)^8 = (\sqrt{2})^8\left(\cos 8\left(-\frac{\pi}{4}\right)+i\cdot\sin 8\left(-\frac{\pi}{4}\right)\right) = 16(\cos(-2\pi)+i\cdot\sin(-2\pi)) = 16(\cos 0+i\cdot\sin 0) = 16$. [4.7].

26. **B** $f\left(\pi,\frac{\pi}{2}\right)=2\pi-\frac{\pi}{2}=\frac{3\pi}{2}$. $g\left(\frac{3\pi}{2}\right)=\sin\frac{3\pi}{2}=-1$. [1.5].

27. **E** Since $|x-y|$ and $|y-x|$ both represent the distance between x and y, the only way this can be satisfied is when they are equal, and they are equal for any values of x and y. [4.3].

28. **B** $S=\frac{n}{2}(6+(n-1)4)=T=\frac{n}{2}(16+(n-1)2)$. Solving for n gives 6. [5.4].

29. **B** $y=(\cos^2 x+\sin^2 x)(\cos^2 x-\sin^2 x)=(1)(\cos 2x)=\cos 2x$. Period $=\frac{2\pi}{2}=\pi$. [3.4].

30. **D** Slope of the line through (x,y) and $(0,0)=\frac{y}{x}$. Slope of the line through (y,cy) and $(0,0)=\frac{cy}{y}=c$. Both slopes are equal, so $c=\frac{y}{x}$. [2.2].

31. **D**

	F	S
P	c	a
M	16	b

32. **B** If (x,y) represents any such point, $|y|$ represents the distance between it and the x-axis. $d=\sqrt{(x-0)^2+(y-0)^2}=\sqrt{x^2+y^2}$. Therefore, $\sqrt{x^2+y^2}=2|y|$. Squaring both sides and simplifying: $x^2-3y^2=0$. [4.1].

33. **D** Let $A=\text{Arctan}\frac{1}{5}$, $B=\text{Arctan}\frac{1}{2}$, and $C=\text{Arctan}\frac{1}{8}$. So $\tan A=\frac{1}{5}$, $\tan B=\frac{1}{2}$, and $\tan C=\frac{1}{8}$. Evaluate $\tan(A+B+C)=\tan[(A+B)+C] = \frac{\tan(A+B)+\tan C}{1-\tan(A+B)\cdot\tan C}$. $\tan(A+B)=\frac{\tan A+\tan B}{1-\tan A\cdot\tan B}=\frac{7}{9}$. So $\tan[(A+B)+C]=\frac{\frac{7}{9}+\frac{1}{8}}{1-\frac{7}{9}\cdot\frac{1}{8}}=1$. Since $\tan(A+B+C)=1$, $A+B+C=\frac{\pi}{4}$. [3.5].

$a+b+c+16=50$, $a+b=26$, $a+c=12$. Subtracting the first two and then the first and third equations gives $c=8$, $b=22$, and $a=4$. Four take Spanish and physics. [5.9].

34. **E** $e^t=1-x$; $y=1+\frac{1}{e^t}$; $y=1+\frac{1}{1-x}=\frac{2-x}{1-x}$. [4.6].

35. **D** Sketch the graphs. It is obvious that $(2,-1),(2,0),(2,1)$ are within the required area. Check $(2,2)$ by substituting into each equation to see if it is in the region. It is not because the second equation gives a value greater than zero. [5.9].

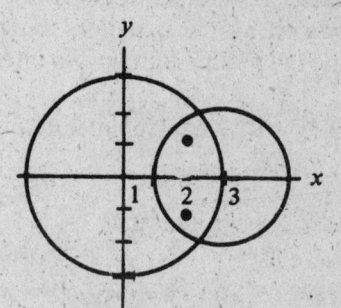

36. **E** Using $\sin^2 x+\cos^2 x=1$, $\sin A=\frac{3}{5}$ implies $\cos A=-\frac{4}{5}$ and $\cos B=\frac{1}{3}$ implies $\sin B=-\frac{2\sqrt{2}}{3}$. $\sin(A+B)=\sin A\cdot\cos B+\cos A\cdot\sin B=\frac{3+8\sqrt{2}}{15}$. [3.5].

37. **E** Substituting the coordinates of the three points into the equation gives $2a=12$, $-3b=12$, and $-4c=12$. $a+b+c=6-4-3=-1$. [5.5].

38. **B** The set of points 13 inches from p is a sphere with center at p and radius 13 inches. The set of points 2 inches from X consists of two planes parallel to X, one above and one below X. Since the radius, 13, is greater than $5+2$, the two sets intersect in two circles. [5.5].

39. **B** $2\#k=\frac{2}{k}-\frac{k}{2}=\frac{4-k^2}{2k}$. $3\#3=\frac{3}{3}-\frac{3}{3}=0$. Therefore, $4-k^2=0$ which implies that $k=\pm 2$. [5.9].

Answer Explanations: Model Test 8

40. **E** By elimination, [A] is a hyperbola and [B] is a parabola. If the last three equations are squared, it follows that [C] is a hyperbola, [D] is a circle, and [E] is an ellipse. [4.1].

41. **D** The domain of the log function is all numbers greater than 0. With a base of 2, the graph is increasing throughout. Therefore $0 < (-x) < 1$. Multiplying through by -1 gives $0 > x > -1$. [2.5, 4.2].

42. **D** $H_1 d_1^2 = H_2 d_2^2$. $d_2 = 3d$. $H_1 d_1^2 = H_2(9d_1^2)$. Therefore, $H_2 = \frac{1}{9}H_1$. [5.6].

43. **B** $(1-2k)x^2 + 8x - 6 = 0$. For equal roots, $b^2 - 4ac = 0$. $b^2 - 4ac = 64 + 24(1-2k) = 0$. $k = \frac{11}{6}$. [2.3].

44. **A** Consider the equation $\frac{x}{x-2} = 5$. $x \neq 2$. $x = 5x - 10$. $x = \frac{5}{2}$. Substituting numbers into the regions of the number line indicated by the equation shows the solution of the inequality contains *no* integers. [2.5].

45. **D** $3x - \frac{7}{x} + 3 = 0$ is equivalent to $3x^2 + 3x - 7 = 0$ and $x \neq 0$. The sum of the roots = $-\frac{b}{a} = \frac{-3}{3} = -1$. [2.3, 4.2].

46. **D** $f(f(x)) = (f(x))^2 - 4 = (x^2 - 4)^2 - 4 = 0$. $(x^2 - 4)^2 = 4$. $x^2 - 4 = \pm 2$. $x^2 = 4 \pm 2 = 6$ or 2. $x = \pm\sqrt{6}$ or $\pm\sqrt{2}$. [1.2, 2.3].

47. **A** Using the definitions of odd and even functions, it is easy to see that (I.) is the only odd function. [1.4].

48. **C** Let $g(x) = r = x + 3$. $x = r - 3$. $f(g(x)) = f(r) = x^2 - 1 = (r-3)^2 - 1 = r^2 - 6r + 8$. [1.2, 2.3].

49. **A** Slope $= \frac{-2-1}{4-2} = -\frac{3}{2}$. $(y-1) = -\frac{3}{2}(x-2)$. $y = -\frac{3}{2}x + 4$. [2.2].

50. **B** Domain is all real numbers except 1 and -2 (both of which make the denominator zero.) [1.1].

MODEL TEST 8

1. **D** $\sec^2 x = 1 + \tan^2 x$ so the equation becomes $1 + \tan^2 x - \tan x - 1 = 0$. $\tan x(\tan x - 1) = 0$. $\tan x = 0$ or $\tan x = 1$. $x = 0°, 180°, 360°$ or $x = 45°, 225°$. $n = 3$. [3.5].

2. **A** $4\pi r^2 = 36\pi$. $r^2 = 9$. $r = 3$. $V = \frac{4}{3}\pi r^3 = 36\pi$. [5.5].

3. **B** Adding the equation gives $3x = 15$. $x = 5$ and $y = 3$. $a - b = 2$. [2.2, 5.9].

4. **A** $h(5) = 1$. $g(1) = 3$. $f^{-1}(x) = x + 1$ so $f^{-1}(3) = 4$. [1.2, 1.3].

5. **C** $x = 3$ is a plane parallel to the yz-plane. $y = 2$ is a plane parallel to the xz-plane. They intersect in a line perpendicular to the xy-plane. [5.5].

6. **E** Period $= \frac{2\pi}{k} = \frac{\pi}{2}$. $k = 4$. Amplitude $= 4$. [3.4].

7. **D** If $P(x)$ represents the polynomial, by Descartes' rule of signs there is one sign change in $P(x)$ which implies one positive real root and one sign change in $P(-x)$ which implies one negative real root. Since there are 4 roots, there must be 2 complex roots also. [2.4].

8. **B** If $P(x)$ represents the polynomial, $P(-k) = k^2 - 3k + k = 0$. $k(k-2) = 0$. $k = 0, 2$. [2.4, 2.3].

9. **D** In one hour, the hour hand moves $\frac{1}{12}$ of the way around the clock, or $\frac{2\pi}{12} = \frac{\pi}{6}$ radians. $\frac{48}{60} \cdot \frac{\pi}{6} = \frac{2\pi}{15}$. [3.2].

10. **D** $\cos^4 x - \sin^4 x = (\cos^2 x + \sin^2 x)(\cos^2 x - \sin^2 x) = (1)(\cos 2x)$. [3.5].

11. **C** If $P(x)$ represents the polynomial, $P(-2)$ represents the remainder when $P(x)$ is divided by $x + 2$. $P(-2) = 20$. [2.4].

12. **B** Let $y = f(x) = x^3 - 4$. To get the inverse, interchange the x and the y and solve for y. $x = y^3 - 4$. $y = \sqrt[3]{x+4}$. [1.3].

13. **C** Using the definition of an odd function, if $f(a) = b$, then $f(-a) = -b$. [1.4].

14. **C** Keeping track of the values of A and x:

Step	1	2	3	4	3	4
A	1		5		12	
x		4		7		10

Print 12. [5.9].

15. **B** Every element of A, (a,b), must be an element of B and every element of B must be an element of C, (a,b,c,d). There are 4 sets which satisfy this condition. They are $\{a,b\}$, $\{a,b,c,d\}$, $\{a,b,c\}$, and $\{a,b,d\}$. [5.9].

16. **B** Counter-examples of [A], [C], and [D] can be found. Choice [E] does not follow if the sets of tourists and vacationers are equal. [5.7].

17. **B** Since $\sin 300° = -\sin 60°$, the answer is [B]. [4.7].

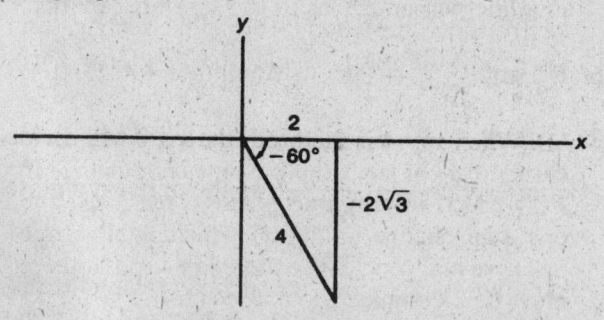

18. **A** Each couple is a unit so there is a circular permutation of four couples $=(4-1)!=6$. [5.1].

19. **A** The first few terms of the sequence are $-0+0-0+0...$. Sum $=0$. [5.4].

20. **E** There are $\binom{5}{3}=10$ ways to get 3 heads, $\binom{5}{4}=5$ ways to get 4 heads, and $\binom{5}{5}=1$ way to get 5 heads. There are $2^5=32$ different ways 5 coins can land. Probability of at least 3 heads $=\frac{16}{32}$. [5.3].

21. **B** Let $A=$ Arcsin x and $B=$ Arccos x. Therefore, $A=2B$. From the diagrams, it is clear that A and B must be complementary. Therefore, $A+B=90$, $2B+B=90$, $B=30$. $x=\cos 30° =\frac{\sqrt{3}}{2}$. [3.6].

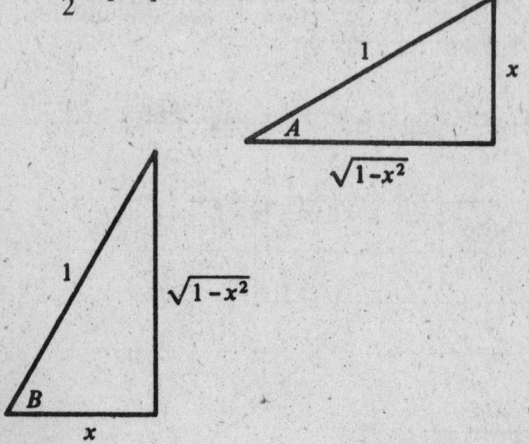

22. **A** A formula for $\tan \frac{x}{2}$. Normal tan period is π.

Period $=\dfrac{\pi}{\frac{1}{2}}=2\pi$. [3.4].

23. **A** Arithmetic series with $t_1=3$, $d=1$, and $S=150$. $150=\frac{n}{2}(6+(n-1)\cdot 1)$. $n=15$. [5.4].

24. **B** If (x,y) represents any of the points, $\sqrt{(x-0)^2+(y-0)^2}=2\sqrt{(x-1)^2+(y-1)^2}$. $x^2+y^2=4(x^2-2x+y^2-2y+2)$. $3x^2+3y^2-8x-8y+8=0$. A circle. [4.1].

25. **C** $g(2,1)=2-1=1$ and $f(2,-1)=2+(-1)=1$. Therefore, $f(1,1)=1+1=2$. [1.5].

26. **B** Complete the square: $(x^2-2x+1)+(y^2-6y+9)=r^2-10+10$. $(x-1)^2+(y-3)^2=r^2$. Center at $(1,3)$.

$r=\dfrac{|5\cdot 1+12\cdot 3-60|}{\sqrt{5^2+12^2}}=\dfrac{|5+36-60|}{13}=\dfrac{19}{13}$.

[4.1, 2.3].

27. **E** If $x+2\geq 0$ the inequality becomes $x+2>x-1$ or $2>-1$ which is always true as long as $x\geq -2$. If $x+2<0$, the inequality becomes $-x-2>x-1$, which becomes $x<-\frac{1}{2}$ when $x<-2$. Therefore, the inequality is true for all values of x. [4.3, 2.5].

28. **B** Probability that item from the red box is defective and item from the blue box is good $=\frac{3}{8}\cdot\frac{3}{5}=\frac{9}{40}$. Probability that item from the red box is good and that item from the blue box is defective $=\frac{5}{8}\cdot\frac{2}{5}=\frac{10}{40}$. Since these are mutually exclusive events the answer is $\frac{9}{40}+\frac{10}{40}=\frac{19}{40}$. [5.3].

29. **B** Complete the squares on both equations to find the coordinates of the centers:

$5(x^2+2x+1)+8(y^2-4y+4)=3+5+32$.
$5(x+1)^2+8(y-2)^2=40$.

Center $(-1,2)$.

$8(x^2-2x+1)-5(y^2-4y+4)=3+8-20$.
$8(x-1)^2-5(y-2)^2=-9$.

Center $(1,2)$. [4.1].

30. **B** $f(x) = x + x(f(x))$.

$f(x) = \dfrac{x}{1-x} \cdot g(x) =$

$x\left(\dfrac{x}{1-x}\right) = \dfrac{x^2}{1-x}$. $\dfrac{f(x)}{g(x)} = \dfrac{\frac{x}{1-x}}{\frac{x^2}{1-x}} = \dfrac{1}{x}$ [1.2].

31. **C** A sphere with equation $(x-2)^2 + (y-3)^2 + (z-4)^2 = 5^2$. [5.5].

32. **B** $\dfrac{F}{Av^2} = K$. First case: $\dfrac{45}{(50)(15)^2} = \dfrac{1}{250} = K$. Second case: $\dfrac{F}{(50)(45)^2} = \dfrac{1}{250}$. $F = \dfrac{(50)(45)^2}{250} = \dfrac{(45)^2}{5}$
$= 9(45) = 405$. $F = 405$. [5.6].

33. **B** Let $r = x - 1$. $x = r + 1$. $g(x-1) = g(r) = x^2 + 2$
$= (r+1)^2 + 2 = r^2 + 2r + 3$. Since $g(r) = r^2 + 2r + 3$,
$g(x) = x^2 + 2x + 3$. [1.2, 2.3].

34. **B** If $F = a + bi$, its conjugate $a - bi$ must be at point B. [4.7].

35. **B** Co-functions of complementary angles are equal. Therefore, $(5n - 30) + 50 = 90$. So $n = 14°$. [3.1].

36. **E** $b^2 - 4ac > 0$. $b^2 - 4ac = 16 - 4k^2 > 0$. $4 > k^2$. So $-2 < k < 2$. However, $k \neq 0$ because if $k = 0$, there would no longer be a quadratic equation. [2.3].

37. **C** f crosses the horizontal axis whenever $2\cos x = 1$ which is when $x = \dfrac{\pi}{3}$ or $\dfrac{5\pi}{3}$. [3.4].

38. **D** $3\#k = \dfrac{3}{k} - \dfrac{k}{3} = \dfrac{9-k^2}{3k}$. $k\#2 = \dfrac{k}{2} - \dfrac{2}{k} = \dfrac{k^2-4}{2k}$.
$\dfrac{9-k^2}{3k} = \dfrac{k^2-4}{2k}$. $18 - 2k^2 = 3k^2 - 12$. $k^2 = 6$.
$k = \pm\sqrt{6}$. [5.9].

39. **D** The domain of the log function is all numbers greater than zero. With a base of $\dfrac{1}{3}$ (being less than 1) the graph is decreasing. Therefore, $0 < (-x) < 1$. Multiplying through by -1 gives $0 > x > -1$. [2.5, 4.2].

40. **B** $f(g(x)) = g(x) + 3 = x^2 + 1$. $g(x) = x^2 - 2$. [1.2].

41. **E** $310° = 250° + 60°$ and $190° = 250° - 60°$.
$\cos 310° = \cos(250° + 60°) = \cos 250°\cos 60° - \sin 250°\sin 60°$. $\cos 190° = \cos(250° - 60°) = \cos 250°\cos 60° + \sin 250°\sin 60°$. Add these two equations to get $\cos 310° + \cos 190° = 2\cos 250°\cos 60° = 2(\cos 250°)\cdot\dfrac{1}{2} = \cos 250°$. The reference angle for $250°$ is $250° - 180° = 70°$ and cosine in the third quadrant is negative. Thus, $\cos 250° = -\cos 70°$. [3.1, 3.5].

42. **C** I. is a hyperbola. It has 2 asymptotes. II. has no vertical asymptote (because there are no values of x that make the denominator equal zero) and one horizontal asymptote. III. has one vertical asymptote and one horizontal asymptote. [4.1, 4.5].

43. **C** Since one of the roots is zero, the product is zero. Since the coefficient of the second term is zero, the sum of the roots is also zero. [2.4, 2.3].

44. **D** When $x^2 - 4$ is factored, the $x + 2$ will divide out leaving $\dfrac{1}{(x-2)(x-2)}$, so the only asymptote is vertical at $x = 2$. [4.5].

45. **D** Find the direction numbers by taking the difference of the corresponding coordinates of the two points: $1, -3, 6$. [D] is the only possibility. [5.5].

46. **E** The middle term is the fourth term: $-\binom{6}{3}(2a^2)^3(a^{1/4})^3 = -\dfrac{6\cdot 5\cdot 4}{3\cdot 2\cdot 1}\cdot 8a^6\cdot a^{3/4}$. The coefficient is -160. [5.2].

47. **E** To find the range, solve for x in terms of y in order to determine the restrictions on y.
$yx^2 - 4y = x^2$. $(y-1)x^2 = 4y$. $x^2 = \dfrac{4y}{y-1}$. $x = \pm\sqrt{\dfrac{4y}{y-1}}$. $\dfrac{4y}{y-1}$ must be greater than or equal to zero. Consider the associated equation $\dfrac{4y}{y-1} = 0$. $y \neq 1$ and $4y = 0$. Substituting numbers into the regions of the number line indicated by the equation shows that y must be greater than one or less than or equal to zero. [2.5, 1.1, 4.5].

48. **D** Using the law of sines, $\dfrac{\sin 45°}{a} = \dfrac{\sin 30°}{8}$
$\dfrac{1}{2}a = 8\dfrac{\sqrt{2}}{2}$. $a = 8\sqrt{2}$. [3.7].

49. **C** From this form of the equation of the hyperbola, $\frac{x^2}{9} - \frac{y^2}{4} = 1$, the equations of the asymptotes can be found from $\frac{x^2}{9} - \frac{y^2}{4} = 0$. Thus, $y = \pm \frac{2}{3}x$. [4.1].

50. **D** $f(i) = 3i^3 - 2i^2 + i - 2 = -3i + 2 + i - 2 = -2i$. [2.4, 4.7].

INDEX

Absolute value, 47, 78
Ambiguous case, 39
Amplitude, 30
Angle
   between 2 lines, 17
   co-function of, 25
   complementary, 26
   quadrantal, 28
Arc length, 27
Area
   of triangle, 41
Arithmetic means, 62, 72
Arithmetic sequence, 61
Asymptote, 42, 50
Average, 72

Binomial Theorem, 57, 79
   integer exponents, 57
      non-integer exponents, 58

Circle, 41, 77
Co-function, 25
Combinations, 56, 78
Complementary angles, 26
Complex numbers, 53
Conic sections, 41
Conjugate
   axis, 42
Conjunction, 70
Constant
   of proportionality, 69
   of variation, 69
Contrapositive, 71
Converse, 70
Cycloid, 52

Degenerate conic, 44
DeMoivre's Theorem, 53, 78
Dependent events, 59
Descartes, Rule of Signs, 20
Determinant, 74, 79
Direction
   cosines, 66
   numbers, 66
Directrix, 42
Discriminant, 19
Disjunction, 70
Distance
   between a point and a line, 17
   between a point and a plane, 66
   between two points, 16
Domain, 11
Dot product, 65
Double angle formulas, 32

Eccentricity, 42
Ellipse, 42, 77
Even functions, 14, 20
Exponents, 45, 78

Factor Theorem, 20
Factorial (!), 55
Focus, 42
Formulas, 76
Frequency, 30
   distribution, 72, 73
Function
   absolute value, 47
   composition of, 12
   continuous, 19
   definition of, 11
   discontinuous, 50
   even, 14, 20
   greatest integer, 49
   inverse, 12, 36
   linear, 16
   multi-variable, 15
   notation, 12
   odd, 14, 20
   periodic, 29
   rational, 49

General Quadratic Formula, 18
Geometric means, 63
Geometric sequence, 62
Greatest Integer Function, 49, 78

Half angle formulas, 32
Harmonic sequence, 74
Homogeneous coordinates, 6
Hyperbola, 42, 77

Identities, 32
Implication, 70
Independent events, 59
Inequalities, 24
Inner product, 65
Inverse, 12, 36, 70

Latus rectum, 42
Law of Cosines, 38
Law of Sines, 38
Limit, 50
Linear functions, 16, 76
Logarithms, 45, 78
Logic, 70

Magnitude, 65
Major axis, 42
Maximum, 18
Mean
   arithmetic, 62, 72
   geometric, 63
Mean proportional, 63
Median, 72
Minimum, 18
Minor axis, 42
Mod, 74
Mode, 72
Mutually exclusive events, 60

Necessary, 70
Negation, 71
Norm, 65

Odd function, 14, 20
Odds, 59
Operations, 73

Parabola, 18, 42, 78
   min-max, 18
Parallel lines, 16
Parameter, 51
Parametric equations, 51
Period, 29
Periodic functions, 29
Permutations, 55, 78
   circular, 56, 78
   repetition, 56, 78
Perpendicular lines, 16
Phase shift, 30
Polar coordinates, 52, 78
Polynomial
   definition, 16
   linear, 16
   quadratic, 18
Probability, 59, 79
   definition, 59
   independent, 59
   mutually exclusive, 60
Pythagorean identities, 32

Quadrantal angles, 28
Quadratic functions, 18, 76

Radian, 27
Range, 11, 72
Rational function, 49
Rational Zero (Root) Theorem, 20
Rectangular hyperbola, 42
Recursion Formula, 61

Relation
   definition, 11
Remainder Theorem, 20
Resultant, 65

Salient point, 111
Sample space, 59
Sequence, 61, 79
   arithmetic, 61, 79
   geometric, 61, 79
   harmonic, 74
Series, 61
Sigma (Σ), 61
Sines, Law of, 38
Slope, 16
Slope-intercept form, 16
Solid of Revolution, 66
Standard position, 25
Statistics, 72
Sufficient, 70
Symmetry
   axis of, 18
Synthetic division, 21

Tautology, 71
Trace, 66
Transformation, 64
Translation, 64
Transverse axis, 42
Triangles, 38
Trigonometric functions, 77

Variation
   combined, 69
   direct, 69
   inverse, 69
Vector, 65, 79
   definition, 65
   dot product, 65
   inner product, 65
   magnitude, 65
   norm, 65
   perpendicular, 65
   resultant, 65
Venn Diagram, 74
Vertex
   ellipse, 43
   hyperbola, 43
   parabola, 18, 42

Zeros
   sum and product, 18, 20

WHENEVER IT'S TEST TIME, THERE'S ONLY ONE PLACE TO TURN FOR TOP SCORES...

BARRON'S: The Test Preparation We Give Is What Keeps Us Famous!

SAT (Scholastic Aptitude Test)
How to Prepare for the College Entrance Examinations (SAT) $9.95, Can. $14.95
Basic Tips on the SAT $4.95, Can. $7.50
Math Workbook for SAT $8.95, Can. $13.50
Verbal Workbook for SAT $8.95, Can. $13.50

14 Days to Higher SAT Scores
(Combines two 90-minute audio cassettes with a 64-page review book) $14.95, Can. $22.50

601 Words You Need to Know for the SAT, PSAT, GRE, State Regents, and other Standardized Tests $6.95, Can. $9.95

How to Prepare for the Advanced Placement Examination in:
American History $8.95, Can. $13.50
Biology $9.95, Can. $14.95
English $9.95, Can. $14.95
Mathematics $9.95, Can. $14.95

PSAT/NMSQT (Preliminary Scholastic Aptitude Test/National Merit Scholarship Qualifying Test
How to Prepare for the PSAT/NMSQT $8.95, Can. $13.50
Basic Tips on the PSAT/NMSQT $3.95, Can. $5.95

ACT (American College Testing Program)
How to Prepare for the American College Testing Program (ACT) $8.95, Can. $13.50
Basic Tips on the ACT $3.95, Can. $5.95

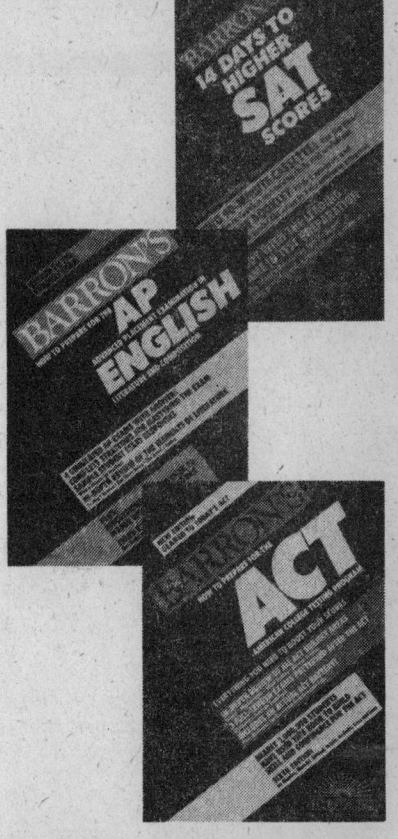

BEAT THE COMPETITION WITH THE COMPUTER BOOST!
All tests and drills on disks featuring color, graphics and sound effects (for Apple & IBM):
Barron's Computer Study Program For The ACT
3 double-sided disks, study guide, user's manual, $79.95, Can. $119.95.
Barron's Computer Study Program For The SAT
6 double-sided disks, user's manual, $49.95, Can. $74.95

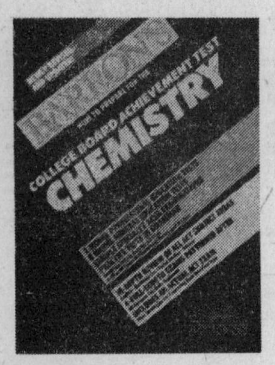

CBAT (College Board Achievement Test) in:
American History/Social Studies $8.95, Can. $13.50
Biology $8.95, Can. $13.50
Chemistry $8.95, Can. $13.50
English $8.95, Can. $13.50
European History and World Cultures $8.95, Can. $13.50
French $8.95, Can. $13.50
German $8.95, Can. $13.50
Latin $6.95, Can. $10.50
Math Level I $8.95, Can. $13.50
Math Level II $8.95, Can. $13.50
Physics $8.95, Can. $13.50
Spanish $8.95, Can. $13.50

CLEP (College Level Exam Programs)
How to Prepare for the College Level Exam Program (CLEP) $8.95, Can. $13.50

BARRON'S EDUCATIONAL SERIES
250 Wireless Boulevard
Hauppauge, New York 11788
In Canada: 195 Allstate Parkway
Markham, Ontario L3R 4T8

Prices subject to change without notice. Books may be purchased at your bookstore, or by mail from Barron's. Enclose check or money order for total amount plus sales tax where applicable and 10% for postage and handling (minimum charge $1.50). All books are paperback editions.